Wetland Plants
for Sustenance of Mankind

Wetland Plants
for Sustenance of Mankind

S.K. Sood

Professor and Dean
Department Life Sciences
Himachal Pradesh University

Shimla - 171005, H.P.

Romita Sharma

Department of Biosciences
Himachal Pradesh University
Shimla – 171005, H.P.

MJP Publishers

Chennai   New Delhi   Tirunelveli

ISBN 978-81-8094-230-3   **MJP Publishers**

All rights reserved   New No. 5 Muthu Kalathy Street,
Printed and bound in India   Triplicane,
Chennai 600 005

MJP 207 © Publishers, 2015

Publisher : C. Janarthanan

This book has been published in good faith that the work of the author is original. All efforts have been taken to make the material error-free. However, the author and publisher disclaim responsibility for any inadvertent errors.

PREFACE

Wetlands are earth's liquid assets and often described as "kidneys of the landscape." These, in fact provide many services and commodities to humanity. Nearly all cultures from ancient times to present day have used wetland resources for their sustenance. However, during the last few decades, the interaction of man with wetlands has been of concern largely due to the rapid population growth accompanied by intensified industrialization, modernization and urbanisation, and needs to be addressed before the knowledge is lost forever. From Indian perspective, wetland services alone are worth Rs 21.45 lakh crores annually, yet wetlands are ignored and reclaimed for all kinds of development projects, consequently accounting for loss of about 40% of its wetlands during 1991-2010. This scenario at all is not good for the health of Indian wetlands. Despite the advantages of wetlands to human beings, enlistment of number of Ramsar sites from India and resurgence in the field of ethnobotany during the past 40 years, no efforts have been made till date to document precious knowledge on the traditional uses of wetlands/ wetland associated plants from India in general and Bilaspur Himalayas in particular – a region with more than 1,016 villages under its 3 Tehsils Ghumarwin, Bilaspur Sadar, Jhanduta and also credited with one of the biggest man-made reservoir on the river Sutlej. Against this backdrop, the present work is a solitary attempt in this direction with a view not only to record their multifaceted usages but also for using them as valuable clues for social forestry endeavours, therapeutic agents, supplementary foods, and sustainable management of species as well as their habitats for the ultimate welfare of mankind. It is expected that this invaluable compendium, the only of its kind available, will prove a milestone in the advancement of knowledge on the subject and certainly become a standard reference for many decades.

Grateful thanks are due to Prof. A.D.N. Bajpayee, Vice-Chancellor, Himachal Pradesh University, Shimla; Prof. T.N. Lakhanpal and Prof. V.K. Mattu for their support and incessant encouragement. We thank Dr Kuldeep Dogra, Dr Suresh Kumar, Mr Ved Prakash, Ms Chandrika Sharma, Mr Rakesh Thakur and Mr Bal Krishan for their ungrudging help in various ways. Thanks are due to authors of books, papers and figures listed in the treatise. Authors are also indebted to the publishers for bringing out the book in the present form. Finally, we would like to thank the members of our families for moral support and unflinching patience during the course of the work.

Suggestions and constructive criticism for further improvement of the treatise will always be welcomed and entertained whole heartedly.

S.K. Sood
Romita Sharma

ABBREVIATIONS

amsl	Above mean sea level		M.P.	Madhya Pradesh
AD	Anno domini		ml	millilitre
Alc.	Alcoholic		N.	North
Alk.	Alkaline		(-) ve	Negative
Aq.	Aqueous		N.W.	North West
Ass.	Assamese		O	Oxygen
B.	Bengal		OAc	Acetoxy
Bo.	Bombay		OH	Hydroxy
Bu	Butyl		OMe	Oxymethyl
C	Carbon		Ori.	Oriya
C.	Central		P	Phosphorus
Ca	Calcium		(+) ve	Positive
cm	Centimeter		P.	Punjab
CNS	Central Nervous System		%	Per cent
d	Dextro		Ph	Phenyl
E.	East		Pl.(s)	Plate (s)
EL	Ethnobotany Laboratory		S.	South
Eng.	English		Sans.	Sanskrit
et al.	(et alia) and other authors		Sp	Specie
etc.	(et cetera) and so on		Spp	Species
Extract(s)	Extract (s)		Sq km	Square Kilometer
Figure. (s)	Figure (s)		Syn.	Synonym
G.	Gujarati		S.W.	South West
HCN	Hydrogen Cyanide		Tam.	Tamil
H.P.	Himachal Pradesh		Tel.	Telugu
''	Inches		U.P	Uttar Pradesh
i.e.	(id est) that is		Var.	Variety
Kan.	Kannada		Vern.	Vernacular
Kash.	Kashmir		viz.	Vide licet (namely)
l	laevo		Vit	Vitamin
m	Meter		Vet	Veterinary
Mal.	Malayalam		Vol.(s)	Volume (s)
Mar.	Marathi		W.	West
mg	Milligram		WTO	World Trade Organisation
Mg	Magnesium			

List of Illustrations

Coloured Photographs

CONTENTS

D

M

N

O

P

R

U

V

W

X

CHAPTER 1

INTRODUCTION

WETLANDS AND IMPORTANCE TO MANKIND

'Wetland' is a generic term that encompasses diverse and heterogenous assemblage of habitats ranging from lakes, estuaries, river flood plains, mangroves, coral reef and other related ecosystems. Abundance of water at least for a part of the year is the single dominant factor. As per the Ramsar (a city in Iran where the first World Convention on Wetlands was held on 2 February, 1971) Convention, Art.1.1 defined wetlands as "areas of marsh, fen, peatland or water, whether natural or artificial, permanent or temporary, with water that is static or flowing, fresh, brackish or salt, including areas of marine water, the depth of which at low tide does not exceed six metres". According to Cowrdin et al., (1979), 'Wetlands are lands, transitional between terrestrial and aquatic ecosystems where the water table is usually at or near the surface, or the land is covered by the shallow water'. Moreover, these are earth's liquid assets and often described as "kidneys" of the landscape (Mitsch & Gosselink, 1986). Broadly speaking, these have been regarded as complex hydrological and biogeochemical systems and recognized as distinctly separate ecosystems between the terrestrial and aquatic ones (Anonymous, 2007). Regarding their importance wetlands are one of the most productive ecosystems of the world, and provide multifaceted valuable services such as recycle nutrients, purify water, attenuate floods, maintain stream flow, recharge ground water, and also serve in providing drinking water, fish, fodder, fuel, wildlife habitat, control rate of run off in urban areas, buffer shorelines against erosion and provide recreation to the society as well as in maintaining the overall cultural, economic and ecological health of the ecosystem (Prasad et al., 2002). At present, an area totalling 145.73 million hectares has been designated under 1,634 wetland sites in the Ramsar List of Wetlands of International importance, amounting to about 6% of the total land mass of the globe. For India, twenty-five sites are the Ramsar sites of global significance, and six are under process of designation. It has been estimated that inland aquatic biodiversity of rivers, lakes, reservoirs and wetlands is very rich in India, and it harbours 15% and 20% of India's floral and faunal diversities, respectively. As is true elsewhere, healthy wetlands are essential in India for sustainable food production and potable water availability for humans and livestock.

Human societies in all parts of the world have used wetland resources for diverse purposes from time immemorial (Seshavatharam, 1990). Infact, the survival of the native wetlands/wetland associated plants and traditional knowledge of native communities living on the fringes of wetlands regarding multifaceted usages of wetland resources for their various basic needs are threatened universally during the last few decades due to the rapid population growth accompanied by intensified industrial, commercial and residential development further leading to pollution of wetlands by domestic, industrial sewerage and agricultural run-offs as fertilizers, insecticides and feedlot wastes, habitat destruction, over-exploitation of resources, unplanned land use and application of inappropriate technologies (Anonymous, 1991; Joshi, 2009); this threat of knowledge loss needs to be addressed urgently before the knowledge is lost forever. It has been seen that of the total annual global ecosystem services which is US $ 33 trillion, the contribution of the wetland ecosystem is about 45% . In India, wetland ecosystem services alone are worth Rs 21.45 lakh crores annually, almost five times more than the Union budget for 2007–2008. Yet, wetlands are ignored and reclaimed for all kinds of development projects, consequently accounting for loss of 38% of its wetlands between 1991-2001 (Vijayan, et.al., 2007).

WETLANDS AND RAMSAR SITES IN HIMACHAL PRADESH

The state of Himachal Pradesh being located between latitudes 30°22′40″N to 33°12′20″N and longitudes 75°45′55″E to 79°04′20″E and having geographical extent of 10.54 per cent of the Himalayan landmass with altitude ranging between 1,450-6,500 m amsl in the lap of north-western Himalayas, has a vast area of wetlands/water bodies rich in extremely important plant resources known for playing a crucial role both in economic and social life of the diverse ethnic populace in 16,997 villages of different sizes covering around 55,678 sq kms under its 12 districts, and also boasts of wide diversity of plant forms of European, Tibetan and Chinese origin in its sub-tropical, temperate, alpine and cold desert regions. In total, 27 natural and 5 man-made wetlands occupying an area of 15 and 712 hectares respectively, covering tropical, subtropical and alpine regions from 1,450-5,093 m amsl have been enlisted from Himachal Pradesh. Five Ramsar sites, viz., Renuka, Pong Dam, Chandratal, Rewalsar and Khajjiar with a total area of 15,736 hectares have been identified so far under the National Programme for Conservation and Management of Wetlands (Anonymous, 2007) whereas benchmark data for others are lacking. Despite the large advantages of wetlands to human beings, resurgence in the field of ethnobotany in India, Himalayas being a rich repository of medicinal and aromatic plants and diverse ethnic communities and changes in sustenance economy, no efforts have been made in the past to document traditional uses of wetlands/wetland associated plants of Himachal Pradesh and, therefore, necessitated the present work.

CURRENT STATUS OF WORK ON INDIAN WETLANDS

Status of Aquatic and Wetland Diversity

The information on aquatic and wetland flora of world is known through the valuable works of Anderson (1919), Arber (2003), Perry (1938), Fassett (1940), Muenscher (1944), Reid (1961), Sculthorpe (1967), Bruggen (1968-1973), Bursch (1971), Aston (1973), Brezy et al. (1973), Beal (1977), Ahmad & Younus (1979), Ancibor (1979), Agami et al. (1986), Beal & Thieret (1986), Tiner (1992), Borman et al., (1997) and Chadde (1998). From Indian perspective, four important publications, viz., Handbook of Common Water and Marsh Plants of India and Burma (Biswas & Calder, 1937), Aquatic Angiosperms (Subramanyam, 1962), Ecology and Management of Aquatic Vegetation in the Indian Subcontinent (Gopal, 1990) and Aquatic and Wetland Plants of India (Cook, 1996) have provided detailed information on their lists and occurrence from various water bodies in India. Thereafter, a number of studies as those of Lal et al. (1997), Chamanlal (1997), Sikarwar & Kaushik (1997), Indra & Venkataraju (1998), Singh & Pandey (1998), Kumar & Banerjee (1999), Roy et al. (1999), Sunil & Sivdasan (2000), Wadoodkhan (2000), Yadav et al. (2000), Dasdas & Singh (2001, 2006), Kothari (2001), Venu et al. (2003), Punekar et al. (2003), Guha & Mondal (2003), Maliya & Singh (2004), Pandey (2004), Barooah & Mahanta (2006), Maliya (2006), Mandal et al. (2006), Narain (2006), Wadoodkhan et al. (2006), Narain & Singh (2008), Rao et al. (2011), Saini et al. (2010) and Udayakumar et al. (2010) have also enriched our knowledge on the subject. Unlike other Indian regions, the literature pertaining to aquatic biodiversity of Himalayas is dismally small. In this regard, contributions of Joshi (1952: Lahul, - H.P.); Kak (1984 a, b; 1990: Kashmir - N.W. Himalaya), Negi et al. (1985: Garhwal Himalaya); Gaur et al. (1993: Dalsera Lake - Garhwal Himalaya); Samant & Pangtey (1994: Kumaon Himalaya); Lal et al. (1997: Uttarkashi, Garhwal Himalaya); Chamanlal (1997: Uttarkashi, Garhwal Himalaya); Singh (2001: Mandi, H.P.) and Srivastava (2003a : Pong Dam Sanctuary, Kangra, H.P.; 2003b : Renuka Lake, Sirmaur, H.P.) are worth mentioning. Moreover, contributions providing overview of Indian wetlands, causes and consequences of wetland losses, economic valuation of wetlands, their management and towards sustainable development and conservation are also there in the literature (Anonymous, 1989; Mahajan, 1989; Nair, 1989; Banerjee & Verma, 1994; Deshmukh & Balaji, 1994; Samant, 1999; Choudhury, 2000; Prasad et al., 2002; Ghosh, 2004; Kumar, 2006).

STATUS OF ETHNOBOTANICAL DIVERSITY

General Human civilization has been influenced both directly and indirectly by plants growing in the vicinity. Even prior to the use of term 'aboriginal botany' by Powers (1874) and 'ethnobotany' by Harshberger (1895), documented knowledge regarding the curative properties of plants can be traced back to the ancient Vedic period (3500-1800 BC) and in other civilizations of the world [Hippocrates (5 BC), Theoprastus (372-287 BC), Pliny (23-79 AD), Dioscorioides (77 AD), Ibn Sina (980-1037), Fuchs (1542), Turner (1551) and Gerard (1597)].

Since the publication of the first book on ethnobotany, An Introduction to Ethnobotany by Faulks (1958), there has been an upsurge of activities in ethnobotanical pursuits all over the World during the last five decades (Towle, 1961; Schultes, 1962, 1967, 1979, 1983, 1986; Hartwell, 1967-71; Martin, 1972; Renfrew, 1973; Craford, 1983; Felger & Moser, 1985; Duke, 1968, 1986; Kindscher, 1987; Schultes & Raffauf, 1990; Taylor, 1990; Turner et al., 1990; Bellany, 1995; Balick & Cox, 1996; Barford & Kvist, 1996; Neuwinger, 1996, 2000; Newall et al., 1996; Redford & Mansaur, 1997; Schultes & Reis, 1997; Brussell, 1998; Schoen & Wynu, 1998; Pieroni, 1999; Gefu et al., 2000; Minnis, 2000; Payee, 2000; Bisset & Vichtl, 2001; Martin et al., 2001; Seitel, 2001; Grotenhermen & Russo, 2002; Lairl, 2002; Idu, 2009) with Mudgal (1987) and Chandra (1990) compiling the diverse ethnobotanical literature existing in various languages of the world. From the view-point of Indian scenario, Janaki Ammal (1955) pioneered the ethnobotanical research whereas Glimpses of Indian Ethnobotany by Jain (1981) is the first book dealing on the subject. Subsequently, the subject matter reviewed in some of the important compilations like Bibliography of Ethnobotany (Jain et al., 1984), Cumulative index to ethnobotany from 1989-2000 with the analysis of regions and the themes covered (Srivastava & Jain, 2000), Indian ethnobotanical literature during last two decades - a graphic view and the future directions (Jain & Srivastava, 2001), a recent detailed one listing 1,600 references of which over 1,250 relate strictly to the ethnobotanical work or very closely allied themes on India and the others on concept, global evolution of the discipline, and methodology of new areas from works of some outstanding foreign ethnobotanists (Jain, 2002) and that on ethnobotanical spectrum of Indian flora: an overview during the past 20 years in the journal ethnobotany (Goel & Tripathy, 2009) have brought together the ethnobotanical literature from disparate sources. In recent times, a number of useful treatises on the subject from Indian perspective have been those of Maheshwari (1992, 2003), Joshi (1995), Arora & Pandey (1996), Rama Rao & Henry (1996), Sinha (1996), Nautiyal et al., (1997), Kaushik (1998), Pandey (1998), Roy et al., (1998), Samant et al., (1998), Singh & Pandey (1998), Dager & Dager (1999), Mudgal et al., (1999), Kaushik & Dhiman (2000), Singh & Kumar (2000a), Sood et al. (2001, 2009a, b, 2010), Parrotta (2001), Kumar (2002), Lalramnghinglova (2003), Singh et al. (2003a, b), Sood & Thakur (2004), Trivedi (2004, 2009, 2010), Viswanathal et al. (2006), Kshirsagar & Singh (2007), Sood & Kaushal (2008), Sood & Sudershana (2008), Sood et al. (2009a, b), etc. and all of these contributed significantly towards our understanding on the various aspects of ethnobotanical science. Barring these, other useful works describing linkages of ethnobotany with the other sciences and disciplines (Manilal, 1989), quality control requirements of medicinal plants used in traditional medicines (Mehrotra, 1990), efforts and concerns about ethnobotanical studies on the Indian plant genetic resources (Arora, 1995), role of ethnobiologists in the traditional knowledge and the sustainable development (Rao, 1996), computing the relative reliability of ethnobotanical claims and ethnobotany in the 21st century in India (Khan, 2001, 2005), ethnobotanical perspective of the Sanskrit plant names (Patil, 2000), Indian plant species used in Tibetan traditional medicines (Jain, 2003; Jain & Goel, 2005), role of traditional and folklore herbals for the development of new drugs (Mehrotra & Mehrotra, 2005), CBD, WTO and Biodiversity Act of India in the ethnobotanical perspective (Pushpangadan & Kumar, 2005), ethnobotany in the new millennium - some thoughts on future direction in Indian ethnobotany (Jain, 2006), scholastic genealogy or mentor student lineages in ethnobotany in India - an appeal for study and record (Jain, 2008), prospects of ethnobotany in 2008 (Shah, 2008), high altitude botanicals in integerative medicine-case studies from North-West Himalaya (Kaul, 2010), ethnobotany in India: some thoughts on future work (Jain, 2011) and Indian Plant genetic resources: role of ethnobotany in determining priorities and issues (Nayar, 2010) have also been done in recent years.

Wetlands Ethnobotanically, very little efforts have been made in the past to document traditional uses of wetlands/wetland associated plants from India; notable ones are those of Srivastava et al., (1987) - Some economically important plants from Beas-Sutlej Link project area in Himachal Pradesh; Subudhi et al. (1992) - Some potential medicinal plants of Mahanadi delta in the state of Orissa; Pareek (1994) - Preliminary ethnobotanical notes on the plants of aquatic habitats of Rajasthan; Das et al. (1996) - Traditional uses of wetland plants of eastern Orissa; Maya et al. (2003) - Economic importance of river vegetation of Kerala; Samanta & Das (2003) - Ethnobotanical studies on *Typha elephantina* Roxb. (Typhaceae) in the southern parts of West Bengal; Nandan & Singh (2004) - Useful macrophytes in Kawvar lake, Noran Bihar; Ravindran et al. (2005) - Ethnomedicinal studies of Pichavaram mangroves of East Coast, Tamil Nadu;

Dash et al. (2007) - Phytotherapeutic uses of mangroves for primary healthcare among the local inhabitants of Bhitarkanika Wildlife Sanctuary, Orissa; Jain et al. (2007) - Edible aquatic biodiversity from the wetlands of Manipur; Jain et al. (2007) – Aquatic/semiaquatic plants used in herbal remedies in the wetlands of Manipur, Northeastern India; Sanilkumar & Thomas (2007) - Indigenous medicinal uses of some macrophytes of the Muriyad wetland in Vembanad - Kol, Ramsar site Kerala; Srivastava & Srivastava (2007) - Diversity and economic importance of wetland flora of Gorakhpur district (U.P.); Pandey & Pandey (2009) - Medicinal value of aquatic and wetland plants of Varanasi district; Kumar & Narain (2010) - Herbal remedies of wetland macrophytes in India; Sarma & Saikia (2010) - Utilization of wetland resources by the rural people of Nagaon district, Assam; Tripathi et al. (2010) - Indigenous semi-aquatic and aquatic angiosperm biodiversity of district Jaunpur, Uttar Pradesh; and Panda & Misra (2011) - Ethnomedicinal survey of some wetland plants of South Orissa and their conservation. Despite exhaustive floristic treatises published by several workers for the state of Himachal Pradesh, e.g., Atkinson (1882), Hooker (1872-1897), Collett (1902), Nair (1977), Chowdhery & Wadhwa (1984), Polunin & Stainton (1987), Stainton (1988), Aswal & Mehrotra (1994), Dhaliwal & Sharma (1999), Singh & Rawat (2000), Singh & Sharma (2006) and a large number of research publications more particularly from its different parts, namely, Chamba (Gupta, 1961, 1964, 1971), Himachal Pradesh (Sarin & Chopra, 1985; Chauhan, 1983, 1989a-c, 1990, 1999, 2003; Singh & Aswal, 1992; Sharma & Sood, 1997; Singh, 1999; Badola, 2001), Kangra Valley (Ahluwalia, 1952; Uniyal & Chauhan, 1971, 1973; Kapur, 1985), Kullu (Rastogi, 1960; Dobriyal et al., 1997), Lahul-Spiti (Rau, 1961; Sarin, 1967; Uniyal et al., 1973; Arora et al., 1980; Chauhan, 1989a-c; Chaurasia et al., 2001), Mandi (Singh, 1993), and Una (Chauhan, 1974) further contibuting to its taxonomical wealth as well as about 30 ethnobotanical reports including 4 books from this region (Uniyal & Chauhan, 1973; Koelz, 1979; Chauhan, 1988, 1996; Kapur, 1993; Singh, 1993, 1999, 2004; Jain & Puri, 1994; Duhoon et al., 1996; Lal et al., 1996; Singh et al., 1998, 2001; Sharma & Rana, 1999; Sood et al., 2001; Negi & Bhalla, 2002; Negi & Subramani, 2002; Chandrasekar & Srivastava, 2003, 2005; Uniyal, 2003; Rana, 2004; Sharma et al., 2004; Singh, 2004a, b; Sood & Thakur, 2004; Sharma & Maheshwari, 2005; Singh & Chauhan, 2005; Verma & Chauhan, 2006, 2007; Uniyal et al., 2006; Kapoor et al., 2010; Parkash & Aggarwal, 2010), it is rather unfortunate that no efforts have been made in the past to document information available on traditional uses of wetlands/wetland associated plants of district Bilaspur of Himachal Pradesh. Against this backdrop, the present work is a solitary attempt in the direction to document the precious knowledge of the local inhabitants in this part of Himalaya on the multifaceted usages of wetland plants with a view not only to conserve it from being lost irreversibly to varied anthropogenic pressures but also for using them as valuable clues for social forestry endeavours, therapeutic agents, supplementary foods, and sustainable management of species as well as their habitats for the ultimate welfare of mankind.

THE STUDY AREA

History and Nomenclature On merger of the territory of Bilaspur into Himachal Pradesh, the district came into being on July 1, 1954. Formerly, it was a princely state and named as 'Kahlur' after Raja Kahal Chand, sixth in the line in the ninth Century a.d. In 1653, Raja Dip Chand (fifteenth of the line) shifted the capital from 'Sunhani' and founded a new one at 'Vyaspur' in honour of sage Vyas, and the present name appears to be its derived form (Mamgain, 1975).

Due to the construction of Bhakra Dam, 256 villages of the district encompassing about 30,000 acres were submerged. As a result, New Bilaspur Township comprising of six sectors underneath with a total area of 8.8 sq km and at a height of 670 m amsl just above the old town of Bilaspur was established.

Geographical Extent Located at an altitude ranging between 290 – 1,980 m at a distance of 93 kms from the state headquarters between 31°12′20″ and 31°35′45″ north latitude and between 76°23′45″ and 76°55′40″ east longitude in the outer hills of Himalayas in the basin of river Sutlej, the district Bilaspur (comparatively the smallest district of H. P.) with an area of about 1,161 sq km is bounded on the north by Mandi and Hamirpur districts, on the west by Hamirpur and Una districts, on the south by Nalagarh area of Solan district, and on the east and north-east by Solan and Mandi districts (Figures. 1-3). As many as about 3.30 lakh people inhabit 1,016 villages in total under its three tehsils, Ghumarwin, Bilaspur Sadar and Jhanduta.

Land, Climate and People The district mostly hilly and undulated, stands divided into two natural parts by the Sutlej river. Geomorphology depicts lesser hills and comparatively wider valleys. The land comprises of small flat strands along the bank of the Sutlej river and other streams. The soil is mostly sandy and intermixed with patches of stiff clay. Mostly, forests are of deciduous type and lie in subtropical zone. Monsoon starts in July and lasts till about the third week of September. October and November constitute the post-monsoon season with an average annual rainfall amounting to 1373.3 mm. The maximum temperature reaches up to 44°– 45°C during summers (March-end of June) and it is an average minimum of 0°–12°C during winters (December–February). Humidities are high especially during the south-west monsoon season whereas the air is comparatively drier in the remaining part of the year.

Predominantly, the district population comprises of Hindus with Brahmins, Rajputs, Kanaits, Rathis and Jhinwars whereas Muslims, Sikhs, Christians and Buddhists are in minority. Caste system prevails. The people are hardworking (Pls. 3J-L; 4B-D; 6D, F), culturally bound, god fearing and superstitious (Plate No. 6A, C). Simplicity and honesty are the hallmarks of their identity. People speak 'Kahluri' (Bilaspuri) which is regarded as an offshoot of Punjabi and the dialects of the erstwhile states of 'Kahlur' and 'Mangal'.

Regarding social life, the joint family system prevails which of late appears to be least preferred due to the impact of modernisation. Typical houses of peasantry are two-storied and made of sun-dried bricks (Plate No. 5D, F) with inmates dwelling in the lower storey and the upper being used as a store room for agricultural produce. Thatched huts (Plate No. 5A, E) and concrete buildings (Plate No. 5B) are also other places of dwellings. Unlike cities, the village schools are small-sized (Plate No. 5C). For clothing, males wear dull coloured 'kurta' (shirt) (Plate No. 4A), 'kapleen' (loin-cloth), white 'pajama' (trouser), 'pahari jootas' (local shoes) and with least preference for turban or 'topi' (skull cap). Women generally wear 'kurti' (a frock reaching to the knees), 'suthan' (trouser), mostly coloured and embroidered 'dupatta' to form the head dress (Plate No. 4D-E; 5F), 'jooti' (local shoes), 'chak' a skull ornament, and 'gajroos' (for wrist) of silver or gold. As elsewhere among the Hindus, a widow puts on white clothes.

Food Habits Maize and rice are the chief staple foods. 'Potande', 'Babru', 'Ankaru', 'Bhalle' and 'Khichari' form the special dishes on festive occasions. Tea is consumed extensively with preference for 'chhach' (butter milk) and 'sharbat' (sweet water) in the morning during summers.

By and large, the rural populace enjoys three meals a day: 'datialu' (morning meal comprising of wheat 'roti' reserved from the evening repast), full meal comprising of maize 'roti', some 'dal' and butter-milk and supper with least preference for rice. The vegetables made of mustard leaves, potato, 'kachalu' (*Arum colocasia*) and 'patrodu' are usually preferred. Because of lesser availability of meat, the fish which is available in abundance is usually consumed throughout the year by certain classes of community. Fruits like mango, 'papita', peach, orange and banana are grown in the district for wider consumption.

RIVER SYSTEM

Sutlej The district has a network of streams that feed the only river Sutlej (Plate No. 2F) of the area. This river enters at village Kasol in the north-east of the district and after coursing for about 90 kms with average width of 45 m at a speed of 8 kms an hour, leaves it at village Neila in the south-west before entering Punjab. Principal tributaries that join along its course are Ali, Gamrola (Plate No. 1D) and Ghambar streams from the south-east and Moni and Sir/Ser streams from the north-west (Figure. 3).

Ali Khad It is a perennial stream that enters the district at villages Kothi Harrar and Manothi (Pargana Bahadurpur). It has its source in the snowless mountains near village Mangu Giyani, Tehsil Arki of Solan district. After coursing for about 35 kms in the district territory, it joins Sutlej river at Berry Ghat, a point in between the hamlets of Kherain and Talwar (Plate No. 1A-C). One of its tributary is named as Tatoh Khad (Plate No. 2L).

Ghamber Khad It is the prettiest Himalayan stream that has its source below Tara Devi in the Shimla hills (Plate No. 1E-F). After entering the district at village Neri of 'Pargana' Ratanpur, it finally joins the Sutlej at villages Dagran and Nerli at a distance of about 13 kms.

Sir, Ser or Seer Khad Taking its rise in Mandi district, it drains the Kot-ki-Dhar and a greater portion of the Ghumarwin tehsil and considered as the largest tributary of the Sutlej river (Plate No. 2H-I). During its course through the district, this is joined by Lanjhta (Plate No. 1K), Lyond (Pl, 1L), Marol (Plate No. 2A), Marsand (Plate No. 2B), Raul (Plate No. 2C) and Sauli (Plate No. 2G).

Suker Khad and Saryali Khad Comparatively, these two khads are small-sized and rise in Hamirpur district (Plate No. 2D, E, J, K). After joining the Sir/Ser or Seer Khad at village Balgar and draining the western portion of the district, it joins the Sutlej at village Seri (16 kms below Bilaspur) and Matla after coursing for about 65 kms in the district.

LAKE

Gobind Sagar Lake Credited as one of the biggest man-made reservoir of water on the river Sutlej in India, it came into existence in the year 1962 in the Shivalik hills of Una and Bilaspur district of Himachal Pradesh (Figure. 4; Plate No. 1G-J). Encompassing a total area of 168 sq kms, its length is about 90 kms in both the districts. Fourteen villages (Dhiungli, Tunglehri, Brahmani Khurd, Tahra No. 27, Bhater Kundu, Chouki, Kashneur, Bedla, Dibru, Chohal, Gahral, Kuthera, Bheri, Kandwari) of the district lie submerged under the water of Gobind Sagar.

Tanks   There are several man-made water tanks. These are Jagatkhana, Kothi Babdi, Kuala, Riwalsar (Plate No. 3A-D), etc., and used for storage and supply of drinking water.

Agriculture   Agriculture, horticulture and animal husbandry are the mainstay of the people of this area. The cultivated lands comprise terraced fields and are usually irrigated by 'kuhls' (irrigation channels). Both traditional and modern agricultural practices are adopted for the cultivation of autumn ('kharif') and spring ('rabi') crops with produce comprising of maize, paddy, ginger, sugarcane, 'kulth', 'mash', and 'oil seeds'. For various agricultural operations, 'hal' (plough), 'deh' (leveller for breaking of clods), 'mesh', (leveller for puddles in paddy fields), 'dendal' (harrow), 'kudal' (hoe), 'karandi' (trowel) and 'pharwa' (spade) are extensively employed. Livestock commonly comprise cattle, buffaloes and cow (Plate No. 3F, G). Local plough is made from the wood of 'Shisham' (*Dalbergia sissoo Roxb.*) whereas for 'Jungda' (Plate No. 4A), (a device for holding together animals during ploughing operations) Salix is extensively employed. In view of the hilly terrain of the district, the irrigation very largely depends on rainfall. Women of the district also lend a helping hand in all agricultural activities besides domestic chores (Plate No. 3H-L; 4E; 6D).

Economy   By and large, the economy is based on agriculture. Some cottage industries like handloom weaving, leather work, carpentry, basketry, pottery, mat making, etc., also contribute to economic upliftment of the rural populace (4B-D, F). As a matter of fact, the economy of the district has received a further boost due to abundant reserves of good qualilty limestone being excavated at Harkhar, Laghat, Panjgino and Kothipura and supplied to nearby Barmana cement plant, and also due to catch of huge fisheries wealth (Plate No. 6E, F) of 'Mahsir' (*Tor pitutora*), 'Gungli' or 'Saloh' (*Sehize-thorase plagiestemus*), 'Gid' or Himalayan Barbel (*Labeo dero*) and other *Labeo* species with annual turn out of more than 1,100 tons.

Fairs and Festivals   Fairs like Nalwari (a state fair held every year in the month of March for trading cattle and lasts for eight days), Baisakhi at Markand Jukhala, Gugga at Gherwin, Naina Devi (three fairs held every year) are important not only from business point of view but also for pilgrimmage interaction and rejoicing, and these also keep people busy throughout the year. As true for Hindus elsewhere, the festivals like Holy, Diwali, Rakshabandhan, etc., are celebrated with great enthusiasm.

Recreational Sports   Gobind Sagar reservoir in the district offers a variety of water sports like swimming, boating (Plate No. 6B), surfing, water-skiing, kayaking, rowing, canoeing, white water river rafting in collaboration with the Directorate of Tourism and Civil Aviation and Directorate of Mountaineering and Allied Sports from August-January. Besides, 'chhinj' (wrestling), a traditional outdoor game of adults and boys is widely prevalent.

Places of Interest   Bhakra Dam (the highest gravity dam in the world, situated in Naina Devi sub-tehsil at a distance of 14 kms from Nangal town), New Bilaspur Town (first planned hill town of the country), Kandrour Bridge (Asia's highest bridge), Namhol (famous for Thakurdwara built by Raja Amar Chand in 1883 A.D. and as main centre of ginger trade), Gobind Sagar lake (a largest man-made reservoir on the river Sutlej) (Figure. 4), Deoth Sidh or Baba Balak Nath temple (40 kms from Bilaspur near Chakmoh), Markandeya shrine (a place about 20 kms from Bilaspur and known for its ancient water spring possessing miraculous healing properties against skin ailments), Shri Naina Devi Ji (a famous temple built by Raja Bir Chand in the 8th century about 15 kms from Ganguwal and about 18 kms from Anandpur Sahib), Vyas Gupha (a place of pilgrimmage where Vyas Rishi of Mahabharta had meditated) and Deoli Fish Farm (Plate No. 3E) are some of the places of interest for tourists and Pilgrimage.

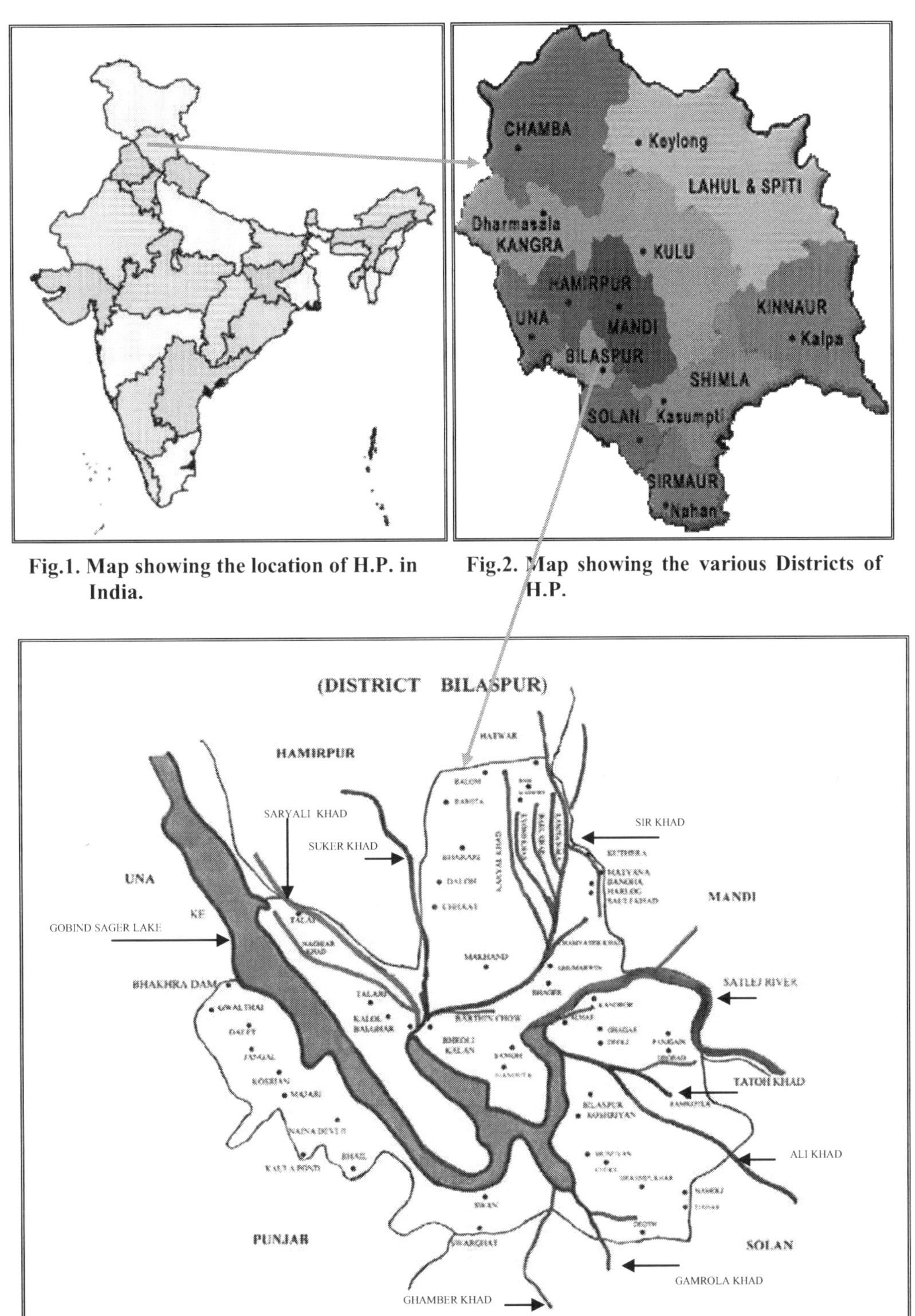

Fig.1. Map showing the location of H.P. in India.

Fig.2. Map showing the various Districts of H.P.

Fig. 3: Map showing wetland areas of District Bilaspur.

Figure 4 Satellite picture showing Gobind Sagar Lake, adjoining villages and landmasses of district Bilaspur

Collection of Ethnobotanical Data

For documenting first hand information pertaining to ethnobotanical explorations on the wetlands/wetland associated plants of district Bilaspur, Himachal pradesh, India, intensive and extensive surveys were undertaken in various wetlands and villages located along the various streams/khads and the river Sutlej traversing its territory in different seasons during the years 2006-2010 (Table I). The field surveys were planned in such a way so as to collect ethnobotanically interesting wetland species either in flowering or fruiting stage. The desired information on ethnobotanical aspects was collected through interviews from knowledgeable people (traditional practitioners, old experienced farmers, family heads, house wives, eminent elderly persons of the community, etc.), as per the methodology outlined by Jain (1987). The data thus collected were verified in villages among the interviewers after showing the same plant sample and even to the same informants on different occasions. Ethnobotanical lore pertaining to wetlands/wetland associated plants was considered valid if at least three informants made similar comments about the uses.

Table I   Flowering and Fruiting Seasons of the Ethnobotanically Important Wetland Plant Species Collected from the Study Area

Name of the Species	Flowering and Fruitng Season
Abelmoschus crinitus Wall.	July-September
Abrus precatorius L.	July-August
Abutilon indicum L. Sweet	July-October
Achyranthes aspera L.	May-October
Acorus calamus L.	June-July

Name of the Species	Flowering and Fruitng Season
Adiantum capillus-veneris L. f.	May-October
Adiantum incisum Forsk.	June-December
Aerva sanguinolenta (L.) Blume	July-October
Aeschynomene aspera L.	October-December
Ageratum conyzoides L.	August-October
Ajuga bracteosa Wall. *ex* Benth.	January-March
Amaranthus gangeticus L.	July-September
Amaranthus paniculatus L.	September-November
Amaranthus tricolor (L.) var. *gangeticus* (L.) Fiori	June-November
Amaranthus viridis L.	May-June
Ampelopteris prolifera (Retz.) Copeland.	August-October
Andrographis paniculata (Burm. f.) Wall. *ex* Nees	October-December
Anisomeles indica (L.) O. Kuntze	September-December
Apluda mutica L.	September-November
Argemone mexicana L.	April-May
Artemisia indica Waldst. & Kit.	August-October
Artemisia scoparia Waldst. & Kit.	August-September
Bacopa monnieri (L.) Pennell	August-October
Barleria cristata L.	July-October
Begonia picta Sm.	August-September
Bidens pilosa L.	June-September
Blyxa auberti Rich.	May-October
Boehmeria platyphylla Don	July-September
Bromus catharticus Vahl	July-October
Bryophyllum calycinum Salisb.	July-November
Calamintha umbrosum (M.B.) Fisch. & C.A. Mey	April-August
Canna indica L.	July-September
Cannabis sativa L.	June-September
Capsella bursa-pastoris (L.) Medik.	April-October
Cardiospermum halicacabum L.	August-March
Cassia absus L.	July-September
Cassia occidentalis L.	May-September
Centella asiatica L.	April-May
Ceratophyllum demersum L.	October-February

Name of the Species	Flowering and Fruitng Season
Cheilanthes bicolor (Roxb.) Fraser-Jenkins.	July-October
Chenopodium album L.	November-April
Chenopodium ambrosiodes L.	June-September
Chenopodium murale L.	January-April
Christella dentata (Forsk.) Brownsey & Jermy.	May-October
Cissampelos pareira L.	April-August
Clematis gouriana Roxb.	June- September
Coix lachryma - jobi L.	November-February
Colocasia antiquorum Schott	June- October
Commelina diffusa Burm. f.	July-September
Commelina paludosa Blume	June-August
Convolvulus arvensis L.	December-March
Conyza bonariensis (L.) Cronquist	June-october
Conyza stricta Willd.	September-October
Costus speciosus (J. Konig) Sm.	July-August
Crotalaria alata Buch.-Ham.	July-November
Crotalaria mysorensis Roth	July-October
Cryptolepis buchanani Roem. & Schult.	April-November
Cucumis pubescens Willd.	May-August
Curcuma longa L.	April-June
Cyathocline purpurea (Buch.- Ham. *ex* D. Don) Kuntze	April-May
Cymbopogon citratus Stapf	June -August
Cymbopogon martinii (Roxb.) Wats.	September-February
Cynodon dactylon (L.) Pers.	August-September
Cyperus compressus L.	August-December
Cyperus distans L.f.	August-December
Cyperus flabelliformis Rottb.	August-December
Cyperus iria L.	July-October
Cyperus rotundus L.	July-October
Datura stramonium L.	June-October
Debregeasia hypoleuca Wedd.	January-April
Dichanthium annulatum (Forssk.) Stapf.	September-December
Dicliptera roxburghiana Nees	October-December
Digitaria griffithii (Hook.f.) Henn.	October-December
Dioscorea belophylla (Prain) Voigt. *ex* Haines.	September-November

Name of the Species	Flowering and Fruitng Season
Dioscorea bulbifera L.	September-February
Echinochloa frumentacea Link	July-February
Eleusine indica Gaertn.	August-November
Emilia sonchifolia (L.) DC. DC.	December-April
Equisetum arvense L.	July-October
Equisetum debile Roxb.	August-December
Eriophorum comosum Wall.	July-September
Eupatorium adenophorum Spreng.	February-May
Euphorbia geniculata Ort.	August-November
Euphorbia helioscopia L.	March-August
Euphorbia hirta L.	March-August
Euphorbia parviflora L.	May-June
Ficus hispida L.f.	March-May
Ficus roxburghii Wall.	April-August
Fragaria indica Andr.	June-August
Fragaria nubicola (Hook.f.) Lindl.*ex* Lacaita	May-October
Fumaria indica (Hausskn.) Pugsley.	February-March
Galium aparine L.	July-August
Geranium nepalense Sweet	June- September
Girardinia heterophylla Decne.	September-October
Gloriosa superba L.	July-September
Gnaphalium pensylvaticum Willd.	December-May
Gomphrena celosioides Mart.	January-April
Hedera helix Clarke	June-November
Hedychium spicatum Buch.-Ham. *ex* Sm.	July-September
Heteropogon contortus (L.) Beauv. *ex* Roem. & Schult.	September-December
Hydrilla verticillata (L.f.) Royle	January-April
Impatiens balsamina L.	July-September
Ipomoea cairica (L.) Sweet	May-July
Ipomoea carnea Jacq.	July- September
Ipomoea muricata (L.)Jacq.	June-August
Ipomoea nil (L.) Roth	June-September
Ipomoea pestigridis L.	September-December
Justicia simplex Don	June-August
Lactuca dissecta Don	May-September

Name of the Species	Flowering and Fruitng Season
Lannea coromandelica (Houtt.) Merr.	March-June
Lantana camara L.	April-June
Lathyrus aphaca L.	April-May
Leea crispa L.	August-September
Lepidagathis cuspidata Nees	December-June
Lespedeza sericea Miq.	November-February
Lindernia ciliata (Colsm.) Pennell	August-February
Macrotyloma unifloruma (Lam.) Verdc.	July-september
Marchantia palmata Nees	January-April
Marsilea minuta L.	November-April
Martynia annua L.	August-November
Medicago denticulata Willd.	February-April
Melothria heterophylla Pittier	May-October
Mentha longifolia L.	November-February
Mentha piperita L.	July-September
Micromeria biflora (Buch.-Ham.) Benth.	January-July
Mirabilis jalapa L.	June-September
Momordica dioica Roxb. *ex* Willd.	July-September
Mucuna pruriens (L.) DC.	July-September
Najas graminea Dd.	August-October
Najas indica (Willd.) Cham.	August-September
Nasturtium officinale R. Br.	April-September
Nerium indicum Mill.	April-June
Ocimum basilicum L.	June-September
Oroxylum indicum Vent.	June-August
Oxalis corniculata L.	April-November
Parthenium hysterophorus L.	August-September
Paspalum distichum L.	August-October
Pennisetum lanatum Klotzsch	June-October
Phragmites karka (Retz.) Trin. *ex* Steud.	May-June
Phyllanthus urinaria L.	July-September
Physalis longifolia Nutt.	July-October
Physalis minima L.	June-August
Plectranthus coetsa Buch.-Ham. *ex* D. Don	October-November
Plumbago zeylanica L.	October-March

Name of the Species	Flowering and Fruitng Season
Poa supina Schrad.	August-September
Pogostemon plectranthoides Desf.	February-April
Polygala arvensis Willd.	July-December
Polygonum barbatum L.	April-July
Polygonum barbatum L. sub sp. *gracile* Dansar.	April-July
Polygonum donii Meisn.	April-July
Polygonum glabrum Willd.	January-September
Polygonum lapathifolium L.	November-February
Polygonum minus Huds.	January-December
Polygonum plebejjum Br. Prodr.	October-April
Polygonum pulchrum Blume	October-March
Polygonum serrulatum Lag.	June-October
Potamogeton crispus L.	December-June
Potamogeton pectinatus L.	October-January
Pteris cretica L.	May-October
Pupalia lappacea Juss.	July-September
Reinwardtia indica Dumort.	February-April
Rhynchoglossum obliquum Blume	July-August
Ricinus communis L.	March-May
Roylea cinerea (D. Don) Baill.	May-October
Rubus ellipticus Sm.	January-May
Ruellia patula Jacq.	January-April
Rumex hastatus D. Don	November-April
Rumex nepalensis Spreng.	May-October
Rungia pectinata (L.) Nees	January-February
Saccharum munja Roxb.	January-April
Saccharum spontanium L.	November-December
Salix oxycarpa Anderss.	April-June
Saussurea heteromella (Don) Hand.-Mazz.	April-June
Selaginella chrysocaulos (Hook. & Grev.) Spring	August-October
Setaria tomentosa Kunth	January-May
Sida acuta Burm.f.	September-December
Silene conoidea L.	March-May
Solanum nigrum L.	January-July
Solanum xanthocarpum Schrad. &Wendl.	March-April

Name of the Species	Flowering and Fruitng Season
Sonchus arvensis L.	May-June
Sonchus asper Hill	March-April
Sorghum halepense (L.) Pers.	July-October
Spilanthes acmella L.var. *oleracea* Jacq.	October-March
Spilanthes paniculata Wall. *ex* DC.	October-March
Stellaria media (L.) Vill.	March-April
Strobilanthes atropurpurens Nees	October-February
Strobilanthes dalhousianus Clarke	July-October
Taraxacum officinale Webb.	March-November
Thalictrum reniforme Wall.	June-August
Thevetia neriifolia Juss. *ex* Steud.	July-September
Trichodesma indicum (L.) Lehm.	September-December
Tridax procumbens L.	July-August
Trifolium resupinatum L.	May-June
Trigonella pubescence Baker	March-May
Uraria picta (Jacq.) DC.	July-November
Urena lobata L.	July-September
Urtica dioica L.	June-October
Vernonia cinerea (L.) Less.	October-March
Veronica anagalis-aquatica L.	November-January
Vicia sativa L.	January-February
Viola pilosa Blume	February-April
Withania somnifera Dunal	September-October
Xanthium strumarium L.	September-November

ENUMERATION

The collected plant specimens were dried, preserved and mounted as per the known herborizing practices outlined by Jain & Rao (1977). Botanical identification of the selected species was first done with the help of regional floras (Chauhan, 1999; Chowdhery & Wadhwa, 1984; Collett, 1902; Dhiman, 1976; Hooker, 1872-1897; Nair, 1977; Polunin & Stainton, 1984, Stainton, 1988) and later carefully matched with the authenticated specimens at the herbaria of Botanical Surveys of India, Northern Circle and Forest Research Institute, Dehradun. Nomenclature of these taxa was confirmed from Bennet (1986) and Wielgorskaya (1995). Excepting for the minor modifications to conform to the present day circumscription, families and genera are delimited mainly after The Flora of British India (Hooker, 1872-1897). Economic valuation of all the presently recorded ethnobotanical species was also carried out to calculate the total importance values (TIV) as per detailed methodology outlined by Belal & Springuel (1996). The nativity of the species has been determined on the basis of Matthew (1969), Maheshwari & Paul (1975), Nayar (1977), Sharma (1984), Hajra & Das (1982), Saxena (1991), Pandey & Parmar (1994), Reddy et al. (2000), Reddy & Raju (2002), Reddy & Reddy (2004), Murthy et al. (2003), Negi & Hajra (2007) and Reddy (2008). For future reference and studies, voucher specimens of afore-mentioned taxa (WL-EBH-1010-1213) have been deposited in the Ethnobotanical Herbarium of Department Biosciences, Himachal Pradesh University, Shimla. Besides, colour photographs of the plant species in their natural habitat were also clicked.

Systematic enumeration of the plants is in alphabetical order of their botanical names, synonyms of the species followed by their respective families. Besides information on English, Sanskrit and Regional Names, distribution, description, reproductive cycle, habitat ecology, material examined, part/s used, folk use/s, active constituents, biological activity and uses in literature has also been furnished and presented, if available. The data gathered were screened with the help of available literature (Ambasta, 1986; Anonymous, 1948-76a; Chopra et al., 1956, 1969; Jain, 1991; Kirtikar & Basu, 1984; Maheshwari & Singh, 1984; Singh et al., 1987; Singh, 1988), besides many other books and articles published in different journals.

CHAPTER 2

WETLAND PLANTS FOR SUSTENANCE OF MANKIND

Abelmoschus Crinitus **Wall. (Plate No. 7A)**

Syn.   *A. racemosus* Wall.; *Bamia cancellata* Wall.; *B. fusiformis* Wall.; *Hibiscus racemosus* Lindl.; *Hibiscus cancellatus* Roxb.

Family   Malvaceae

Vern.   Beuli

English, Hindi and Regional Names

    Eng.   Long haired abelmoschus;

    Hindi   Bankapas;

    B.   Bankapas, Birkapas.

Distribution   Mostly in Himalaya, N.W. India, W. Bengal.

Description   Annual herbs upto 1m tall with fusiform roots and cordate leaves which are obtusely 5-angled and crenately toothed. Stipules linear. Peduncles much shorter than the petiole, racemose at the ends of the branches each with 2 bracts at the base. Bracteoles linear-setaceous. Flowers yellow with a purple centre. Anthers scattered. Capsule ovoid, furrowed. Seeds numerous, reniform.

Flowering & Fruiting   July-September.

Habitat Ecology   Grassy meadows, Open slopes; **Papral Khad**-740 m.

Material Examined   EBH-WL-1010; 22.07.2008.

Parts Used	Whole Plant. Root. Tubers.
Folk Uses	2-5 g of powdered plant given twice a day for 3-5 days to check **dysentery**. Herb also yields a fibre employed for **ropes** and **cordage**. Roots **edible** and its paste heals fresh **cuts**. Paste of tubers crushed in butter milk with seeds of *Crocus sativus* L., *Cucumis melo* L., *Sorghum bicolor* L. given orally as well as its decoction sprayed over the sheep and goats head for **sexual arousal**.
Biological Activity	Plant exhibits antimalarial and antimicrobial activity.
Uses in Literature	Known to be used in India as an aphrodisiac, edible and for diarrhoea and dysentery (seeds, tubers) **(Bhogaonkar & Kanerkar, 2007; Jain, 1964, 1991; Jain & Tarafder, 1970; Pal & Srivastava, 1976; Pande et al., 2006; Reddy & Raju, 2000; Saxena, 1986; Sinha, 1996; Sood & Thakur, 2004; Trivedi et al., 1985).**

Abrus precatorius L. (Plate No. 7B)

Syn.   *A. laevigatus* Meyen; *A. melanospermus* Hassk.; *A. acutifolius* Blume; *A. pausifolius* Desv.; *A. minor* Desv.

Family   Fabaceae

Vern.	Raten

English, Hindi, Sanskrit and Regional Names

Eng.	Crab'eye, Indian liquorice, Jeriquity, Papernoster pea, Rosary pea, Red bead vine, Weather plant;
Hindi	Chirmiti, Gnunghachi;
Sans.	Angaravallari, Aruna, Bhilabhushna, Chatak, Gunja, Kanchi, Rakta, Ratika;
Ass.	Gunja, Krishanala, Kakachinchi;
B.	Chunhati, Gunj, Kawet;
G.	Chanoti, Gunja;
Kan.	Guengungi;
Kash.	Shangir;
Mal.	Irattimadhuran, Shekkunni;
Mar.	Chanoti;
Oriya	Kainsho, Kotibepolo;
P.	Labri, Ratak;
Tam.	Adisamiyai, Atti, Ghurie-ghenza;

Tel.	Kukkutamu, Sinnaguruginja.
Distribution	Exposed areas throughout India below 1,200 m.
Description	A copiously-branched climber with slender branches and 3-4 long leaves. Leaflets 24-30. Racemes more numerously flowered on longer peduncles. Calyx and corolla similar. Pods incurved with smooth valves.
Flowering & Fruiting	July-August.
Habitat Ecology	Moist slopes, common; **Karyal Khad** - 615 m.
Material Examined	EBH-WL-1015; 20.07.2008.
Parts Used	Roots. Leaves. Seeds.
Folk Uses	Chewed leaves applied on **cuts** and **wounds**; also 2-3 tsp of its juice taken twice daily for 3 days to cure **cough**. Powdered roots used as **contraceptive**. 20 ml decoction of its roots and leaves good against **fever** and **headaches**. Paste of seeds applied locally in **sciatica**, **stiffness** of shoulder joints and **paralysis**. 1-2 g powdered seeds given with fodder as remedy for curing **anthrax**. Bruised seeds used criminally for **poisoning** cattle and **homicidal** purposes.
Chemical Constituents	Chief ingredients are hypophorine, trigonelline, methyle ester of N, N-di Me-tryptophan methocation and precatorine (leaves), glychyrrhizin, pinitol, crystalline compound, flavonoids, triterpene glycosides-abrusosides A-D; 4,5,7-trihydroxyflavon and taxifolin-3-glucoside (leaves, roots), precol, glycyrrhizin, abrasion, precasene, isoflavion-quinones I, I I, I I I, (roots), abrin A and B, abrincin, abridin, abric acid, hypophorine, alainine, serine, abralin, abrine and abranin (seed), xylolucosy ledelphinidin and p-coumarylgalloyl-glucosldelphnidin (seed coat), galactose, arabinose and xylose (leaves, stems, roots, seeds).
Biological Activity	Abortifacient, alexeteric, analgesic, antiestrogenic, antifertility activity, antitumourous, anthelmintic, aphrodisiac, cerebrotonic, cicatrizant, CNS-depressant, spasmogenic activities found to be (+)ve.
Uses in Literature	Used earlier as blood purifier, cytotoxic, diuretic, emetic, purgative, tonic, uterotonic, vermifuge; and for abortion, conception, contraception, homicidal purposes, impotency, leprosy, leucorrhoea, migraine, nervous disorders, night blindness, poisoning cattle, removing maggots in wounds, sore mouth, skin diseases, stomach pain, syphilis and white leucorrhoea **(Ambasta, 1986; Asolkar et al. 1992; Balu et al., 2000; Bhat et al., 2002; Bhogaonkar & Kanerkar, 2007; Biswas et al., 2010; Chauhan, 1999; Dager & Dager, 1996; Dwarakan & Alagesaboopathi, 2000; Deshmukh et al., 2000; Duke et al., 2002; Goel & Rajendran, 2000; Henry, 1999; Hosagoudar & Henry, 1996a; Jadhav, 2009; Jain, 1991, 1996; Jain, et al., 1991, 2010; Kaushik & Dhiman, 2000; Khan &**

Khanum, 2000; Khare & Khare, 2005; Kumar, 2002; Lindley, 1981; Maheshwari et al., 1996; Mandal & Basu, 1996; Masih, 2003; Meena & Yadav, 2010; Mishra, 2008; Nag et al., 2009; Noumi, 2010; Painuli & Maheshwari, 1996; Pande et al., 2006; Parabia & Pathak, 2007; Parkash & Aggarwal, 2010; Prajapati et al., 2006; Rana et al., 2003; Rao & Henry, 1995; Reddy & Raju, 2000; Retnam & Martin, 2006; Samwatsar & Diwanji, 2000; Sarkar et al., 2000; Saren et al., 2000; Satapathy, 2008; Sharma, 1996; Sharma & Malhotra, 1984; Saini, 1996b; Singh, 2000; Singh & Pandey, 1996; Singh et al., 1996a; Siwakoti & Siwakoti, 2000; Siwakoti & Varma, 1996; Sood & Thakur, 2004; Thakor, 2009).

Abutilon indicum L. Sweet (Plate No. 7C)

Syn.   *A. asiaticum* (L.) Sw.; *A. guineense* (Schumach.) Baker f. & Exell; *Sida guineensis* Schumach.

Family   Malvaceae

Vern.	Jangli bhindi.

English, Hindi, Sanskrit and Regional Names

Eng.	Country mallow;
Hindi	Jhampi, Khangani, Kanghi;
Sans.	Atibala;
B.	Petari;
Kan.	Tutti;
Mal.	Velluram;
Tam.	Perum tutti;
Tel.	Adavibanda, Botlabenda, Tutti.
Distribution	Sub-Himalayan tract and hills upto 1,200 m and hotter parts of India.
Description	Softly tomentose perennials up to 3 m. Stems round, frequently tinged with purple. Leaves ovate to orbicular-cordate. Flowers solitary on jointed peduncles, orange-yellow. Capsules hispid, awns erect. Seeds 3-5, reniform, black or dark brown.
Flowering & Fruiting	July-October.
Habitat Ecology	Found near water; **Chamyater Khad** - 613 m.
Material Examined	EBH-WL-1022; 18.08.2009.
Parts Used	Roots. Leaves.

Folk Uses	About 10 g crushed leaves prescribed for post-delivery complications, especially to get rid off remains of **placenta** after birth; its decoction also considered good against **toothache, tender gums** and **inflammation of bladder**. Roots used in **jaundice**.
Chemical Constituents	Found to possess mucilage, tannin, organic acid, traces of asparagin and ash containing alkaline sulphates, chlorides, magnesium phosphate and calcium carbonate **(leaves)** and asparagin **(roots)**.
Biological Activity	Plant shows analgesic, hepatoprotective and hypoglycaemic activity.
Uses in Literature	Reported earlier in Indian literature as astringent, demulcent, expectorant, diuretic, laxative; and for boils, chest infection, chronic cystitis, fever, fomenting painful parts of body, gleet and gonorrhoea, piles, ulcers and urethritis **(Ansarali & Sivadasan, 2009; Asolkar et al., 1992; Bhatt et al., 1999; Bhattacharyya, 1996; Biswas et al., 2010; Chatterjee & Pakrashi, 1997; Chopra et al., 1956; Kaushik & Dhiman, 2000; Lindley, 1981; Mishra, 2008; Parrotta, 2001; Retnam & Martin, 2006; Rout & Panda, 2010; Saini, 1996a; Satapathy, 2008; Sharma, 1996; Upadhye et al., 1994, Yadav et al., 2010).**

Achyranthes aspera L. (Plate No. 7D)

Syn.   *A. lanceolata* Wall.; *A. wightiana* Wall.; *Digera muricata* Mart.

Family   Amaranthaceae.

Vern.	Puthkanda, Lathjeera.

English, Hindi, Sanskrit and Regional Names

Eng.	Prickly chaff flower;
Hindi	Adhajhara, Chirchira, Lathjira, Orga;
Sans.	Adhoganta, Adhavashalya, Aghamargava, Aghata, Apamarga, Apangaka, Dhamargava;
Ass.	Apang;
B.	Apang, Chirchiti;
Bo.	Aghada;
G.	Aghedo;
Kan.	Uttarane;
Mal.	Katalati;
Mar.	Aghadha, Aghara;
Oriya	Apamaranga, Apamargo;
P.	Kutri;

Tam.	Chirukadaladi, Nayurivi;
Tel.	Apamarganu, Uttarn.
Distribution	An abundant weed throughout India.
Description	Erect stiff herbs with branched stem and thick, pubescent leaves. Petiole short. Flowers in slender or panicled spikes, pink or greenish white. Fruits deflexed.
Flowering & Fruiting	May-October.
Habitat Ecology	Waste places, roadsides, near water sources; **Chamyater Khad** - 613 m.
Material Examined	EBH-WL-1023; 12.09 2009.
Parts Used	Bark. Leaves.
Folk Uses	Young leaves cooked as **spinach**. Also, its poultice prepared in goat's milk recommended for healing **limb fracture** of domestic cattle. Likewise, paste of its leaves prepared with that of *Coccinia grandis* (L.) Voigt considered good for all types of **wounds**. Leaf juice as such or its root grounded with that of *Pergularia daemia* (Forsk.) Chiov and *Calotropis gigantea* (L.) Br. applied to heal maggot **wounds**. Decoction of root given twice daily for three days for **menstrual pain**. Bark good against **constipation**
Chemical Constituents	Important constituents are ecdysterone (polypodine A), two oleanolic acid based saponins, ecdysone **(roots)**, ecdysterone **(leaves)**, glycosides, saponins, amino acids, fatty acids, hormone cellysone and aliphatic dihyroxyketone **(plant)**.
Biological Activity	Saponins reported to increase tone of hypodynamic heart and force of contraction of failing papillary muscle.
Uses in Literature	Reported so far for bleeding after termination of foetus, bleeding piles, blood dysentery, chicken pox, cholera, conjunctivitis, cough, cuts and wounds, deafness, dermal diseases, diarrhoea, earache, fever, foot cracks, gastric troubles, haemorrhoids, headache, insect bites, leucorrhoea, malaria, measles, migraine, night blindness, piles, relieving pain during delivery, removal of spines, rheumatism, scorpion and snake bites, teeth infection, tumours, ulcers and urinary troubles **(Ahmed & Borthakur, 2005; Ambasta, 1986; Asolkar et al., 1992; Bhogaonkar & Ahmed, 2007; Bhogaonkar & Kanerkar, 2007; Biswas et al., 2010; Chauhan, 1999; Dwarakan & Ansari, 1996; Duke et al., 2002; Gupta, 1997; Hosagoudar & Henry, 1996a; Jadhav, 2009; Jain, 1991; Jain & Singh, 1997; Jain et al., 1997, 2010; Jha et al., 1996; Kamble et al., 2010; Kapur & Singh, 1996; Kaushik & Dhiman, 2000; Khan & Khanum, 2005; Lalramnghinglova, 2003; Maheshwari et al., 1996; Manandhar, 1996a; Martin & Retnam, 2005; Meena & Yadav, 2010; Mishra, 2008;**

Pande et al., 2006; Parabia & Pathak, 2007; Parrotta, 2001; Prajapati et al., 2006; Rana et al., 2003; Ramdas et al., 2000; Rao & Henry, 1995; Rastogi & Mehrotra, 1991; Retnam & Martin, 2006; Rout & Panda, 2010; Saini, 1996a; Samwatsar & Diwanji, 1996; Satapathy, 2008; Sharma, 1996; Shukla & Verma, 1996; Singh & Pandey, 1996; Singh & Rao, 2003; Solanki et al., 2007; Sood et al., 2009b; Swami & Gupta, 1996; Thakor, 2009; Yadav et al., 2010).

Acorus calamus L. (Plate No. 7E)

Syn.   *A. belangeri* Schott; *A. cassia* Bertol.; *A. griffiyhii* Schott; *A. nilaghirensis* Schott; *A. casia* Bertol.

Family   Araceae.

Vern.	Barae.

English, Hindi, Sanskrit and Regional Names

Eng.	Calamus, Sweet flag;
Hindi	Bach, Ghar, Ghorbach;
Sans.	Bhadra, Galani, Jalaja, Shadgrantha, Ugra, Vgragandhadha, Vacha;
B.	Bach;
G.	Gandhilovaj, Godavaj, Vekhand;
Kan.	Baji;
Mal.	Vayampu;
P.	Bariboj, Warch;
Tam.	Vashambu;
Tel.	Vadaja, Vasa, Wasa.
Distribution	Throughout India.
Description	Aromatic marshy herbs having creeping rootstock with leaves possessing stout midrib of wavy margins and equitant bases. Spathe in continuation of foliaceous peduncle. Spadix sessile, flowers bisexual, greenish yellow. Perianth concave. Stamens 6, filiform, linear filament and reniform anther. Ovary conical. Berries oblong.
Flowering & Fruiting	June-July.
Habitat Ecology	Frequent along banks of lakes and in marshy places; **Raul Khad** - 746 m.
Material Examined	EBH-WL-1024; 08.07.2008.
Parts Used	Root. Rhizome.

Folk Uses

Root paste applied on **cuts** and **toothache**; also its paste mixed with honey given to check **stammering** in children (½ teaspoon in the morning for 20 days). Paste of rhizome in mustard oil makes a good lotion for body **massage** in children and in adults to relieve **body pain**. Also, its paste in combination with equal quantity of mint and roasted fruits of *Terminalia chebula* given to cure throat **infection**. Smoke of burnt rhizome good for patients suffering from **epilepsy** and **hysteria**.

Chemical Constituents

Active properties due to acoradin 2,4,5-trimethoxybenzaldehyde, 2,5 dimethoxybenzoquinone, galangin and sitosterol **(rhizome)**, β-asarone, myristic (1-3), palmitic (18.2), palmitoleic (16.4), stearic (7.3), oleic (29.1), linoleic (24.5) and arachidic (3.3%) acids **(essential oil rhizome)**, vit-c, luteolin-6- and 7-diglucoside **(leaves, rhizome)**, calamarone, calamendiol and isocalamendiol **(roots)**.

Biological Activity

Rhizomes CNS depressant and spasmogenic.

Uses in Literature

Recorded in India as analgesic, anthelmintic, antibacterial, antigonadotropic, antipyretic, antiseptic, antispasmodic, anxiolytic, aphrodisiac, carcinogenic, carminative, CNS- sedative, diaphoretic, emmenagogue, expectorant, insecticidal, insectifuge, hallucinogen, hepatosis, hypotensive, larvicidal, laxative, negative inotropic, neurotonic, ovicide, tonic and tranquilizer; and for asthma, body ache, bronchitis, cholera, cold, cough, cuts, dysentery, epilepsy, fever, malaria, pain in neck, paralysis, pthisis, skin diseases, snake bite, stomach ache, teething in children, tuberculosis, tumour, typhoid, ulcer, uterosis, vaginosis, varicosis, vertigo, water retention and wound **(Asolkar et al., 1992; Biswas & Calder, 1984; Biswas et al., 2010; Chak & Sharma, 1965; Dam & Hajra, 1997; Dandiya & Sharma, 1961, 1962; Duke et al., 2002; Gupta, 1997; Henry, 1999; Islam, 1996; Jain, 1991, 1996; Kaushik & Dhiman, 1984; Kaul, 1997; Khan, 2003; Kharkongor & Joseph, 1997; Kumar, 2002; Kumar & Narain, 2010; Lindley, 1981; Malla & Chhetri, 2009; Martin & Retnam, 2005; Nath & Bordoloi, 1989; Panda, 1996; Pande et al., 2006; Parrotta, 2001; Prajapati et al., 2006; Rana et al., 2003; Rastogi & Mehrotra, 1995b; Saini, 1996b; Saren et al., 2000; Sharma, 2000; Singh, 2000, 2003c; Singh et al., 1996b; Siwakoti & Siwakoti, 1999; Siwakoti & Varma, 1996; Sood & Thakur, 2004).**

Adiantum capillus-veneris L. f. (Plate No. 7F)

Syn.   *A. wattii* Baker

Family   Adiantaceae.

Vern.	Vikrantaa.

English, Hindi, Sanskrit and Regional Names

Eng.	Maiden hair fern, Rock fern, Venus-hair fern;
Hindi	Hansraj, Mubaraka, Pursha;
Sans.	Astmabayda;
B.	Hangsapadi;
G.	Hanspadi;
Kan.	Hansraj;
Kash.	Dumtuli, Geutheer.
P.	Buikallan;
Tam.	Sirupoolai.
Distribution	Mainly in W. Himalaya; ascending upto 2,400 m.
Description	Delicate ferns having 10-23 cm long, suberect, glabrous stipes and bi-pinnate fronds with short terminal pinna. Lamina deltoid, ovate, 2-3 pinnate. Pinnae widely spaced, 8-13 pairs, alternate, petiolate; venation cyclopteridian. Sori borne at the roundish sinuses of the crenations, obreniform, 2mm in diameter. Sporangia shortly-stalked, globose. Spores tetrahedral, hyaline, 36-50 μm in diameter.
Flowering & Fruiting	May-October.
Habitat Ecology	Walls and rocks along damp places, especially along streams; **Kothi Khad** - 810 m; **Chamyater Khad** - 613 m; **Raul Khad** - 646 m; **Sir Khad** - 620 m.
Material Examined	EBH-WL-1014; 02.07.2008.
Parts Used	Whole Herb. Fronds.
Folk Uses	15-25 ml decoction of the fern good against **cough**, shortness of **breath, jaundice**, diseases of the **spleen, kidney stones** (twice daily till cure). Paste of fronds good for chronic **rheumatism** and **gouty** swellings. Fern also used as a **flavouring agent** for tobacco and as a substitute for **tea**.
Chemical Constituents	Chief constituents of **fern** are flavonoids (rutin, isoquercitin), terpenoides (adiantone), tannins, mucilage, kaempferol, quercetol, astragalin,

lutcolol, rutin, triterpenoid, isoquercitrin, nicotiflorin, querciturone, narinyenin, hesperidin sulphuretin and genisterin.

Uses in Literature   Reportedly discovered in India as antibacterial, anticancerous, antifungal, anti-implantation, antipyretic, aphrodisiac, astringent, demulcent, depuative, diaphoretic, diuretic, emetic, emmenagogue, emollient, expectorant, hypoglycaemic, laxative, stimulant and tonic; and for asthma, bee and scorpion sting, breast cancer, centipede bite, constipation, cough, cytosis, dropsy, dysmenorrhoea, eye ailments, hard tumours, spleen, liver and kidney stone, liver and other viscera, menstrual regularization, mouth ulcers, pectoral, respirosis, rheumatism, rhinosis, rickets, sclerosis, snakebite and wound healing **(Ambasta, 1986; Asolkar et al., 1992; Brickell, 1996; Cauis, 2003; Chandra, 1997; Chatterjee & Pakrashi, 1997; Dixit & Singh, 2004; Dixit & Vohra, 1984; Duke et al., 2002; Guha Bakshi et al., 1999; Jain, 1991; Kaul, 1997; Kaushik & Dhiman, 2000; Khare, 1996; Khullar, 1984, 1994; Lalramnghinglova, 1996; Lindley, 1981; Manandhar, 1996b; May, 1978; Pande et al., 2006; Prajapati et al., 2006; Seth et al., 2002; Seth & Kumar, 2007; Singh, 2003b; Maheshwari et al., 1996; Singh & Viswanathan, 1996; Singh et al., 2007; Wangchuk et al., 2008).**

Adiantum incisum Forssk. (Plate No. 8A)

Syn.   *A. caudatum* L

Family   Adiantaceae

Vern.	Morpanki.

Hindi, Sanskrit and Regional Names

Hindi	Marshikha;
Sans.	Mayarshika, Nilkanthashikha;
B.	Mayurshikha;
P.	Adhsarita-ki-jhari; Gunkeri, Kanghai;
Tam.	Myle kondai.
Distribution	W. Himalaya, Tropical and S. Africa, Arabia.
Description	Leafy ferns with 5-10 cm long, dark chestnut-brown, tufted stipes. Fronds simply pinnate, often elongated and rooted at the apex. Sori borne at the edge of the lobes. Rhizome short and clothed with thin, brown, concolorous scales. Stipes proximate, 1.0-5 cm long, brown, hairy. Spores brown, 21-28x24-53µm, nonperinate, smooth.
Flowering & Fruiting	June-December.
Habitat Ecology	Shady places, rock-crevices, slopes and walls at elevation of 300-1,200m; **Panyala Khad** - 810m.

Material Examined	EBH-WL-1025; 02.07.2008.
Parts Used	Whole Plant. Fronds.
Folk Uses	Paste of frond good against **skin diseases**. 2-3 powdered plant given with luke warm water, once daily for **blood pressure**.
Chemical Constituents	**Plant** yields 1b-hentriacontanone, hentriaconetone, β-sitosterol, fernene, adiantone, isoadiantone steroids, triterpenoids and flavonoids.
Biological Activity	Et.OH (50%) extract of plant - hypoglycaemic and spasmogenic.
Uses in Literature	Plant useful in India for bone fracture, cough, diabetes, eye diseases, mouth blisters, pneumonia; and as antibacterial, anticancer, antipyretic, antitussive, aromatic, astringent, diuretic, emetic, febrifuge, hypoglycaemic and tonic **(Ambasta, 1986; Asolkar et al., 1992; Cauis, 2003; Chopra, 1933; Dixit & Singh, 2004; Guha Bakshi et al., 1999; Jain, 1991; Kaul, 1997; Khan & Khanum, 2005; Khullar, 1994; Lalramnghinglova, 1996; Lindley, 1981; Manandhar, 1996b; Pande et al., 2006; Rastogi & Mehrotra, 1995a; Seth et al., 2002; Seth & Kumar, 2007; Singh & Viswanathan, 1996; Singh et al., 2007; Subramaniam, 2000; Wangchuk et al., 2008).**

Aerva sanguinolenta (L.) Blume (Plate No. 8B)

Syn. *A. scandeans* (Roxb.) *Moq.; Achyranthes sanguinoenta* L.

Family   Amaranthaceae.

Vern.	Sufed-phulia.
Regional Name	
B.	Nuriya;
Distribution	Found throughout India; upto 1,800m in Himalaya.
Description	Erect or rambling perennial undershrubs with long grey tomentose branches and elliptic, acute, mucronate leaves. Flowers bisexual, silvery white, oblong, sessile. Fruits dry, acute. Seeds lenticular, black shining.
Flowering & Fruiting	July-October.
Habitat Ecology	Wastelands; **Matwana Khad** - 610m.
Material Examined	EBH-WL-1012; 03.10.2008.
Part Used	Twig.
Folk Use	A garland of twig of the plant around the neck of sick cattle believed to cure **digestive disorders**.

Uses in Literature	Discovered earlier for digestive disorders, dysentery; and as diuretic and demulcent **(Ambasta, 1986; Asolkar et al., 1992; Chandra, 1997; Jain, 1991; Kaushik & Dhiman, 2000; Maheshwari & Singh, 1992; Pande et al., 2006; Rana *et al.*, 2003; Retnam & Martin, 2006).**

Aeschynomene aspera L. (Plate No. 8C)
Family   Fabaceae

Vern.	Solu.

English, Hindi and Regional Names

Eng.	The sola plant;
Hindi	Sola;
B.	Phul-sola; Shola;
Kan.	Alaginagida, Bendu, Kerebondu, Tankali;
Mal.	Kadessum;
Tam.	Attuneddi;
Tel.	Neerujeelu-gabendu.
Distribution	Common in Bengal extending to the Malay Islands.
Description	Sparsely branched tall erect robust shrubs. Racemes corymbose, 2-6 flowered, axillary or terminal. Flowers pale yellow. Bracts ovate-cordate, pedicels and peduncles covered with sparse bristles. Pods smooth.
Flowering & Fruiting	October-December.
Habitat Ecology	Common in moist places; **Lyond Khad -** 617m.
Material Examined	EBH-WL-1026; 10.11.2008.
Part Used	Leaves.
Folk Use	Leaves **edible**.
Uses in Literature	Commercially important due to its use for shola hats and other articles manufactured from soft pith of the plant **(Biswas & Calder, 1984; Chandra, 1997; Das et al., 1996; Jain, 1991; Kumar & Narain, 2010).**

Ageratum conyzoides L. (Plate No. 8D)

Syn.   *A. cordifolium* Roxb

Family   Asteraceae

Vern.	Ukal Booty, Phulnu.

English, Hindi, Sanskrit and Regional Names

Eng.	Goat weed, White weed;
Hindi	Visadodi;
Sans.	Visamustin;
B.	Dochunty;
G.	Ajgandha, Gandharisedardi, Mankdamari;
Kan.	Uralgidda;
Mal.	Appa, Muryampacha, Siangit, Tahiayam, Tambak jantan;
Mar.	Ghaneraosadi;
Oriya	Boksunga, Poksunga;
Tam.	Pumpillu.
Distribution	Throughout India.
Description	Erect annual herbs possessing petioled ovate, alternate leaves and terminal corymb heads. Achenes black; pappus scales 5 awned, florets pale blue.
Flowering & Fruiting	August-October.
Habitat Ecology	Edges of cultivated areas, meadows, waste places, prominent along irrigated channels; **Sir Khad -** 620m.
Material Examined	EBH-WL-1135; 12.09.2008.
Parts Used	Whole Plant. Leaves.
Folk Uses	Snuff prepared by mixing its powdered leaf with black pepper checks **headache**. Poultice of leaves applied on **boils** and its juice good against **cuts, wounds, blisters, burns** and **bleeding of nose**. Latex used in ophthalmic **infections**. Plant extract used as **eye lotion**.
Chemical Constituents	Constituents isolated from different **plant parts** are ageratochromene, γ-cadinene, 6-demethyoxyageratochromene, β-caryophyllene, d-α-pinene, ocimene, d-cardinene, eugenol, methyleegenol, d-α-pinene, ocimene, d-calanene, methyeugenol in 7-methoxy-2,2-dimethyl-chromene, ageratochromene dimmer **(essential oil)**, saturated al-

iphatic hydrocarbons, α-spinasterol, stigmast-7-en-3ol, quercetin, kaempferol, fumarix, caffeic acids **(leaves)**, phenol and essential oil **(flower buds)**.

Biological Activity   Antibacterial and antifungal activities found to be positive.

Uses in Literature   Reported to be used in India as anthelmintic, antilithic, antiseptic, colic, diuretic, edible, emetic, flatulent, haemostatic, insecticidal, shampoo, symbolic; and for bed wetting, bleeding, body swellings, boils, burns, cancer, cuts, dysmenorrhoea, fever, gastrointestinal disorder, leprosy, leprous sores, piles, typhoid, ringworm, scabies, skin diseases, snake bite, uterine disorders and wounds **(Asolkar et al., 1992; Baruah & Sarma 1984; Bhalla et al., 1996; Biswas et al., 2010; Chaudhury & Neogi, 2000; Deokota & Chhetri, 2009; Hajra & Baishya, 1997; Idu et al., 2008; Jain, 1981, 1991; Jain et al., 2010; Jamir et al., 2008; Joshi, 2008; Kaushik & Dhiman, 2000; Kharkongor & Joseph, 1997; Kumar, 1996, 2002; Kumar & Jain, 1998; Lalramnghinglova, 1996; Mudgal et al., 1999; Pande et al., 2006; Parrotta, 2001; Patil et al., 2007; Prajapati et al., 2006; Rana et al., 2003; Rastogi & Mehrotra, 1995b; Rout & Panda, 2010; Sahu et al., 2009; Samwatsar & Diwanji, 1996; Sasidharan & Swarupanadan, 2000; Singh & Srivastava, 2000; Sinha, 1986; Siwakoti & Varma, 1996).**

Ajuga bracteosa **Wall.** *ex Benth.* **(Plate No. 8E)**

Syn.   *A. remota Benth*

Family   Lamiaceae

Vern.   Neelkanthi.

English, Sanskrit and Regional Names

    Eng.   Bungle;

    Sans.   Nilkanthi;

    Kash.   Jan-1-adam;

    P.   Khurbanti.

Distribution   W. Himalaya: Kashmir – Nepal. W. Bengal, Punjab, Gangetic Plains.

Description   Softly pubescent annual herbs with stem or branches arising from the rootstock. Leaves stout; lower petioled and upper ones sessile. Corolla lilac. Nutlets ellipsoidal.

Flowering & Fruiting   January-March.

Habitat Ecology   Exposed slopes, grasslands; **Karyal Khad -** 615m.

Material Examined   EBH-WL-1136; 05.02. 2009.

Part Used	Leaves.
Folk Uses	Plant juice mixed with juice of *Centella asiatica* (1:1 ratio) given to women in **gonorrhoea** and in **intermittent fever** accompanied with **shivering**. 15-20ml decoction of powdered dried leaves given twice daily as **blood purifier** and for **stomach pain**. Dried leaves immersed in mustard oil good against **sores**.
Chemical Constituents	These are triacontanyl hexacosanol, β-sitosterol and its glucoside tetracosanoic acid, ecdysteroids **(aerial parts)**, ceryl alcohol, γ-sitosterol, cerotic and palmitic acids alongwith a glucoside constituents **(leaves)**.
Biological Activity	Plant exhibits antiinflammatory, antiplasmodial and antipyretic activities.
Uses in Literature	Discovered earlier in Indian litrature as aromatic, astringent, ammenorrhoea, blood purifier, diuretic, febrifuge, stimulant, tonic; and for boils, burns, fever, gout, lice killer, rheumatism, scabies, swelling and worms **(Ambasta, 1986; Ansari, 1991, 1997; Arya & Goel, 2000; Arya & Parkash, 2000; Chauhan, 1999; Jain, 1991; Kaul, 1997; Kapur & Singh, 1996; Kapur et al., 1996; Kirtikar & Basu, 1984; Pande et al., 2006; Prajapati et al., 2006; Rana et al., 2003; Rastogi & Mehrotra, 1995a; Semwal et al., 2010; Singh & Rao, 2003).**

Amaranthus gangeticus L. (Plate No. 8F)

Syn. *A. melancholicus L.; A. oleraceus* Willd

Family Amaranthaceae

Vern.	Lal sag.

English, Hindi, Sanskrit and Regional Names

Eng.	Love-lies-bleeding, Red cocks-comb;
Hindi	Lalnattiya, Rajkili;
Sans.	Marishal;
B.	Banspatanatiya;
G.	Abdaudambho.
Distribution	Throughout India, Sri Lanka Tropical Asia, Africa & America.
Description	Erect herbs having stout stem possessing grooved branching and rhomboid-ovate, lanceolate leaves. Flowers clustered in the axils and forming a long terminal. Bracteoles subulate. Perianth long. Stamens 3. Capsules long, ovoid, rugose. Seeds lenticular, smooth, shining, black.

Flowering & Fruiting	July-September.
Habitat Ecology	Near water sources; **Sunhani Khad** - 625m.
Material Examined	EBH-WL-1099; 12.09.2008.
Parts Used	Whole Plant. Leaves.
Folk Uses	Plant cooked and eaten like **spinach** at the seedling stage. Poultice of leaves applied in **ulcerated** conditions of throat and mouth.
Chemical Constituents	Isolation of amaranthin, isoamaranthine, o-β-D-glucopyranosides of betanidine and isobetanidine **(leaves)** characterized.
Uses in Literature	Known to be useful as antipyretic, astringent, emetic, emmenagogue, expectorant; and for biliousness, burning sensations, diarrhoea, dysentery, fleshy tumours, haemorrhages, inflammations, liver complaints, menorrhagia, suppuration, toothache and vulnerary **(Guha Bakshi et al., 1999; Jha et al., 1996; Kirtikar & Basu, 1984; Rastogi & Mehrotra, 1995a; Uphof, 2001).**

Amaranthus paniculatus L. (Plate No. 9A)

Syn.   *A. speciosus* Sims; *A. sanguineus L.; A. stictus* Willd.; *A. frumentaceus* Ham.; *A. farinaceus* Herb.; *A. anacardaha* Ham.; *A. flavus var. bracteatus* L.

Family   Amaranthaceae

Vern.	Chollayi.
Hindi, Sanskrit and Regional Names	
Hindi	Chaulai, Chu, Chumarsa, Ganhar;
Sans.	Rahadrei, Rajagiri, Rajashkini;
B.	Bathu, Chuko, Natya;
G.	Chuko, Rajgaro;
Kan.	Kire, Soppu;
Kash.	Bhutanatroz;
Mar.	Rajagira;
Tam.	Pungi kirai.
Distribution	Throughout India.
Description	Tall robust annuals with striated stem and long-petioled, elliptic or ovate-lanceolate, acuminate leaves. Spikes sub-erect, red. Seeds yellowish - white or pitchy black.
Flowering & Fruiting	September-November.

Habitat Ecology	Commonly a weed in maize fields and moist places; **Marsand Khad** - 600 m; **Auhr Khad** - 610 m.
Material Examined	EBH-WL-1207; 15.09.2007.
Parts Used	Aerial Parts. Leaves. Seeds.
Folk Uses	Tender shoots and leaves used as **vegetable**. Seeds consumed as **food** by the poor; its flour especially eaten during **fast days**. 6g powdered seeds in divided dose given with cold water thrice a day for 5 days to check **diarrhoea**.
Chemical Constituents	These are rutin **(aerial parts, inflorescence)** linoleic, oleic, palmitic, vernolic, stearic, sterculic and malvalic acids **(seed oil)**, α-, β- and γ-amylases **(seeds)** and oxalic acid **(leaves)**.
Uses in Literature	Known in India for abscess, boils, cereals, gonorrhoea, piles, purifying blood, sores and as diuretic, laxative, vegetable and antidote for snakebites **(Agarwal, 2003; Ambasta, 1986; Arora & Pandey, 1996; Chopra et al., 1956; Guha Bakshi et al., 1999; Kirtikar & Basu, 1984; Mao, 1993; Pande et al., 2006; Pandey & Parmar, 1994; Rana et al., 2003; Roy et al., 1998; Samant et al., 2001a; Sharma & Rana, 2005; Sood & Thakur, 2004; Sood et al., 2001).**

Amaranthus tricolor (L.) var. *gangeticus* (L.) Fiori (Plate No. 9B)

Syn.   *A. tricolor* L.; *A. lanceolatus* Roxb.; *A. tristis* L.; *A. oleraceus* Roxb.; *A. polygamous* Roxb.; *A. lividus* Roxb.; *A. amboinicus* Herb.; *A. inamoenus* Willd.; *A. melacholicus* L.

Family   Amaranthaceae

Vern.	Lal Sag.

English, Hindi, Sanskrit and Regional Names

Eng.	Chinese spinach, Tampala;
Hindi	Lal sag;
Sans.	Tanduliya;
B.	Dengo, Lalshak;
Tam.	Serikkeerai;
Tel.	Chirakura.
Distribution	Throughout India and Sri Lanka; cultivated grounds.
Description	Stout, erect, leafy green herbs. Leaves variable, lanceolate, ovate. Bracts awned. Sepals long. Seeds lenticular, pitch-black.
Flowering & Fruiting	June-November.

Habitat Ecology	Plains, moist grounds, river banks; **Sir Khad -** 620m.
Material Examined	EBH-WL-1011; 03.08.2007.
Parts Used	Root. Leaves. Seeds.
Folk Uses	Leaves used as **vegetable** and **salad**. Fresh extract of root used in acute **diarrhoea** along with rice water. Seeds used in **leucorrhoea**.
Chemical Constituents	Fatty oils, protein, fat fibre, carbohydrate, minerals, calcium, phosphorus, iron, thiamine, riboflavin, niacin, vitamin, energy, carotene **(leaf, seed)**, arginine, cystine, histidine, leucine, isoleucine, lycine, methionine, phenylalanine, threonine **(leaves)** have been isolated.
Biological Activity	Plant exhibits antioxidant activities.
Uses in Literature	Used earlier in India as antidote, astringent, cooling agent; and for bowel, diarrhoea, dysentery, haemorrage, leucorrhoea, mouth sores, toothache, ulcerated throat and vomiting **(Guha Bakshi et al., 1999; Jain, 1991; Pande et al., 2006; Prajapati et al., 2006; Rao & Henry, 1995; Retnam & Martin, 2006).**

Amaranthus viridis L. (Plate No. 9C)

Syn.   *A. polystachus* L.; *A. fasciatus* Roxb.; *A. gracilis* Desf.; *Albersia caudatus* Boiss.; *Chenopodium caudatum* Jacq.; *Euxolus caudatus* Moq.

Family   Amaranthaceae

Vern.	Cholai.

English, Hindi, Sanskrit and Regional Names

Eng.	Green or wild amaranth, Green flowered;
Hindi	Chaulai;
Sans.	Tanduliya, vishaghna;
B.	Bannote;
G.	Dhinmado;
Kan.	Chilikirae soppy, Daglisoppu;
Mal.	Bayam munyit;
Mar.	Lhanamat;
Tam.	Keerai-kuppai, Sinna-kerrai;
Tel.	Chailaka thotakura.
Distribution	Waste places throughout India.

Description	Erect, glabrous annuals possessing long-petioled, ovate leaves. Clusters lax or in slender branches. Flowers long, bracts acute membranous. Seeds black, obtuse.
Flowering & Fruiting	May-June.
Habitat Ecology	Frequent on wastelands, cultivated fields, roadsides; **Karyal Khad** - 615 m; **Suker Khad** - 860 m; **Gamrola Khad** - 528 m.
Material Examined	EBH-WL-1027; 06.05. 2007.
Part Used	Whole Plant.
Folk Uses	Plant cooked as a **pot herb** and tastes like spinach. Also, used as a cattle **fodder**.
Chemical Constituents	Found to contain amasterol **(roots)**, saponin, spinosterol (24-ethyl-22-dehydrolathosterol) as major component along with 24-methyllathosterol, 24-ethyllathosterol, 24-mehtyl-22-dehydrolathosterol, 24-ethyl-cholesterol and 24-ethyl-22dehydrocholesterol **(plant)**.
Biological Activity	Plant shows juvenomimetic activity. Amasterol inhibited seed germination and growth of seedlings in lettuce and *Helminthosporium oryzi*.
Uses in Literature	Reported in India as appetizer and cooling medicine in snake bite; and for abortion, centipede bites, constipation, dental caries, diarrhoea, dizziness and kidney stones **(Ambasta, 1986; Asolkar et al., 1992; Banerjee & Ghora, 1996; Chopra et al., 1956; Das, 2000; Guha Bakshi et al., 1999; Jain, 1991; Jain et al., 2010; Jha et al., 1996; Joshi, 1995, 2008; Kaushik & Dhiman, 2000; Khan & Khanum, 2005; Kumar & Narain, 2010; Lalramnghinglova, 1996; Maheshwari et al., 1996; Rana et al., 2003; Rastogi & Mehrotra, 1993; Retnam & Martin, 2006; Saini, 1996b; Singh et al., 1996a; Sood & Thakur, 2004; Uphof, 2001).**

Ampelopteris prolifera (Retz.) Copeland. (Plate No. 9D)

Syn.   *Goniopteris prolifera* Presl; *Hemionitis prolifera* Retz.; *Nephrodium proliferum* Keys.

Family   Thelypteridaceae

Vern.	Sena.
Distribution	Throughout India.
Description	Ferns upto 120 cm long with fibrous roots and creeping rhizomes possessing ovate, dentate scales. Frond stipes dark coloured, sparsely scaly throughout and with prominently grooved rachis bearing numerous vegetative buds. Pinnae usually 10 in pairs or numerous, al-

ternate. Veins upto 10 pairs. Sori superficial and along veins. Spores brown, 24-25x38-42 µm, perinate.

Flowering & Fruiting	August-October.
Habitat Ecology	Plains, foot hills, along river banks and water courses; **Karyal Khad** - 615 m; **Sir Khad** - 620 m.
Material Examined	EBH-WL-1208; 12.08. 2008.
Parts Used	Roots. Leaves. Fronds.
Folk Uses	Fern planted as **ornamental** plant in gardens. Young tip of fronds eaten as a **vegetable** and also **pickled**. Coconut oil boiled with its leaves considered good against various **skin ailments**. Paste of roots applied as poultice for curing **eczema**.
Chemical Constituents	Active principles due to proteins, steroids, triterpenoids, flavones, flavonoids, sugars **(fronds)**.
Biological Activity	Antiviral and antibacterial found to be positive.
Uses in Literature	Reported to be described for aperient, eczema and as purgative **(Ambasta, 1986; Dixit and Vohra, 1984; Kaur, 1989; Khullar, 2000; Seth & Kumar, 2007; Singh, 2003b; Singh & Viswanathan, 1996; Singh et al., 2007; Wangchuk et al., 2008)**.

Andrographis paniculata (Burm. f.) Wall. ex Nees (Plate No. 9E)

Syn. *Justicia paniculata* Burm

Family Acanthaceae

Vern.	Kalmegh.

English, Hindi, Sanskrit and Regional Names

Eng.	Creat, Kalmegh, Kirayat;
Hindi	Charayetah, Kiryat, Mahatila;
Sans.	Bhuinmba, Kirata;
B.	Kalmegh;
G.	Kariyate, Kiryata, Kiriyati;
Kan.	Nelaberu;
Mal.	Kiriyattu, Nelaveppu;
Mar.	Olenkirayat;
Oriya	Bhuinimba;
Tam.	Nilavembu, Shiratkuchi;

Tel.	Nelavembu.
Distribution	Throughout India (Lucknow–Assam); sometimes cultivated.
Description	Erect annual herbs with quadrangular stems and entire leaves. Racemes axillary, pedicels pubescent. Inflorescence sympodial. Corolla white. Capsules glabrous. Seeds rugose, deep brown.
Flowering & Fruiting	October-December.
Habitat Ecology	Near water sources; **Panyala Khad -** 810m; **Chamyater Khad** - 613 m.
Material Examined	EBH-WL-1016; 18.08. 2009.
Parts Used	Whole Plant. Roots. Leaves.
Folk Uses	Powdered plant mixed with 'sarson' oil applied in **itching**. Whole plant extract mixed with goat's milk and egg albumen administered for **bone fracture**. Poultice of leaves applied on the spot of snake and scorpion **bite**. Leaf juice along with cardamom, cinnamom and cloves given orally for **bowel complaints** and **appetite**. 5-10ml leaf extract given thrice daily till cure for **dysentery, liver complaints** and **jaundice**. Young leaves added in **food** preparations.
Chemical Constituents	Active properties due to stereostructure of a diterpene glucoside-neo-andro-grapholide, 14-deoxy-11-oxoandograolide, 14-deoxy-11,12, didehydroandrographolide and 14-deoxyandrographolide, a mixture of four bitter substances, caffeic, chlorogenic and dicaffeoylquinic acids **(leaves)**, andrographin, panicolin, apigenin 4,7-dimethyle ether and mono-o-methyleothin **(roots)**.
Biological Activity	Plant shows antibacterial, anticancer, antidiarrhoeal, antimicrobial, cardiovascular and choleratic activities.
Uses in Literature	Discovered earlier as abortifacient, adaptogen, adrenocortical stimulant, alternative, analgesic, anthelmintic, antiaggregant, antiandrogenic, antiatherosclerotic, antibacterial, antibiotic, antifertility, antiinflammatory, antidiabetic, antifungal, antiserotonin, antispermatogenic, antityphoid, cholinergic, febrifuge, hepatotective, hypotensive, tonic and vermicidel; and for asthma, bite, body pain, cold, cuts, cough, debility, diarrhoea, dysentery, dyspepsia, eczema, fever, fractures, gastric troubles, heat diseases, influenza, intestinal worms, jaundice, leprosy, malaria, paralysis, piles, scabies, sexual weakness in males, skin diseases, snakebite, sneezing, stomach pain, tuberculosis, ulcers, whooping cough and wounds **(Alagesaboopathi & Balu, 2000; Balu et al., 2000; Banerjee & Ghora, 1996; Biswas et al., 2010; Chandra, 1997; Choudhury et al., 2008; Das, 2000; Duke et al., 2002; Dwarkan & Alagesaboopathi, 2000; Girach et al., 2000; Goel & Rajendran, 2000; Gogoi et al., 2003; Goud et al., 2000; Gupta, 1997; Henry, 1999; Jain, 1991, 1996, 2010; Khan & Khanum, 2005; Kumar, 2002; Lindley, 1981; Mahato & Mahato, 1996; Narain, 2006; Pal,**

2000; Pande et al., 2006; Pandey et al., 1996, 2000; Parrotta, 2001; Prajapati et al. 2006; Rahman, 2000; Rao & Henry, 1995; Rao et al., 2000; Retnam & Martin, 2006; Rosakutty et al., 1999; Rout & Panda, 2010; Saini, 1996b; Saren et al., 2000; Satapathy, 2008; Satapathy & Brahmam, 2000; Thomas & Britto, 2000; Yasodamma et al., 2009).

Anisomeles indica (L.) O. Kuntze (Plate No. 9F)

Syn. *A. ovata* R. Br.; *Nepeta indica* L

Family Lamiaceae.

Vern.	Basinga.
Hindi and Regional Names	
Hindi	Kalabhangra;
Bo.	Gobura, Gopali;
G.	Chodharo;
Kan.	Mangamari soppu;
Mal.	Chedayan;
Oriya	Bhutamari;
Tam.	Chedayan, Erumattai.
Distribution	Tropical Asia, upto 1,800 m.
Description	Strongly scented herbs upto 180 cm tall. Leaves coarsely crenate, accuminate. Flowers purplish. Nutlets bearing ellipsoid and compressed seeds.
Flowering & Fruiting	September-December.
Habitat Ecology	Wastelands; **Lyond Khad** - 617m.
Material Examined	EBH-WL-1021; 16.10.2008.
Part Used	Leaves.
Folk Uses	1-2 g powdered shade dried leaves given along with cow's milk twice a day before sunrise and after sunset for a period of 2-3 weeks for treating **rickets** in children. Smoke of burnt plant used as a mosquito **repellent**. Decoction of leaf prescribed for **gastrointestinal** disorders in animals.
Chemical Constituents	These are n-hentriacontane, glutinone, glutinol, friedelin ovatodiolide, anisomelic acid, betulin, methyl p-hydroxycinnamate, β-sitosterol and its glucoside **(aerial parts)**, stigmasterol, β-sitosterol, paraffins and fatty acids **(root)**, diterpenes, ovatodiolide and its derivatives

(leaves) and terpene hydrocarbon, citral and geranic acid **(flowers)**, β-sitosterol, letulinic acid, ovatodiolide, anisomalic acid **(plant)**.

Biological Activity	Et OH (50%) extract of herb hypothermic.
Uses in Literature	Found to be useful earlier as an asringent, carminative, tonic; and for abdominal pain, catarrh, cold, dyspepsia, eczema, flavouring purposes, intermittent fevers, rheumatism, skin problems, snakebites and toothaches **(Anonymous, 1948-76a; Chandra, 1997; Chopra et al., 1956; Dhyani & Sharma, 1987; Guha Bakshi et al., 1999; Jadhav, 2009; Jain, 1991; Joshi, 2008; Lindley, 1981; Pande et al., 2006; Parrotta, 2001; Rana et al., 2003; Retnam & Martin, 2006; Saini, 1996a; Singh, 1999; Siwakoti & Varma, 1996; Uphof, 2001; Yoganarasimhan, 1996).**

Apluda mutica (L.) (Plate No. 10A)

Syn.   *A. gigantea* Spreng.; *Andropogon glucus* Retz.; *Calaminea gigantea* Beauv.

Family   Poaceae

Vern.	Chofki basar.
Distribution	Throughout India; ascending Himalaya upto 2,700m.
Description	Perennial leafy grasses having thick branches, flexous stem and wiry roots. Leaves subbifarious, linear-lanceolate, cuspidately acuminate, glaucous. Spathes about 10 cm long, cymbiform, thin. Panicle usually lax. Spikes green. Spikelets with 1 terminal fertile floret and 2 lateral staminate. Rachilla glabrous. Glumes 2, unequal. Lemma awn twisted, margins thin. Palea membranous. Stamens 1. Style 2-fid. Stigma 2. Fruit caryopsis, ellipsoid.
Flowering & Fruiting	September-November.
Habitat Ecology	Common throughout the area; **Karyal Khad** - 615 m; **Sir Khad** - 620 m.
Material Examined	EBH-WL-1028; 07.08. 2008.
Part Used	Plant.
Folk Use	Plant paste mixed in small-sized wheat dough given three to four days as a remedy for animals suffering from **tongue** and **mouth sores**.
Uses in Literature	Repeatedly described earlier as a good fodder and for thatching purposes **(Bor, 1973; Duthie, 1888; Jain, 1991; Rana et al., 2003; Rao & Henry, 1995; Roy, 1984; Tarafder et al., 1997).**

Argemone mexicana L. (Plate No. 10B)
Family   Papaveraceae

Vern.	Chooly, Badi-Kandayi.

English, Hindi, Sanskrit and Regional Names

Eng.	Mexican poppy, Prickly poppy, Yellow mexican poppy;
Hindi	Bharbhand, Biladhutura, Brahmadundi, Brahmi, Satiyanashi, Ujarkanta, Shialkanta;
Sans.	Brahmadandi, Himadugdha, Hemashikha, Hemavati, Hemarha, Patuparni, Pitapushpa, Rukmini, Suverna;
B.	Baroshial kanta, Shialkantatha, Shialkantata, Siakanta;
Kan.	Arasina, Ummatta, Datturigida;
Mal.	Brahmadanti;
Mar.	Darusi, Firangidhotra, Kantedhotra, Kantedhotrey, Pinraladhotra;
Oriya	Kanta kusum;
P.	Bhatkateya Bhatmil, Bherband, Kandiari, Katsi, Satyanasa, Sialkanta;
Tam.	Bramadandi, Kurrukukkum.
Tel.	Brahmadandicettu;
Distribution	Naturalised throughout India.
Description	Erect robust annual herbs with sessile, variegated leaves and bright yellow flowers. Juice yellow. Capsules short, many seeded dehiscing at the top.
Flowering & Fruiting	April-May.
Habitat Ecology	Moist places, along nallahs; **Papral Khad -** 740 m; **Gobind Sagar Lake -** 490 m.
Material Examined	EBH-WL-1017; 10.04.2008.
Parts Used	Roots. Seeds.
Folk Uses	Paste of roots applied as poultice for chronic **skin diseases**. Pounded seeds mixed with oil applied in **eczema** and **itching**.
Chemical Constituents	Found to contain hydroxy, epoxy and keto fatty acids, myristic, palmitic, oleic and linolenic acids, 9- and 11-oxo-octacosanoic and 11-oxo-triacontanoic acids **(seed oil)** and four alkaloids identified as heletrine **(leaves, roots)**.

Biological Activity	Analgesic, antibacterial, antiinflammatory, antiseptic, antitrypanosomic, antiviral, carminative, contraceptive, demulcent, depurative, diaphoretic, diuretic and embryotoxic activities confirmed.
Uses in Literature	Known to be useful as emmenagogue, hallucinogen, haemostat, psoriasis, nauseant, purgative, vermifuge; and for boils, conjunctivitis, cutaneous troubles, eye diseases, fever, gonorrhoea, intestinal worms, jaundice, leprotic spots, leprous skin, leucoderma, leucorrhoea, malaria, menstrual disorders, night blindness, piles, pyorrhoea, rat bite, respiration, rheumatism, scabies, skin diseases, spermatorrhoea, syphilis, toothache, ulcers and wound **(Ansarali & Sivadasan, 2009; Bhogaonkar & Ahmed, 2007; Duke et al., 2002; Jadhav, 2009; Jain, 1991, 1996; Jain et al., 2010; Joshi, 2008; Kaul, 1997; Kaushik & Dhiman, 2000; Khan & Khanum, 2005; Khanna et al., 1996; Kumar, 1996, 2002; Lindley, 1981; Masih, 2003; Painuli & Maheshwari, 1996; Pande et al., 2006; Patil, 2009; Patil et al., 2007; Prajapati et al., 2006; Prakasha et al., 2010; Rana et al., 2003; Rao & Henry, 1995; Rastogi & Mehrotra, 1991; Retnam & Martin, 2006; Sahu et al., 2009; Saini, 1996a; Samwatsar & Diwanji, 1996; Singh & Pandey, 1996; Siwakoti & Varma, 1996; Swami & Gupta, 1996; Thakor, 2009; Yadav et al., 2010; Yasodamma et al., 2009).**

Artemisia indica Waldst. & Kit. (Plate No. 10C)

Syn.   *A. dubia* Wall.; *A. myriantha* Wall.; *A. paniculata* Roxb.; *A. lavandulaefolia* DC.; *A. parviflora* Wight; *A. vulgaris* L.

Family   Asteraceae

Vern.	Charmaar.

English, Hindi, Sanskrit and Regional Names

Eng.	Fleabane, Indian wormwood, Motherwort, Mugwort;
Hindi	Dona, Gathivana, Majtari, Mastaru, Nugduna;
Sans.	Barha, Bahikusum, Barhipushpa, Grantthika, Granthiparna, Granthiparnaka, Guchhaka, Guthaka, Kakapushpa, Kukura, Nagadamani, Nilapushpa, Saraparni, Shukla, Shukabarha, Tailaparnaka, Vanyadamanaka;
Mal.	Appa, Damanakam, Kattuchatti, Makkippu, Mashipatri, Nilampala;
P.	Afsuntin, Banjiru, Buimadaran, Chambra, Puujan, Tarkha, Tataur;
Tam.	Mashibathiri, Tirunama;
Tel.	Davanamu, Mashipatri.
Distribution	Mountainous districts of India; upto 4,000m in the W. Himalaya.

Description	Hoary pubescent or tomentose small shrubs with paniculatly branched leafy stems and large, ovate leaves. Corollas glabrous. Achenes oblong-ellipsoid, minute.
Flowering & Fruiting	August-October.
Habitat Ecology	Near ponds and lakes; **Raul Khad -** 646m; **Gobind Sagar Lake -** 490m.
Material Examined	EBH-WL-1018; 11.10. 2007.
Part Used	Leaves.
Folk Uses	Fresh juice of leaves cures **itching** in eyes occurring during summers; also applied to the head of young children for prevention of **convulsions**. Poultice of boiled young leaves also considered good against **headache**.
Chemical Constituents	Terpinnenol-4, β-caryophyllene, artemisia alcohol, linalool, cineol, camphor, borneol, cucalyptol **(aerial parts)** have been isolated.
Uses in Literature	Reported in India as an abortifacient, alexiteric, allergenic, analgesic, anthelmintic, antidote, antipileptic, antiseptic, antispasmodic, appetiser, choleretic, deobstruent, diaphoretic, immunostimulant, insecticidal, insectifuge, larvicidel, nervine stimulant, pungent, sedative and stomachic; and for asthma, foul ulcers, hysteria, itching, nervous and spasmodic affections, measles, scorpion-bite, skin diseases and snake-bite **(Bhattacharyya, 1996; Deokota & Chhetri, 2009; Duke et al., 2002; Joshi, 2008; Kirtikar & Basu, 1984; Lalramnghinglova, 2003; Lindley, 1981; Malla & Chhetri, 2009; Panda, 1996; Pande et al., 2006; Rastogi & Mehrotra, 1995b; Sharma & Rana, 1999; Uphof, 2001).**

Artemisia scoparia **Waldst. & Kit. (Plate No. 10D)**

Syn.   *A. elegans* Roxb.*; A. trichophylla* Wall

Family   Asteraceae

Vern.	Churisaroj.
Regional Names	
P.	Biur, Dona, Durunga, Jhan, Lasaj, Marua, Pilajau.
Distribution	Upper Gangetic Plains, W. Himalaya: Kashmir - Lahul (1,700-2,350m).
Description	Glabrous hoary annuals with perennial rootstock. Radical leaves petioled, broadly ovate; cauline filiform. Heads minute. Achenes minute, ellipsoid.
Flowering & Fruiting	August-September.
Habitat Ecology	Near water sources; **Gobind Sagar Lake -** 490m.

Material Examined	EBH-WL-1019; 04.08.2009.
Parts Used	Whole Plant. Leaves.
Folk Uses	Powdered leaves (2-4 g) prescribed twice daily for 3 days with cold water to relieve **headache**. Smoke considered good for **burns**. Infusion of plant given as a **purgative**. Plant juice used as ear drops for checking **pains**.
Chemical Constituents	Active properties associated with β-sitosterol, oxalic acid, a lactone **(root)**, capillin, 1-ph-2, 4-hexadiyne-1-ol, vanillin, scoparone, flavonoids **(leaves)**, capillaxin, coumarine and scopanone **(flower heads)**.
Biological Activity	Hypotensive, tranquilizing and sleep inducing (6-7-di-Meo-coumarin) and unrolithicases activities confirmed.
Uses in Literature	Useful for promoting secretion of biles and bile salt **(Hooker, 1872-97; Kirtikar & Basu, 1984; Guha Bakshi et al., 1999; Kala, 2003; Rastogi & Mehrotra, 1995b; Sood & Thakur, 2004; Watt, 1972)**.

🅑

Bacopa monnieri (L.) Pennell (Plate No. 10E)

Syn. *B. monnieria* (L.) *Wettst.; Gratiola monniera L.; Herpestis monniera* (L.)
H. B. & K.; *Lysimachia monnieria L.; Monniera cuneifolia* Michx.

Family Scrophulariaceae

Vern.	Jalnema.

English, Hindi, Sanskrit and Regional Names

Eng.	Bacopa, Thyme leaved gratiola, Waterhyssop;
Hindi	Brahmi, Jalnim, Safedchanmi;
Sans.	Brahmi, Sarasvati;
B.	Adabirni, Brihmi sak, Jalnimba;
Kan.	Nirbrahmi;
Mal.	Brahmi, Nirbrahmi;
Tam.	Nirpirami, Piramiyapundu;
Tel.	Sambranicettu.

Distribution	Wet places throughout India upto 1,200m.
Description	A glabrous prostrate or creeping, juicy annual herb having simple branches and opposite decussate, sessile, fleshy, obscurely veined, entire, punctuate leaves and pale blue flowers. Fruits ovoid, acute. Seeds minute, numerous, striated.
Flowering & Fruiting	August-October.
Habitat Ecology	Damp and wet areas throughout India; **Karyal Khad -** 615 m; **Sir Khad** - 620 m.
Material Examined	EBH-WL-1222, 04.04.2009.
Parts Used	Whole Plant. Leaves.
Folk Uses	Paste of leaves applied on head against **aches**. Poultice of boiled plant applied on chest for acute **bronchitis** and **cough** in children.

Chemical Constituents	Contains apigenin-7-glucuronids and luteolin - 7- glucuronide **(leaves)**.
Biological Activity	Antianxiety property +ve.
Uses in Literature	So far recorded as an adaptogen, analgesic, anticancer, anticonvulsant, antitoxidant, antitumour, anxiolytic, aperient, catarrah, convulsant, diuretic, emetic, expectorant, hypotensive, hypertensive, laxative and neurotonic; and for asthma, bilious disorder, bronchitis, diabetes, diarrhoea, epilepsy, fever, hoarseness, impotency, inflammation, insanity, insomnia, leprosy, lethargy, leukoderma, memory development, malarial fever, rheumatism, scabies, splenomegaly, stammering and swollen painful joints **(Agarwal, 1986; Ambasta, 1986; Ansarali & Sivadasan, 2009; Asolkar et al., 1992; Banerjee & Ghora, 1996; Chopra et al., 1956; Das, 2000; Duke et al., 2002; Gogoi et al., 2003; Goud et al., 2000; Jain, 1991; Kaushik & Dhiman, 2000; Khan & Khanum, 2005; Kumar & Narain, 2010; Nadkarni, 1976; Pande et al., 2006; Parrotta, 2001; Patil, 2009; Prajapati et al., 2006; Rana et al., 2000; Rastogi & Mehrotra, 1991; Retnam & Martin, 2006; Rosakutty et al., 2000; Rout & Panda, 2010; Singh & Kumar, 2000b; Singh & Rao, 2003; Sudhakar & Vedavathy, 2000; Upadhye et al., 1994, Yoganarasimhan, 1996).**

Barleria cristata L. (Plate No. 10F)

Syn.   *B. noctiflora* L.; *Barleria-canthus noctiflora* Oerst.; *B. ciliata* Roxb.; *B. dichotoma* Roxb.

Family   Acanthaceae

Vern.	Morni.

English, Hindi, Sanskrit and Regional Names

Eng.	Bluebell barleria, Crested puple;
Hindi	Tadrelu;
Sans.	Jhinti, Jhintica, Kurabaka, Raktapushpa;
Ass.	Jhinli;
B.	Jhanti, Sadajati;
Ori.	Koileka;
P.	Tadrelu;
Tam.	Nilachemmuli, Udamulli;
Tel.	Ettapulapeddagoranta.

Distribution	Throughout India.

Description	Small-sized prickly undershrubs with pubescent branches and obtuse, acute, pubescent leaves. Bracteoles with simple spines or denticulate. Corolla purplish, pink, pubescent, round-ovate. Capsule 4-seeded.
Flowering & Fruiting	July-October.
Habitat Ecology	Open slopes, very common in moist places; **Gobind Sagar Lake** - 490 m; **Sir Khad** - 620 m.
Material Examined	EBH-WL-1020; 02.10.2008.
Parts Used	Leaves. Flowers.
Folk Uses	Flowers offered to **appease** various deities. Paste of leaves applied thrice daily on **cuts, wounds** and curing **toothache**.
Chemical Constituents	Apigenin, naringenin and apigenin glucuronide, anthraquinones, barlacristone and cristabarlone isolated from **flowers**.
Biological Activity	Hypoglycaemic, spasmolytic and CNS depressant activity in mice +ve. EtOH (50%) extract of plant spasmogenic.
Uses in Literature	Considered useful for anaemia, asthma, body pains and swellings, headache, inflammation, snake–bites, toothache and in drinks **(Biswas et al., 2010; Jain, 1991; Khan & Khanum, 2005; Pande et al., 2006; Parrotta, 2001; Singh, 2003a; Singh & Srivastava, 2000; Sood & Thakur, 2004; Watt, 1972).**

Begonia picta Sm. (Plate No. 11A)

Syn.   *B. echinale* Royle; *B. erosa* Wall

Family   Begoniaceae

Vern.	Khattu.
Distribution	Himalaya: 700-2,000m (H.P. – Bhutan).
Description	Small, erect herbs with ovate acuminate leaves, glabrous pubescent stem and flowers in terminal cymes. Rootstock of one or few tubers and inflorescence scarcely exceeding the leaves which are cordated and serrated. Stipules lanceolate, hairy. Peduncle pubescent. Bracts oblong, persistent. Petals 2, small. Capsule with one wing much longer. Seeds light brown, shortly ellipsoid.
Flowering & Fruiting	August-September.
Habitat Ecology	Moist slopes; **Raul Khad** - 646m.
Material Examined	EBH-WL-1021; 25.08.2007.
Parts Used	Whole Plant. Stem.

Folk Uses	Poultice of plant effective for relieving **body pains** and **heat stroke**. Peeled leaf stalks and stems **pickled** and eaten as **salad**.
Uses in Literature	Known earlier as edible (leaves), colic; for curing bristles on tongue and mouth ulcers and dysentery **(Ambasta, 1986; Jain, 1991; Jain et al., 1991; Kumar, 2002; Pande et al., 2006; Ranjan, 2000; Rao & Henry, 1995; Rao & Jamir, 1982; Sood & Thakur, 2004).**

Bidens pilosa L. (Plate No. 11B)

Syn.   *B. bipinnata* Wall.; *B. chinensis* Willd.; *B. leucantha* Willd.; *B. odorata* Cav.; *B. pilosa* var. *bimucronata* (Turcz.) Sch. Bip.; *B. pilosa* var. *minor* (Blume) Sherff.; *B. tripartita* Wall.

Family   Asteraceae

Vern.	Lamb.
English, Hindi and Regional Names	
Eng.	Spanish needle;
Hindi	Phutium, Chirchitha;
G.	Phutium, Samarakokadi;
P.	Chirchitta.
Distribution	Throughout India.
Description	Glabrous, pilose or pubescent erect herbs with variable leaves and heads on long stout peduncles. Achenes black with short stout awns.
Flowering & Fruiting	June-September.
Habitat Ecology	Wastelands, shady slopes, damp places near fresh water; **Sir Khad - 620 m; Koshriyan Khad -** 520 m; **Barthin Khad** - 585 m.
Material Examined	EBH-WL-1184; 18.08.2007.
Parts Used	Aerial Parts. Leaves. Flowers.
Folk Uses	Aromatic infusion of plant good for **cough**. Leaf juice **antiseptic** and used as drops for **eye** and **ear complaints**. Powdered leaves (5 g) mixed with white pepper (4:1 ratio) prescribed thrice a day for 5 days for **tonsils, sore throat** and **cough**. Paste of aerial plant parts in combination with *Taraxacum officinale* and a pinch of alum used for **mouth ulcers**. Grounded dried flower buds mixed with alcohol used as a mouthwash in **toothache**.
Chemical Constituents	Reported to contain a polyacetylene glycoside – bipannatpolacetyleoside and hyperoside, isookanin-7-o-D-glucoside, okanin and maritimein **(leaves)**.

Uses in Literature	Used in India as antibacterial, antidiabetic, antiseptic, antispasmodic, appetizer, astringent, diuretic, emmenagogue, expectorant, eye and ear drops, fistulae, fungicide, galactagogue in veterinary medicine, haemostatic, hypoglycaemic, parasiticide, phototoxic, protisticide, rich source of vitamins C, stimulant, styptic, substitute for tea and symbolic to ward off evil spirits, tonic and vermifuge; and for asthma, counteracting peptic ulceration, cough, cuts, diarrhoea, eye sores, headache, kidney problems, leprosy, pectorial, phthisis, pustules, skin diseases, snake bite, sores, swollen glands, toothache, treat bladder, tumours, ulcerative colitis, ulcers, vulnerary and wounds **(Ambasta, 1986; Arora & Pandey, 1996; Bennet et al., 1991; Chandra, 1997; Chhetri, 2005; Chopra et al., 1956; Chowdhery, 1996; Devi, 2003; Duke et al., 2002; Jain, 1991; Kirtikar & Basu, 1984; Pande et al., 2006; Parrotta, 2001; Prajapati et al., 2006; Rajendran et al. 2002; Rana et al., 2003; Rastogi & Mehrotra, 1991; Siwakoti & Varma, 1996; Sood & Thakur, 2004; Subramaniam, 2000).**

Blyxa auberti **Rich. (Plate No. 11C)**
Family Hydrocharitaceae

Vern.	Jaladhru.
Description	A caulescent glabrous herb having radical, grass-like leaves. Flowers bisexual, solitary spathe 10 cm long, lobes obtuse; stamens 3, ovary spathe, fruit 6 cm long. Seeds elliptic.
Flowering & Fruiting	May-October.
Habitat Ecology	In ponds; **Sauli Khad -** 765 m.
Material Examined	EBH-WL-1279; 10.02 2009.
Part Used	Leaves.
Folk Use	15-20 ml juice of leaves given twice daily till cure for **stomach ache**.

Boehmeria platyphylla **Don (Plate No. 11D)**

Syn. *B. macrophylla* Hornem.; *B. macrostachya* Willd.; *B. scabrella* Wedd.; *Urtica macrostachya* Wall

Family Urticaceae

Vern.	Chamrala, Paryoon.
Hindi Name	
Hindi	Gargela.

Distribution	Tropical and Subtropical Himalaya from Shimla eastwards.
Description	Shrubby. Branches soft, glabrous or substrigose. Leaves large triangular, coarsely toothed and pubescent. Fruits compressed or angled.
Flowering & Fruiting	July-September.
Habitat Ecology	Rocky slopes, Moist places; **Kothi Khad** - 810 m; **Lyond Khad** - 617 m.
Material Examined	EBH-WL-1137; 28.08.2008.
Parts Used	Whole Plant. Young Leaves. Bark.
Folk Uses	Juice of plant as well as plant paste along with a pinch of black pepper applied as an emollient for quick **healing of cuts** and **wounds**. Poultice of bark paste good for healing **fractured bones**.
Chemical Constituents	Properties associated due to 5α- stigmastan-3,6-dione, pamolic acid, oleanolic acid, lupeal, β-sitosterol **(leaves, twigs)**, 3,4-di-OME-w-(2-piperidyl) acetophenone, cryptopleurine and secophenanthroquinolizidine **(plant)**.
Biological Activity	Hypotensive activity and effect on CNS found to be +ve.
Uses in Literature	Known to be used for cuts, diarrhoea (vet), dysentery, eczema, fibre for cordage and fishing lines, stomach diseases, wounds; and as fodder and galactagogue **(Ambasta, 1986; Asolkar et al., 1992; Deokota & Chhetri, 2009; Jain, 1991; Jain et al., 1991; Kharkongor & Joseph, 1997; Manandhar, 1987; Pande et al., 2006; Rao, 1996; Rao & Henry, 1995; Rastogi & Mehrotra, 1991; Sinha, 1986; Sood & Thakur, 2004; Sood et al., 2009b; Uniyal & Chauhan, 1973).**

Bromus catharticus Vahl (Plate No. 11E)

Syn.   *B. annuus* Herb.; *B. anatolicus* Boiss.; *B. arvensis* Duthie; *B. commulatus* M. Bieb.; *B. japonicus* Thunb.; *B. kochii* C. C. Gmel.; *B. mollis* Duthie; *B. multiflorus* Hort.; *B. pendulus* Schk.; *B. phygius* Boiss.; *B. polymorphus* Hoh. Enum.; *B. secalinus* M. Bieb.; *B. squarrosus* Herb.; *B. unilateralis* Schur *L.C.* Wheeler; *B. velutinus* Nocc.; *B. villiferus* Steud. Syme

Family   Poaceae

Vern.	Jangdu.
English Name	
Eng.	Prarie grass.
Distribution	W. Himalaya: 2,000-4,650 m; Europe, W. & N. Asia, China, Japan.
Description	Erect annuals upto 60 cm with glabrous stem and linear, flaccid, hairy leaves. Sheaths rarely all glabrous or hairy; ligule shoot ovate, toothed.

Panicle narrow or open, rachis glabrous; branches solitary, slender, flexuous. Spikelets densely, pale green. Rachilla scabrid, lanceolate; awn intra-apical. Anthers small. Grains linear-oblong.

Flowering & Fruiting	July-October.
Habitat Ecology	Moist places; **Landy Khad -** 615 m; **Lyond Khad** - 617 m; **Karyal Khad** - 615 m.
Material Examined	EBH-WL-1138; 22.09.2007.
Part Used	Whole plant.
Folk Use	Used extensively as a winter **forage** grass.
Chemical Constituents	Found to contain nitrates, Si as SiO^2, 45% ash and presence of transaconitic acid **(grass)**.
Uses in Literature	Decoction of fruits taken as a diuretic and pectoral **(Asolkar et al., 1992; Caius, 2003; Uphof, 2001; Ward, 1980).**

Bryophyllum calycinum Salisb. (Plate No. 11F)

Syn.   *B. pinnatum* (Lamk.) Kurz; *Cotyledon rhizophylla* Roxb.; *C. pinnata* Lamk.

Family   Crassulaceae

Vern.	Patharchatta.
Hindi and Regional Names	
Hindi	Zakham-haiyat;
B.	Koppata;
G.	Ghayamari;
Tel.	Sima-jamudu.
Distribution	Tropical plains of India; throughout tropics of the world.
Description	Tall erect glabrous perennial herbs. Leaves opposite, crenate. Flowers large, pendent, in spreading panicles with opposite branches. Calyx with a long inflated tube; lobes 4, short, valvate. Corolla campanulate. Stamens 8. Carpels 4, free or connate at the base. Follicles 4, many-seeded.
Flowering & Fruiting	July-November.
Habitat Ecology	Throughout moist parts; **Karyal Khad** - 615 m; **Koshriyan Khad** - 520 m; **Sir Khad** - 620 m.
Material Examined	EBH-WL-1166; 25.08.2007.

Part Used	Leaves.
Folk Uses	Slightly toasted leaves applied as poultice to heal **wounds, bruises, boils** and **bites** of venomous insects and for removing **kidney stones**.
Chemical Constituents	**Leaves** contain malic, isocitric and citric acids, p-coumaric, ferulic syringic, caffeic and p-hydroxybenzoic acids, flavonoids (quercetin, kaemferol), n- hentriacontane, n-tri- triacontane, alpha- and beta-amyrin and sitosterol.
Biological Activity	Leaves exhibit styptic, antiseptic and astringent properties.
Uses in Literature	Earlier used as analgesic, antiaggregant, antibacterial, anticancer, antidemic, antiinflammatory, antiplaque, antiprostagladin, antiseptic, antispasmodic, antitussive, choleretic, cicatrizant, cytotoxic, diuretic, emollient, expectorant, fungicide, styptic; and for abrasions, bruises, boils, centurions, diabetes, diarrhoea, dysentery, insect bites, preventing discoluration and wounds **(Choudhury et al., 2008; Duke et al., 2002; Jha et al., 1996; Khare, 2004; Pande et al., 2006; Rastogi & Mehrotra, 1995b; Rout & Panda, 2010; Singh et al., 1996a).**

Calamintha umbrosum (M.B.) Fisch. & C.A. Mey (Plate No. 12A)

Syn. *C. clinopodium* var. *umbrosa* Hook.; *C. repens* Benth.; *C. nepalensis* Fisch.; *Clinopodium repens* Roxb.; *Thymus repens* Don Prodr

Family Lamiaceae

Vern.	Jangli-tulsi.

English, Hindi and Regional Names

Eng.	Wild basil;
Hindi	Ban tulsi;
B.	Ban tulsi;
Mar.	Karimthumba.
Distribution	Temperate Himalaya; (Kashmir – Bhutan; 1,350-4,000m).
Description	Procumbent hairy perennial herbs possessing globose lax or dense-fid whorls. Bracts short. Flowers purple. Nutlets minute, subglobose, smooth, dry.
Flowering & Fruiting	April-August.
Habitat Ecology	Near ponds; **Ali Khad -** 580m; **Ghamber Khad** - 523m.
Material Examined	EBH-WL-1185; 10.05.2008.
Part Used	Whole Plant.
Folk Use	Yellowish-brown **dye** extracted from the herb.
Chemical Constituents	Stachyrose, cis-piperitenone oxide (63.05) and piperitenone oxide (17.98%) identified in **essential oil**.
Biological Activity	C.N.S. depressant and sedative in mice.
Uses in Literature	Used as stomachic, styptic and emmenagogue **(Guha Bakshi et al., 1999; Hooker, 1872-97; Rana et al., 2003; Rastogi & Mehrotra, 1995b; Uphof, 2001).**

Canna indica L. (Plate No. 12B)

Syn. *C. chinensis* Willd.; *C. indica* var. *orientalis* Rose

Family Cannaceae

Vern.	Sudershan.

English, Hindi, Sanskrit and Regional Names

Eng.	Indian bead, Indian reed, Indian shot;
Hindi	Sabbajaya, Sarvajjaya;
Sans.	Devakili, Kamkshi, Krishnatamara, Sarvajaya, Shilarambha, Vanakadali;
B.	Kamakshi, Sarbajaya;
G.	Akalabera;
Mal.	Kattuvala;
Mar.	Devakeli;
Oriya	Sorobojoya;
P.	Hakik;
Tam.	Kalvalai, Puvalai;
Tel.	Krishnatamara.

Distribution	Throughout India.
Description	Perennial rhizomatous herbs possessing tuberous rootstocks, leafy stem, oblong, acute leaves and terminal simple flowers. Bracts ovate, green. Sepals small, lanceolate. Corolla greenish. Fruit globose, echinate, indehiscent. Seeds black, globose.
Flowering & Fruiting	July-September.
Habitat Ecology	Oftenly found near moist places; **Lyond Khad** - 617 m; **Papral Khad** - 740 m.
Material Examined	EBH-WL-1139; 29.09.2007.
Parts Used	Roots. Stem. Leaves. Flowers. Seeds.
Folk Uses	Infusion of young leaves and inflorescence taken to cure **stomach pain**. Extract of root given twice a day for **urinary troubles**. Roots yield fibres for **cordages**. Pieces of 100 g of its stem and root stalks, boiled with rice water and pepper given as **antidote** to cattle for eating poisonous grasses. Juice of seed used as **ear drops**. Decoction of roots with 25 g of ginger and 21 black peppers boiled in rice water given to cattle for **poisonous fodder**.

Uses in Literature	Used in India as an antidote to poisonous fodder, febrifuge, hallucinogen, hypertensive, immunodepressant, inebriants, intoxicant, laxative and ornaments; and for dropsy, dyeing, headache, meal plates, necklaces, rope twine, sack clothes, swelling of abdomen in cattle, worshipping, wrapping up of goods, edible purposes and thatching houses **(Ambasta, 1986; Bajpayee & Dixit, 1996; Duke et al., 2002; Jain, 1991; Lalramnghinglova, 1996; Pande et al., 2006; Rama Rao & Henry, 1996; Ranjan, 2000; Siwakoti & Siwakoti, 2000; Subramaniam, 2000; Watt, 1972).**

Cannabis sativa L. (Plate No. 12C)

Syn.   *C. indica* Lamk.; *C. indica* var. *indica* (Lamk.) Wehmer

Family   Cannabinaceae

Vern.	Bhang.

English, Hindi, Sanskrit and Regional Names

Eng.	True hemp, Soft hemp, Indian hemp;
Hindi	Bhang, Charas, Ganja, Gur, Kinnab, Siddhi;
Sans.	Bhanga, Chapola, Ganjika, Indrasana, Jaya, Ujaya, Vijaya Vrijpatta;
B.	Bhang, Ganja;
Kan.	Bhangi-gida;
Mal.	Kancavu, Sivamuli;
Mar.	Bhangajhada;
Tam.	Ganea, Bhangi;
Tel.	Ganjayi.
Distribution	Throughout India; wild in N.W. Himalaya. Cultivated in temperate and tropical regions.
Description	Erect tall annual herbs with alternate leaves and lateral stipules. Flowers axillary green. Achenes compressed, crustaceous. Seeds flattened, pendulous.
Flowering & Fruiting	June-September.
Habitat Ecology	Wastelands, moist places; **Gobind Sagar Lake** - 490 m; **Sir Khad** - 620 m.
Material Examined	EBH-WL-1140; 16.08.2008.
Parts Used	Stem. Leaves.

Folk Uses

Infusion of young leaves and inflorescence taken to cure **stomach pain** and **flatulence**. Poultice of the young leaves used for healing **cuts** and **wounds**. Stem fibres used for making **ropes**. Leaves and top of female plants used as **intoxicant**. Seeds and leaves mixed with powder of gram (*Cicer arictinum*) and spices used for making **'Pakoras'**.

Chemical Constituents

Active constituents are friedelin, epifriedelinol and N-(p-hydroxy-β-pheylethyl) –p-hydroxy-trans-cinnamamide, cannabidivarin and tetrahydro cannabivarin, 1-dehydrotetrahydrocannabinol, cannabidol, cannabinol, 3-one, stigmast-4, 22-dien-3-one, stigmast-5-en-3β-ol-7, O-glucosides friedelin, epifriedelinol, β-sitosterol, carvone, dihydrocarvone **(roots)**, ferrloyltyraimne, p-coumaroyal tyramine and O-glucoside of apigenol, O-glycosides of vitamin, isovitexin and orientin glycosides **(leaves)**.

Biological Activity

Resins produced inhibitory action on respiratory and musculotropic action; also CVS and CNS active.

Uses in Literature

Reported earlier as abortifacient, alternative, analgesic, aphrodisiac, anesthetic, antiasthmatic, antibacterial, anticonvulsant, antidote, anti-inflammatory, antipyretic, antiseptic, antiserotonin, astringent, bhang ganja (dried flowering tops of female plants), bronchodilator, cardiotonic, carminative, cataleptic, cataract, charas resinous exudation (leaves), cholagogue, CNS stimulant, convulsant, demulcent, diaphoretic, diuretic, narcotic, sedative, soporific, spasmolytic, stomachic, tonic; and for asthma, convulsions, cough, dropsy, headache, hemp seed oil, inflamed eyes, insanity, leucorrhoea, malaria, melancholy, nose-bleeding, otalgia, palpitation, piles, restlessness, thirst, paints, varnishes and soaps **(Agarwal, 2003; Ambasta, 1986; Bhogaonkar & Kanerkar, 2007; Bose, 1984; Chandra, 1997; Chauhan, 1999; Chevallier, 1996; Dey & Bahadur, 1973; Duke et al., 2002; Harborne & Baxter, 2001; Henry, 1999; Jadhav, 2009; Jain, 1991; Jamir et al., 2008; Joshi, 1995, 2008; Kaul, 1997; Kaushik & Dhiman, 2000; Khan & Khanum 2005; Khare, 2004; Kirtikar & Basu, 1984; Kumar, 2002; Kumar & Pullaiah, 2003; Lindley, 1981; Malla & Chhetri, 2009; Negi et al., 1999; Noumi, 2010; Pande et al., 2006; Prajapati et al., 2006; Rana et al., 2003; Ranjan, 2000; Ranjan, et al., 2000; Retnam & Martin, 2006; Saini, 1996a; Sarin, 1990; Semwal et al., 2010; Shah, 1997; Sharma & Rana, 2005; Singh, 2000; Singh et al., 1996a; Siwakoti & Siwakoti, 2000; Siwakoti & Varma, 1996; Sood & Thakur, 2004; Yadav et al., 2010).**

Capsella bursa-pastoris (L.) Medik. (Plate No. 12D)

Syn.   *C. bursa* Raf.; *C. pastoris* Rupr

Family   Brassicaceae

Vern.	Seksi.
English Names	
Eng.	Shepherd'spurse, St. James wort.
Distribution	Throughout Temperate India.
Description	Annual herbs with forked hair. Leaves pinnatifid, lanceolate, lobed at the base. Flowers small, white, racemed. Pods flat, triangular.
Flowering & Fruiting	April-October.
Habitat Ecology	Grasslands; **Barthin Khad** - 585m; **Suker Khad** - 860m.
Material Examined	EBH-WL-1169; 22.04.2008.
Part Used	Aerial Parts.
Folk Uses	Aerial parts used as a **vegetable**. Decoction of plant drunk to lower **blood pressure**.
Chemical Constituents	Essential oil from **aerial parts** contain camphor, hespiridin, rutin, luteolin-7-rutinoside, quercetin-3- rutinoside, luteolin-7-galactoside, β-sitosterol, sinigrin, and α-methylsulfinyldecyl glucosinolates.
Biological Activity	Plant shows antibacterial, antithrombotic and anti-HIV virus activities.
Uses in Literature	Known so far in India for chest pain, cystitis, diarrhoea, dropsy, dysentery, eye ailments, haematuria, kidney disorders, menorrhagia, stomach ache; and as antiemetic, antiseptic, edible (vegetable, salad, pickled fruits), CNS depressant, diuretic, haemostatic, hypertensive, larvicidal, laxative, muscarinic, myostimulant, negative chronotropic, oxytocic, positive inotropic, rubefacient, urinary, uterocontractant, vasodilator, vermifuge, substitute for ergot and stimulant (seeds) **(Ambasta, 1986; Ballabh & Chaurasia, 2006; Chandra, 1997; Chandersekhar & Srivastava, 2003; Chevallier, 1996; Das & Agarwal, 1991; Dey & Bahadur, 1973; Duke et al., 2002; Harborne & Baxter, 2001; Jain, 1991; Jain & Goel, 2005; Joshi, 2008; Kala & Ravat, 2001; Khare, 2004; Kirtikar & Basu, 1984; Kumar, 2002; Lal et al., 1996; Panda, 1996; Prajapati et al., 2006; Rana et al., 2003; Roy et al., 1998; Samant et al., 2001a; Sharma, 2003; Singh et al., 2001; Singh & Kumar, 2000b; Srivastava et al., 2003; Wangchuk et al., 2008).**

Cardiospermum halicacabum L. (Plate No. 12E)
Family Sapindaceae

Vern.	Fuka-fucha.

English, Hindi, Sanskrit and Regional Names

Eng.	Balon vine, Heart pea;
Hindi	Kanphuti, Kapalphoti,
Sans.	Sakralata, Indravalli;
Kan.	Kanakayya;
Mal.	Ulinna;
Tam.	Mudukkottan, Modikkottan;
Tel.	Vekkudutiga.
Distribution	Throughout India. Most tropical and subtropical countries.
Description	Annuals with furrowed branches, deltoid or ovate leaves and deeply cut leaflets. Flowers white. Fruits wide, broadly pyriform. Capsule papery, green. Seeds black, opaque, 4.
Flowering & Fruiting	August-March.
Habitat Ecology	Fairly common in shady places; **Raul Khad -** 646m.
Material Examined	EBH-WL-1170; 25.03.2008.
Parts Used	Whole Plant. Leaves. Seeds.
Folk Uses	Poultice of plant paste used for curing **swellings** in leg. Powdered dried plant mixed with mustard oil applied to cure **sores** and **wounds**. Leaves fried in castor oil taken everyday orally for fever. Seeds eaten as **tonic**.
Chemical Constituents	These are **seed oil** 19.5-23.5%, fatty acid, (+) pinitol, 7-o-glucuronides of apigenin, chrysoeriol and luteolin **(leaves)**.
Biological Activity	Alcoholic and aqueous extrs. prevented cellular and extracellular injuries by stabilising lysosomal membrane and enzyme leakage.
Uses in Literature	Known so far for body pain, hydrocele, nervous diseases, rheumatism, stiffness of the limbs, snakebite, swellings and tumour **(Balasubramaniam & Prasad, 1996; Bhattacharyya, 1996; Biswas et al., 2010; Girach et al., 1999; Hosagoudar & Henry, 1996a; Jain, 1991; Jain et al., 2010; Kaushik & Dhiman, 2000; Khare & Khare, 2000; Painuli & Maheshwari, 1996; Patil, 2009; Prajapati et al., 2006; Rana et al., 2003; Rao & Henry, 1995; Retnam & Martin, 2006; Rosakutty et al.,**

1999; Samwatsar & Diwanji, 1999; Shukla & Verma, 1996; Siwakoti & Siwakoti, 2000; Siwakoti & Varma, 1996; Thomas & Britto, 1999; Yasodamma et al., 2009).

Cassia absus L. (Plate No. 12F)

Syn.   *C. coccinea* Wall.; *C. exigua* Roxb.; *Senna absus* Roxb.; *S. exigua* Roxb

Family   Fabaceae

Vern.	Chakshu.

Hindi, Sanskrit and Regional Names

Hindi	Banar, Chaksi, Chakut, Chakshu;
Sans.	Arangakulithika, Chakshushya, Drikaprasada, Kulmasha, Kulathika, Kumbhkarna, Lochanathhita;
G.	Chimah;
Kan.	Kuti;
Mal.	Karinkolla;
Tam.	Mulaippal-virai;
Tel.	Chanupala vittulu.
Distribution	Foot of W. Himalaya.
Description	An erect annual upto 60 cm high with stem and leaves clothed with grey bristly viscose hairs. Leaves long-petioled; leaflets oblong, very oblique, long. Stipules persistent. Racemes narrow. Corolla reddish yellow. Pods flat, coarsely hairy. Seeds thin hairy.
Flowering & Fruiting	July-September.
Habitat Ecology	Moist places, waste places; **Karyal Khad** - 615 m; **Panyala Khad** - 810 m.
Material Examined	EBH-WL-1133; 28.09.2008.
Parts Used	Leaves. Seeds.
Folk Uses	Leaf ash used as healing agent for **cuts** and **wounds**. Seeds used in **ophthalmic** and **skin troubles**. Decoction of leaves given twice daily for **cough**. Decoction of seeds used in **eye diseases**. Extract of seed used as **blood purifier**.
Chemical Constituents	Active constituents isolated are palmitic, gentisic, 5-O-D- glucopyranosyl gentisic acids, ethyl-α-D-galactopyranoside, apigenin, luteolin, hydnocarpin, aloe-emodin **(roots)**, quercetin, rutin **(leaves)**, chaksine, isochaksine **(roots, leaves)**, β-sitosterol-β-glycosides, raffinose and a D-galacto-D-mannan **(seed)**.

Biological Activity	Diuretic activity confirmed.
Uses in Literature	Useful as astringent, cathartic, diuretic and stimulant; and for constipation, cough, ringworm and skin diseases **(Ambasta, 1986; Deshmukh et al., 2000; Jain, 1991; Khan & Khanum, 2005; Khare, 2004; Lindley, 1981; Muratkar & Rothe, 2000; Parrotta, 2001; Prajapati et al., 2006; Sharma, 2000; Singh, 2000).**

Cassia occidentalis L. (Plate No. 13A)

Syn. *C. foetida* Pers.; *C. sophera* Wall.; *Senna occidentalis* Roxb

Family Fabaceae

Vern.	Badi-aeluan.

English, Hindi, Sanskrit and Regional Names

Eng.	Fetid cassia, Negro coffee, Rubbish cassia, Stinking weed, Coffee-senna;
Hindi	Barikasaumdi, Chakundra, Karaumdi, Kasunda;
Sans.	Arimarda, Dipana, Kanaka, Kasarmardah; Karkasha, Karamardah, Vimarda;
B.	Kalkashunda;
G.	Kasodari; Kasundari;
H.	Kasunda;
Kan.	Doddagace;
Kash.	Badde haedma;
Mal.	Ponnaviram, Ponnariviram;
Mar.	Hikal;
Oriya	Karundri;
Tam.	Ponnavirai, Peravirai, Nattam takarai;
Tel.	Karinda.

Distribution	Common throughout India; scattered in Himalaya.
Description	A diffuse subglabrous undershrub possessing long leaves and pubescent, foetid, glaucous leaflets. Racemes short-peduncled, few-flowered, corymbose. Axillary bracts thin, caducous. Pods recurved, subcompressed. Seeds 15-30, dark olive-green, ovoid and hard.
Flowering & Fruiting	May-September.
Habitat Ecology	Growing on wastelands after the rains; **Ali Khad** -580 m; **Auhr Khad** - 610 m; **Sir Khad** - 620 m; **Suker Khad** - 860 m.

Material Examined	EBH-WL-1223; 19.05.2008.
Parts Used	Roots. Leaves. Seeds.

Folk Uses   10-15 ml decoction of root taken orally once daily in the morning for **stomach disorders** and **dysentery**. Roasted seeds used as a substitute for **tea**. Paste of fresh leaflets warmed in groundnut oil applied for healing **cuts** and **wounds** and suppuration of **boils**. Poultice of crushed leaves with egg albumen and goat's milk also bandaged for healing **bone fracture**.

Chemical Constituents   Reported to contain new xanthone-cassiollin-along with emodin, ph scion, chrysophanol, α-3-sitosterol, methylmorphaline, phytosterol **(seeds)**, physcion-β-D-glucopyranoside **(flowers, seeds)**, ph scion, emodin, β-sitosterol, a phytosterol **(flowers, roots)**, 1,4,5-tr hydroxy anthraquinone derivatives-islandicin, helminthosporin, xanthorin, 1,4,5-trihydroxyanthraquinone derivatives-islandicin, helminthosporin, xanthorin, 1,4,5-trihydroxy-7-methoxy-3-metylanthraquinone (II **(seeds)**, dianthronic heteroside, matteucinol -7- rhamnoids, jaceidin -7- rhmnoside, chrysohenol and a bianthraquinone **(leaves)**.

Biological Activity   Seeds exhibit +ve diuretic activity.

Uses in Literature   Useful in India as anthelmintic, antiemetic, diuretic, febrifuge, laxative, substitute for coffee and tonic; and for body swellings, boils, burns, chest pain, cold, cough, cutaneous diseases, diabetes, diarrhoea, eczema, eye infection, fever, filariasis, headache, itch, jaundice, mental disorders, paralysis, rotten teeth, scabies, scorpion sting, skin diseases, snake bite, stomach ache, urinary troubles and whooping cough **(Ambasta, 1986; Balasubramanian & Prasad, 1996; Barua et al., 2000; Bhogaonkar & Ahmed, 2007; Biswas et al., 2010; Bora, 2000; Chandra, 1997; Chaudhury & Neogi, 2000; Chauhan, 1999; Deshmukh et al., 2000; Girach, 1992; Goel & Rajendran, 2000; Jain, 1991; Khan & Khanum, 2005; Kumar, 2002; Lindley, 1981; Maheshwari et al., 1996; Mishra, 2008; Parrotta, 2001; Prajapati et al., 2006; Rao & Henry, 1995; Rao et al., 2000; Rastogi & Mehrotra, 1995b; Reddy & Raju, 2000; Samwatsar & Diwanji, 1996, 2000; Sharma, 2000; Singh, 2000; Singh et al., 2000, 2001; Sinha, et al., 1996; Sood & Thakur, 2004; Swami & Gupta, 1996).**

Centella asiatica L. (Plate No. 13B)

Syn.   *Hydrocotyle asiatica* L.; *H. wightiana* Wall.; *H. lurida* Hance

Family   Apiaceae

Vern.   Brahmi.

English, Hindi, Sanskrit and Regional Names

Eng.	Indian pennywort, Thick leaved pennywort;
Hindi	Brahmamanduki, Khulakhudi;
Sans.	Bhekapanni, Bheki, Divya, Manduki, Manduka brahmi, Mandukaparnika, Mandukaparni, Mahaushadi, Supriya, Tvasthi;
Ass.	Manimurni;
Kan.	Urage, Von-de-laga;
Mal.	Kodagam, Muthal;
Mar.	Brahmi;
Tam.	Babbassa, Vallari;
Tel.	Bokkudu.

Distribution   Throughout India from Himalaya–Sri Lanka. Tropical and Subtropical regions.

Description   Prostrate herbs with glabrous, pubescent petioled, orbicular - reniform leaves, short peduncle and small, ovate bracts embracing the flowers. Umbel 6- flowerd. Carpels oblong, subcylindric. Fruits orbicular, reticulated. Seeds compressed laterally.

Flowering & Fruiting   April-May.

Habitat Ecology   Prominent along water channels in fields; **Karyal Khad** - 615 m; **Kothi Khad -** 810 m.

Material Examined   EBH-WL-1211; 10.04.2008.

Parts Used   Whole Plant. Leaves.

Folk Uses   2-3 g powdered plant taken on empty stomach with milk for enhancing **memory**, improving quality of **voice**, and as cooling **drink**. Leaf extract taken in the morning for **expelling worms**; its poultice good against **skin ailments**. Plant fed to cattle as **fodder**.

Chemical Constituents   3-glucosylquercetin, 3-glucosyl kaempferol, 7-glucosyl kampferol isolated from **(leaves)**.

Biological Activity   Antiprotozoal and spasmolytic activities confirmed.

Uses in Literature   Known in India for boils, cholera, cough, diarrhoea, dysentery, eczema, eye complaints, fever, gastric disorders, headache, insanity, leprosy, leucorrhoea, liver complaints, memory, mouth sores, nervine disorders, respiratory disorders, scabies, skin diseases, syphilis, tubercles, tumour, scorpion bite, urinary complaints, vegetable, wounds; and as an anthelmintic, blood purifier, brain tonic, colic pain, coolant, diuretic, emmenagogue, galactagogue (vet), insecticidal, local drink, pscyoactive drug, postnatal tonic and stomachic **(Agrawal, 2003; Ambasta, 1986; Ansarali**

& Sivadasan, 2009; Arora & Pandey, 1996; Augustine & Sivadasan, 2005; Banerjee & Bhattacharyya, 1996; Bennet et al., 1991; Biswas et al., 2010; Bondya et al., 2006; Borthakur et al., 2004; Chakraborty et al., 2003; Chauhan, 1999; Chevallier, 1996; Choudhury et al., 1996; 2008; Chowdhary, 1996; Das, 1999; Das & Sharma, 2003; Dey & Bahadur, 1973; Devi, 2003; Duke et al., 2002; Ghora, 1996; Gogoi & Borthakur, 2001; Gogoi et al., 2003; Goud & Pullaiah, 1996; Harborne & Baxter, 2001; Henry et al., 1996; Jain, 1991, 1996; Jamir, 1995; Jamir et al., 2008; Jha & Verma, 1996; Karnick, 1994; Katewa & Arora, 1997; Kayal et al, 2003; Kaushik & Dhiman, 2000; Khare, 2004; Kharkongor & Joseph, 1997; Kirtikar & Basu, 1984; Kshirsagar & Singh, 2000; Kumar, 2002; Kumar & Narain, 2010; Kurian, 1999; Lalramnghinglova, 1999, 2002, 2003; Maheshwari et al., 1996; Mao, 1993; Murthy et al., 2003; Negi et al., 1999; Panda, 1996; Pande et al., 2006; Pandey & Parmar, 1994; Prajapati et al., 2006; Prakasha et al., 2010; Prasad, 1996; Rai & Upadhay, 1996; Rajendran & Sikarwar, 2003; Rana et al., 2003; Ranjan, 2000, 2003; Rawat & Chowdhary, 1998; Rao & Henry, 1995; Reddi et al., 2005; Retnam & Martin, 2006; Said, 1997; Saklani & Jain, 1994; Samant et al., 2001a; Sarma et al., 2002; Satpathy & Brahmam, 1996; Semwal et al., 2010; Sen & Batra, 1997; Sharma & Rana, 2000; Sharma & Sood, 1997; Singh, 1999, 2000, 2003c; Singh & Kumar, 2000b; Singh & Pandey, 1996; Sinha, 1996; Siwakoti & Varma, 1996; Sood & Thakur, 2004; Subramaniam, 2000; Sudhakar & Vedavathy, 2000; Thomas & Britto, 2003; Tiwari & Tiwari, 1996; Upadhyay & Singh, 2005; Verma et al., 1995; Viswanathan, 1997; Yasodamma et al., 2009; Watt, 1972).

Ceratophyllum demersum L. (Plate No. 13C)
Family Ceratophyllaceae

Vern.	Jalaj, Shaival.

English, Hindi, Sanskrit and Regional Names

Eng.	Coontail, Hornwort;
Hindi	Sivara;
Sans.	Shivala;
B.	Sheoyala;
Tel.	Nasu.
Distribution	Ponds and lakes all over India from temperate to tropics.
Description	Herbs having 1-3 cm long internodes, 4-9 dichotomously forked leaves in whorl and axillary, solitary, minute, sessile flowers. Nutlets ellipsoid or ovoid, tuberculate.

Flowering & Fruiting	October-February.
Habitat Ecology	Common in stagnant water; **Sauli Khad** - 765 m; **Sir Khad** - 620 m.
Material Examined	EBH-WL-1020; 25.01.2009.
Part Used	Whole Plant.
Folk Uses	Plant used as a **green manure**; also source of food for **fish** and **ducks**. Its juice mixed with sesamum oil prescribed for **discoloured skin**. Also, 10-15 ml deoction of the plant given twice daily for 7-10 days for **biliousness** and **ulceration**.
Chemical Constituents	**Herbs** rich in proteins (24.6%), calcium and magnesium.
Biological Activity	Exihibits +ve purgative, antibilious and antimicrobial properties.
Uses in Literature	Prescribed for cardiac affections, giddiness, haemothermia, intrinsic haemorrhages, leucorrhoea, morbid thirst, rheumatism spermaturia, venereal diseases and as a biofertilizer **(Ambasta, 1986; Biswas & Calder, 1984; Das et al., 1996; Joshi, 2008; Khan & Khanum, 2005; Kumar & Narain, 2010).**

Cheilanthes bicolor (Roxb.) Fraser-Jenkins. (Plate No. 13D)

Syn.   *Pteris bicolor* Roxb

Family   Sinopteridaceae

Vern.	Shopa.
English and Hindi Names	
Eng.	Silver fern;
Hindi	Shankhapuhpi.
Distribution	Throughout the W. Himalaya; 600-1,500 m
Description	Rhizome short, erect with scaly dark-brown apex and stipe of variable length. Lamina 20-30 cm long, dark-brownish-black, thick, glossy, 1-2 pinnate. Sori indusiate, marginal; indusia light-brown, broad. Spores dark-brown, globose to tetrahedral, perinate.
Flowering & Fruiting	July-October.
Habitat Ecology	Near water channels; **Kothi Khad -** 810 m; **Tatoh Khad** - 710 m.
Material Examined	EBH-WL-1168; 10.08.2008.
Parts Used	Roots. Rhizome. Fronds.
Folk Uses	Paste of its roots and rhizomes along with black pepper (*Piper nigrum*) applied locally for treating **cuts** and **wounds**.

Uses in Literature	Fronds used in India for seasonal cold fever **(Asolkar et al., 1992; Khullar, 1994; Seth & Kumar, 2007; Singh et al., 2007; Vasudeva, 1999).**

Chenopodium album L. (Plate No. 13E)

Syn.   *C. giganticum* Don; *C. lacinatum* L.; *C. nepalense* Hort.; *C. purpurascens* Ham.; *C. viride* L.; *C. vulpinum* Wall

Family   henopodiaceae

Vern.	Bathu.

English, Hindi, Sanskrit and Regional Names

Eng.	Bacon weed, Dirty dick, Lamb quarters, White goose foot, Wild spinach;
Hindi	Bathua, Bethu sag, Khartua sag;
Sans.	Agralohika, Cheilika, Mridapatri, Shakarata, Vastaki;
B.	Chandan betu, Bethusag;
Kan.	Hancike;
Mal.	Vastuccira;
Tam.	Pasupukkirai;
Tel.	Pappukura.
Distribution	Tropical-Temperate Himalaya (Kashmir–Sikkim).
Description	Erect herbs having angled stem which is often striped green and red. Leaves extremely variable with long petiole. Flowers minute in axillary clusters or cymes. Seeds rarely vertical.
Flowering & Fruiting	November-April.
Habitat Ecology	Common in cultivated fields; **Saryali Khad** - 670 m; **Sir Khad -** 620 m.
Material Examined	EBH-WL-1124; 18.03.2009.
Parts Used	Whole Plant. Leaves.
Folk Uses	Paste made from its leaves applied on **wounds** and **sores** of cattle with favourable results. Leaves used as **vegetable**. 10-15 ml of plant juice taken orally (once daily for atleast 15 days) against **joint pains** and as a health **tonic**.
Chemical Constituents	β-ecdysone, polysone and polypodine B isolated from **leaves**.
Uses in Literature	Used as a acrid, anthelmintic, aphrodisiac tonic, carminative, diuretic, field indicator for magnesium, laxative, sweet pot herb; and for cardiac disorders and general debility, dyspepsia, flatulence, haemorrhoids,

helminthiasis, oleaginous, ophthalmopathy, pharyngopathy, seminal weakness, strangury and splenopathy **(Ambasta, 1986; Arora & Pandey, 1996; Asolkar et al., 1992; Bajpayee & Dixit, 1996; Banerjee & Ghora, 1996; Chaudhary & Neogi, 2000; Chaurasia et al., 2000; Chopra et al., 2005; Jain, 1991; Jha et al., 1996; Joshi, 2008; Kshirsagar & Singh, 2007; Pande et al., 2006; Parrotta, 2001; Prajapati et al., 2006; Rana et al., 2003; Sharma, 1996; Singh, 2000; Yadav et al., 2010).**

Chenopodium ambrosiodes L. (Plate No. 13F)

Family Chenopodiaceae

Vern.	Bathu.
English, Hindi, Sanskrit and Regional Names	
Eng.	Wormseed;
Hindi	Khatua;
Sans.	Kshetravastuka;
Kan.	Kaaduvoma;
Tam.	Kattasambadan.
Distribution	Throughout India.
Description	Aromatic glandular tall herbs with flowers clustered in slender, axillary and terminal spikes which are simple or panicled. Seeds dark brown, small with obtuse margins.
Flowering & Fruiting	June-September.
Habitat Ecology	Meadows near water channels; **Gobind Sagar Lake** - 490 m; **Sir Khad** - 620 m.
Material Examined	EBH-WL-1213; 14.08.2008.
Part Used	Aerial Parts.
Folk Use	Aerial parts used as a **vegetable**.
Chemical Constituents	Isolated components characterized are chenopodosides A and B **(aerial parts)**.
Uses in Literature	Reported so far as an abortifacient, amoebicide, analgesic, anthelmintic, antiasthmatic, antibacterial, antifeedant, antimalarial, antipyretic, antiseptic, antispasmodic, antitussive, antiulcer, carminative, cholera, colic, contraceptive, decongestant, diaphoretic, diuretic, fungicide, poison, edible (aerial parts), emmenagogue, paralytic, snake repellant, sedative; and for amoebic dysentery, appendicitis, asthma,

cough, cramps, diarrhoea, opthalmia, panacea, tuberculosis, wounds and intestinal parasites **(Abraham, 1997; Ambasta, 1986; Chevallier, 1996; Dey & Bahadur, 1973; Duke et al., 2002; Hosagoudar & Henry, 1996b; Jain, 1991; Kapur & Singh, 1996; Kumar, 2002; Mahato & Chaudhary, 2005; Nunez & Castro, 1995; Pande et al., 2006; Parrotta, 2001; Prajapati et al., 2006; Rana et al., 2003; Saklani & Jain, 1994; Sharma, 2003; Sharma & Rana, 2005; Singh & Kumar, 2000b; Sood & Thakur, 2004; Sood et al., 2001, 2009b).**

Chenopodium murale L. (Plate No. 14A)

Syn.   *C. gandhium* Ham.; *C. hookerianum* Moq

Family   Chenopodiaceae

Vern.	Baj-bhanga.
Distribution	Upper Gangetic Valley and Punjab.
Description	Glabrous erect herbs having bright-green rhombic or deltoid-ovate leaves with acute sides lobed and sharply toothed base. Sepals obtusely keeled, incurved in fruits. Seeds opaque.
Flowering & Fruiting	January-April.
Habitat Ecology	As a weed in fields; Near **Sir Khad -** 620 m.
Material Examined	EBH-WL-1167; 22.03.2009.
Part Used	Leaves.
Folk Uses	Leaves used as **pot herb** and considered rich source of iron; also its extract (5-10 ml, twice daily for 15 days) prescribed as a **blood purifier**.
Uses in Literature	Leaves used as vegetable **(Hooker, 1872 - 97; Jain, 1991; Patil et al., 2007; Kapur et al., 1996; Saini, 1996b).**

Christella dentata (Forsk.) Brownsey & Jermy. (Plate No. 14B)

Syn.   *Thelypteris dentata* (Forsk.) S.D. Jones

Family   Thelypteridaceae

Vern.	Kathu.
Distribution	Throughout India.
Description	Ferns upto 90 cm in size with short, stout rhizome having scaly apex which is brown, linear with entire margins. Fronds short, abrupt, attenuated. Stipes light brown, scaly and hairy. Lamina herbaceous,

unipinnate. Pinnae sessile, 15-25 pairs; apex acuminate, lobed, acroscopic, veins 6-9 pairs. Sori in 4-6 pairs, indusium reniform, green, hairy. Spores dark brown or black, kidney-shaped, perisporiate or perinate.

Flowering & Fruiting	May-October.
Habitat Ecology	Ravine channels, moist shady places, rarely along roadside slopes; **Sauli Khad -** 765 m; **Sir Khad -** 620 m; **Chamyater Khad -** 613 m.
Material Examined	EBH-WL-1190; 22.09.2008.
Part Used	Frond. Leaves.
Folk Uses	Extract of leaves mixed with honey taken after meal by people suffering from **tuberculosis**.
Chemical Constituents	Fern contains caffeic acid and p-coumaric acid.
Uses in Literature	Antibacterial agent **(Pande et al., 2006; Singh, 2003b; Yasodamma et al., 2009)**.

Cissampelos pareira L. (Plate No. 14C)

Syn. *C. caapeba* L.; *C. convolvulacea* Willd.; *C. discolor* Buch.; *C. diversa* Miers; *C. elata* Miers; *C. eriantha* Miers; *C. grallatoria* Miers; *C. hernandifolia* Wall.; *C. hirsuta* DC.; *C. orbiculata* Buch.; *C. obtecta* Wall.; *C. subpeltata* Thwaites; *Cocculus orbiculatus* DC.; *C. villous* Wall.; *C. membranaceus* Buch.; *Menispermum orbiculatum* L.

Family Menispermaceae

Vern.	Bhatindu.

English, Hindi, Sanskrit and Regional Names

Eng.	False pareira baruva, Ice vine, Pareira browa, Velvet-leaf;
Hindi	Akauadi, Dakhnirbissi, Harjeuri, Pari;
Sans.	Ambasthha, Ambashthika, Avidhakarni, Brihattica, Chichinnveshika, Devi, Shishi
B.	Akanadi, Nemuka, Nimuka, Tejomalla;
Bo.	Pahadmul;
Goa.	Parayel;
Kan.	Pahadmul;
Mal.	Kattuvalli, Patuvalli;
Mar.	Padavalli, Padavel, Pahadvel, Paharmul, Paharvel;
Oriya	Okanabindhi;

P.	Bat, Batindu path, Bel, Haiyat, Katori, Parbik, Pataki, Pilajur, Sakkhmi, Sucum-yeat, Tikri;
Tam.	Appata, Pomuskite, Pun mushtie, Punaittitta, Puttutiruppi, Sina, Titta, Tuvan, Tuvigaba, Vata-tirupie;
Tel.	Adivibankatige, Nemuka, Nirmuka, Pata, Tejomala.
Distribution	Tropical and Subtropical India; Cosmopolitan in warm regions.
Description	Leafy climbing shrubs with peltate, obtuse mucronate leaves and rarely glabrous branchlets. Male cymes axillary, superposed. Decompound bracts minute. Female racemes axillary, densely imbricate, petioled.
Flowering & Fruiting	April-August.
Habitat Ecology	Rocky moist slopes; moist places and fields near **Raul Khad -** 646 m.
Material Examined	EBH-WL-1141; 10.04.2009.
Parts Used	Whole Plant. Roots. Leaves.
Folk Uses	Sun-dried, pea-sized tablets of the plant paste prescribed thrice daily for seven days for **malaria**. Extract of leaves used for external application in **itch, sores** and **sinuses**. Leaf juice in combination with jaggery and a pinch of sugar useful against **chronic dysentery**. Fresh leaves mixed with wheat dough fed to livestock for **stomach disorders**. Leaves also given to animals to increase the **quantity of milk**. Root paste applied externally in **snakebite**. 15-20ml decoction of roots good against **stomach ache** and **cough** (thrice daily for 3 days).
Chemical Constituents	These are cycleanine, (-) bebearine, hayatinin, hayatin, (+) quercitol **(leaves)**, hyatin, picrate, chloroplatriate, methiocide, methochloride, eleven quaternary alkaloids like menismine, cissamine, hayatidine, hayatinine, pareirine d-isochodrodendrine, 1-berbeerine etc. **(root, root bark)**, cissampareine and dl-curine dimethiodide **(plant)**.
Biological Activity	CNS depressant, hypoglycaemic activity confirmed.
Uses in Literature	Earlier reported as alexeteric, analgesic, antiabortive, antiinflammatory, antifertility, antipyretic, astringent, colic, diuretic, stomachic, tonic and local drinks; and for abscess, bleeding piles, body ache, boils, bronchitis, burns, carbuncle, catarrhal infection of bladder, chest pain, child birth, cold, constipation, cough, cystis, despepsia, diabetes eye complaints, fever, gastric complaints, headache, indigestion, jaundice, leprosy, leucoderma, loose motion, malaria, menstruation, jaundice, kidney stones, peptic ulcers, scorpion bites, skin diseases, smallpox, snakebite, sores, spleen enlargement, stomach pains, toothache, urinary complaints, venereal complaints, wounds and fermentation **(Ambasta, 1986; Deokota & Chhetri, 2009; Dwarakan & Ansari,**

1996; Duke et al., 2002; Girach et al., 2000; Goel & Rajendran, 2000; Gogoi et al., 2003; Hosagoudar & Henry, 1996a; Islam, 1996; Jain, 1991, 1996; Jain & Singh, 1997; Jain et al., 1991; Kaushik & Dhiman, 2000; Khan & Khanum, 2005; Kumar, 2002; Lindley, 1981; Maheshwari et al., 1996; Manandhar, 1996a; Masih, 2003; Painuli & Maheshwari, 1996; Pande et al., 2006; Rama Rao et al., 2008; Rana et al., 2000, 2003; Rao & Henry, 1995; Rastogi & Mehrotra, 1991; Reddy & Raju, 2000; Retnam & Martin, 2006; Saini, 1996a; Saren et al., 2000; Sarkar et al., 2000; Satapathy, 2008; Satapathy & Brahmam, 2000; Sen & Behera, 2009; Singh, 2000; Siwakoti & Siwakoti, 2000; Siwakoti & Varma, 1996; Sood et al., 2009b; Tarafder & Chaudhuri, 1997; Thakor, 2009).

Clematis gouriana Roxb. (Plate No. 14D)

Syn.   *C. canna* Wall.; *C. javana* DC

Family   Ranunculaceae

Vern.	Bakerbel.
Distribution	Hill districts of W. Himalaya: 300-1,000 m.
Description	An extensive pubescent climber possessing pinnate leaves and membranous, ovate-oblong, acuminate leaflets. Flowers yellowish. Sepals obovate, revolute. Achenes hairy.
Flowering & Fruiting	June- September.
Habitat Ecology	Near Water Sources; **Ali Khad** - 610 m; **Marsand Khad -** 610 m.
Material Examined	EBH-WL-1001; 22.07. 2008.
Part Used	Leaves.
Folk Use	Paste of leaves applied on forehead to relieve **headache**.
Biological Activity	Analgesic, anthelmintic, antibacterial, antimicrobial and anticonceptive activities confirmed.
Uses in Literature	Earlier known for fever and muscular skeletal disorder **(Hooker, 1872-97; Kirtikar & Basu, 1984)**.

Coix lachryma - jobi L. (Plate No. 14E)

Syn.   *C. agrestis* Lour.; *C. arundinaceae* Lamk.; *C. exaltata* Jacq.; *C. ovata* Stokes; *C. pendula* Salisb.; *C. puellarum* Balans.; *C. stigmatosa* Koch; *Lithagrostis lacryma*-jobi Jacq

Family   Poaceae

Vern.	Bajayanti-mala.

English, Hindi, Sanskrit and Regional Names

Eng.	Job's tears;
Hindi	Baru, Dabir, Gaduta, Garahadua, Gargaridhan, Garun, Gulbigadi, Gurlu, Kaiya, Kasei, Sankhru, Sankru;
Sans.	Gavedhu, Gavedhuka, Gavedu, Gojivha, Gundraguttha, Jargadi;
Ass.	Koamonee, Sohru;
B.	Gurgur, Kunch;
G.	Kasai, Ranzonndle;
Kan.	Majutti;
Mal.	Jilaibatu, Mulaitiksu;
P.	Sanklu;
Tam.	Kattu-kundamani;
Tel.	Adavi guruginja.
Distribution	Throughout the hotter and damper parts of India; wild or cultivated.
Description	Tall annual grasses upto 1m; stem stout, leafy, rooting at the leaf node. Leaves wavy; sheaths smooth. Ligule very short, glabrous. Spikes suberect, peduncled. Lower lemma empty, upper lemma as long as spikelet. Female palea a little shorter. Glumes chartaeous. Caryopsis ovoid or globose.
Flowering & Fruiting	November-February.
Habitat Ecology	Found growing in wet places and standing water; **Karyal Khad -** 615 m; **Sauli Khad -** 765 m; **Sir Khad -** 620 m.
Material Examined	EBH-WL-1142; 26.02.2009.
Parts Used	Roots. Stem. Leaves. Fruits. Seeds.
Folk Uses	Foliage (leaves, stem) provides a useful **fodder**. 1-2g powdered roots taken with honey twice daily for 5 days for **menstrual disorders**. Grains eaten by people for **purification** of blood. Fruits grounded with

water drunk regularly for a long time to cure **tuberculosis**. Hardened involucres called, 'job's tear,' used as **ornaments**.

Chemical Constituents   Found to possess 1,4-benzoxazine derives, benzoxazolinones, benzoxazolimenes **(roots)**, coixenolide, palmilolenic and vauenic acids **(seeds)**.

Biological Activity   Aqueous extract found to lower B.P. in rabbits.

Uses in Literature   So far known in India as anticancer agent, bitter, blood purifier, cathartic, depurative, diuretic, tonic; and for anthrax, catarrhal affections, chest complaints, diabetes, dysentery, fever, inflammation of urinary passages, menstrual complaints, reducing weight of the body and smallpox **(Asolkar et al., 1992; Biswas et al., 2010; Bor, 1973; Cauis, 2003; Duthie, 1888; Jain, 1991; Kumar, 2002; Kumar & Narain, 2010; Mehra, 1982; Mudaliyar, 1921; Prajapati et al., 2006; Roy, 1984; Singh & Pandey, 1996; Siwakoti & Siwakoti, 1999; Siwakoti & Varma, 1996).**

Colocasia antiquorum **Schott (Plate No. 14F)**

Syn.   *Arum colocasia* L.; *A. nymphaeifolium* Roxb.; *Caladium esculentum* Vent.; *C. nymphaeifolium* Vent.; *Colocasia acris* Schott; *C. fontanesii* Schott; *C. nymphaeifolia.* Kunth; *C. pruinipes* Koch

Family   Araceae

Vern.   Jangli Kachalu.

English, Hindi, Sanskrit and Regional Names

Eng.	Cocoyam, Dasheen, Eddo, Taro;
Hindi	Arvi, Kachalu, Ghuiya;
Sans.	Kachwi, Kachu;
B.	Kachu;
Kan.	Kacchi, Shamagaddu;
Mal.	Shembu;
Mar.	Alu;
Oriya	Saru;
Tel.	Chamadumpa, Chamagadda.

Distribution   Moist and dry places, throughout hotter parts of India; wild or cultivated.

Description   Coarse herbs having dark green, heart-shaped peltate leaves. Peduncles solitary. Inflorescence spathe 20-35 cm, tube green, blade orange.

Spathe erect, pale yellow. Spadix small. Flowers pistillate, pea green. Ovaries 1-locular with 36-67 ovules. Sterile flowers white to pale yellow; staminate flowers and sterile tip pale orange. Stamens 3-6, connate. Fruit orange coloured.

Flowering & Fruiting	June-October.
Habitat Ecology	Commonly forming groups or rows along the margin of lakes, ponds or on wet soil; **Ali Khad -** 580 m.
Material Examined	EBH-WL-1147; 28.05.2008.
Parts Used	Leaves. Tubers.
Folk Uses	Tubers used as a **vegetable** and for making **pickle**. Leaves used for making **'Patrodu'** (a delicacy fairly relished by hill folks).
Chemical Constituents	**Tubers** yield phytoalexin-9,12,13-trihydroxy-10 (E) – octadecenoic acid, -14α-methyl - 5α- cholesta – 9, 24-dien-3β, 7α-diol (I), 14α-ethyl-24- methylene-5α- cholesta-9,24 – dien-3α, 7α- diol (II) nonacosane, cyanidin-3-glucoside, β- sitosterol and stigmasterol.
Uses in Literature	Known as vegetable, rubefacient, stimulant; and for alopaecia, body-ache, buboes, congestion of portal system, earache, flatulence, gravies, haemorrhage, inflamed glands, insect stings, otorrhoea, production of industrial alcohol, puddings, piles and soups **(Agarwal, 2003; Ambasta, 1986; Augustine & Sivadasan, 2005; Bandya et al., 1990; Banerjee & Ghora, 1996; Biswas & Calder, 1984; Chopra et al., 1956; Chowdhery, 1996; Chun-Linn & Jieru, 1995; Das, 1999; Dey & Bahadur, 1973; Dwivedi, 2003; Huidrom, 1996; Islam, 1996; Jain, 1991; Jha et al., 1996; Kapoor et al., 2010; Kulkarni & Kumbhjokar, 1996; Kumar & Narain, 2010; Lal et al., 1996; Prajapati et al., 2006; Rana et al., 2003; Saini, 1996a; Samwastsar & Diwanji, 1996; Solanki et al., 2007; Sood & Prakash, 2007).**

Commelina diffusa Burm. f. (Plate No. 15A)

Syn.   *C. agraria* Kunth; *C. caespitosa* Roxb.; *C. communis* Watt; *C. deficiens* Flor.; *C. longicaulis* Jacq.; *C. salicifolia* Bojer

Family   Commelinaceae

Vern.	Kapala.
Distribution	Throughout the hotter parts of India.
Description	Herbs with creeping stem rooting at the nodes and glabrous ciliate leaves. Spathes acute, base rounded. Peduncles small with 2 cymes. Petals blue or white. Capsule oblong with tuberculate seeds.
Flowering & Fruiting	July-September.

Habitat Ecology	Wet forests, drains, swampy areas and other wet places; **Karyal Khad** - 615 m; **Kothi Khad -** 810 m; **Sir Khad** - 620 m.
Material Examined	EBH-WL-1148; 16.08.2007.
Parts Used	Leaves. Seeds.
Folk Uses	Leaves used as a **vegetable** and **fodder**. Seeds eaten in times of **scarcity**.
Chemical Constituents	Acylated anthocyanins, zwitterionic anthocyanins, alkaloids flavonoid glycoside flavocommelin, saponins, tannins, coumarins, lectin **(leaves)** have been isolated.
Uses in Literature	Reported earlier as antidote; and for biliousness, body swellings, dysentery, fever, fractured bones and rashes **(Bhogaonkar & Ahmed, 2007; Jain, 1991; Pande et al., 2006; Prajapati et al., 2006).**

Commelina paludosa Blume (Plate No. 15B)

Syn.   *C. benghalensis* Wall.; *C. persicariaefolia* Wight; *C. polyspatha* Wight; *C. maculata* Edgew.; *C. semi-ovata* Ham.; *C. communis* Roxb.; *C. donii* Dietr.; *C. obliqua* Ham. *ex* Don

Family   Commelinaceae

Vern.	Chura.
Hindi, Sanskrit and Regional Names	
Hindi	Kanchara;
Sans.	Kanchata;
B.	Jatakan-chura;
Mar.	Kena;
Tam.	Kanavazhai, Kanangakarai;
Tel.	Vennadevikura.
Distribution	Throughout India; upto 1,800 m in the Himalaya.
Description	Erect branched herbs upto 90 cm having glabrous leaves with diffuse stem and acuminate, glabrous, rounded base. Spathes ovate–cordate, peduncled. Capsule oblong, obtuse. Seeds ellipsoid, compressed, terete smooth.
Flowering & Fruiting	June-August.
Habitat Ecology	Edges of cultivation, wastelands moist slopes; **Karyal Khad -** 615 m; **Sir Khad** - 620 m.
Material Examined	EBH-WL-1148; 04.08.2007.

Part Used	Leaves.
Folk Use	Leaves used as a **pot herb** in times of scarcity.
Uses in Literature	Herb reported to be used in India for body and foot sores, burns, bruises, eyelid infection, leprosy, piles; and as demulcent, emollient, laxative, refrigerent and tonic **(Ambasta, 1986; Chandra, 1997; Chopra et al., 2005; Devi, 2003; Dwivedi, 2003; Jain, 1991; Jha et al., 1996; Khan & Khanum, 2005; Kharkongor & Joseph, 1997; Painuli & Maheshwari, 1996; Pande et al., 2006; Rana et al., 2003; Rao & Henry, 1995).**

Convolvulus arvensis L. (Plate No. 15C)

Syn.   *C. chinensis* Ker. ; *C. divaricatus* Wall.; *C. maleolmii* Roxb

Family   Convolvulaceae

Vern.	Dudhua-bel.

English, Hindi, Sanskrit and Regional Names

Eng.	Deer foot bindweed, Field bindweed, Small binweed;
Hindi	Beri, Haranpadi, Harinpadi, Hiranpaddi, Prasarna;
Sans.	Bhadhbadali, Bhadravala, Gondol, Rajbala;
B.	Gandhbadali, Gondol;
G.	Nari, Veladi;
Mar.	Chandvel, Haranpag;
P.	Harinpadi, Hiranpadi.
Distribution	W. India. All temperate and subtropical regions.
Description	Twining glabrous pubescent herbs. Leaves ovate-cordate, petiole small. Corolla glabrous, funnel- shaped, pink. Capsule glabrous, glabrous. Seed subtrigonous, reddish-brown.
Flowering & Fruiting	December-March.
Habitat Ecology	Frequent in wastelands, dry slopes, agricultural fields; **Sir Khad -** 620 m.
Material Examined	EBH-WL-1150; 06.03.2009.
Parts Used	Roots. Whole Plant. Leaves.
Folk Uses	2-3 g powdered plant given with water to restore normal function of the body and for improving body **immunity** against diseases, twice a day for 8-10 days. Decoction of roots given orally as **tonic.**
Chemical Constituents	Constituents identified are α-amyrin, campesterol, stigmasterol, β-sitosterol, n-alkanes and n-alkanols **(aerial parts)**, cuscohygrine and coumarins **(roots)**, and β-me-escoletin **(all parts of plants).**

Biological Activity	Plant extract hypotensive, anticonvulsant, mascarinic and nicotinic.
Uses in Literature	Known to be used as fodder, purgative, stimulant, substitute for detergent; and for confusion, debility, duodenal ulcers, dysentery, gastric, insanity, insomnia, psychoneurosis, spermatorrhoea and uterine bleeding **(Anonymous, 1948-76a; Dwivedi, 2003; Jain, 1991; Kapur & Singh, 1996; Kaul, 1997; Khan & Khanum, 2005; Khare, 2004; Painuli & Maheshwari, 1996; Pande et al., 2006; Parrotta, 2001; Patil et al., 2007; Rana et al., 2003; Rastogi & Mehrotra, 1995b).**

Conyza bonariensis (L.) Cronquist (Plate No. 15D)

Syn.   *Erigeron bonariensis* (L.) Cronquist

Family   Asteraceae.

Vern.	Shesherda.
Distribution	Garden weed throughout plains of India.
Description	Perennial herbs with erect stems having stiff hair. Leaves narrow, lanceolate, green in colour, coarsely toothed, covered with fine hair. Flowers numerous, arranged in pyramidal panicles. Fruits cypsela.
Flowering & Fruiting	June-October.
Habitat Ecology	Moist shady places; **Sauli Khad -** 765 m.
Material Examined	EBH-WL-1144; 22.09.2008.
Part Used	Whole Herb.
Folk Uses	15-20 ml decoction of the plant (twice daily for 7 days) given in mouth and throat **diseases**. Plant used as an occasional **fish poison**.
Chemical Constituents	These are polyphenol, polyphenolic acids, caffeic major several flavones, butenalide-(RS) 3-methoxy – 6- oxodec – 2- en-4-olide (1), trans-lachnophyllum lactone, Cis- lachnophyllum methyl ester and germacrane alcohol **(whole herb)**, caryophyllene oxide (22.6%), spathulenol (15.1%), limonene (6.7%) **(leaf oil)**, Geranyl acetone (25.3%), germacrene D (14.0%), limanene (7.1%), bicyclogermacrene (4.7%) **(stem oil)**, trans-α-bergamotene (24.3%), (E) –β-ocimene (14.6%), zingiberene (13.8%), limanene (7.6%), γ-cadinene (5.6%), β-caryophyllene (5.6%), matricaria ester (16.0%) **(flower head oil)**, (Z)-nerolidol (19.9%), nonane (7.8%) and β-sesquiphellandrene (5.0%) **(root oil)**.
Uses in Literature	Known so far as anthelmintic, antidiarrhoeal, anti-rheumatic, anti-ulcer gastric, diuretic, febrifuge, insecticide; and for bronchitis, diarrhoea, dysentery, gonorrhoea, healing, heart burning sensations, liv-

er, ulcers and urinary tract infections **(Jain, 1991; Kaul, 1997; Keller & Ghillean, 2008; Pande et al., 2006; Rastogi & Mehrotra, 1995a).**

Conyza stricta **Willd. (Plate No. 15E)**

Syn.   *C. absinthifolia* DC.; *C. pinnatifida* Roxb.; *Blumea trisulca* DC.; *Baecharis trifurcata* Trel

Family   Asteraceae

Vern.	Hodari.
Distribution	Subtropical Himalaya from Kashmir eastwards, ascending to 1,700m.
Description	Corymbosely branched, pubescent leafy herbs with narrowly linear, toothed leaves. Heads numerous, peduncled. Achenes pubescent, pappus reddish.
Flowering & Fruiting	September-October.
Habitat Ecology	Grassy localties; **Sir Khad** - 620 m; **Sunhani Khad -** 620 m.
Material Examined	EBH-WL-1145; 18.09.2008.
Part Used	Whole Plant.
Folk Use	Poultice of plant good against **bone fracture**.
Chemical Constituents	Conyzic acid and strictic acid have been isolated **(whole plant)**.
Use in Literature	Reported as a medicinal plant **(Jain, 1991; Pande et al., 2006; Rana et al., 2003; Rastogi & Mehrotra, 1991).**

Costus speciosus **(J. Konig) Sm. (Plate No. 15F)**

Syn.   *C. speciosa* var. *nepalensis* (Rose) Baker; *C. arabicus* Jacq.; *Hellenia grandiflora* Retz.; *Banksia speciosa* Koenig

Family   Zingiberaceae

Vern.	Kedae-ki-challi.

English, Hindi, Sanskrit and Regional Names

Eng.	Elegant grass, Srilanka calumba root, Spiral ginger;
Hindi	Ken, Keu, Kemuk, Kust;
Sans.	Bramhatirtha, Kashmira, Kushtha, Kusthabeda, Padmakarna, Padmapatra, Padmapunya, Paushkara, Padmapatra, Sagara, Subandhu, Kemuka;
Ass.	Jamlakhuti;

B.	Keu;
Bo.	Gudurichakada, Kemuka;
G.	Pakaramula;
Kan.	Chengalvakoshtu;
Mal.	Anakkuva, Anappu, Channa, Channakuva, Kottam, Marujanna, Narikkurampu, Patimukam;
Mar.	Penva, Pinnaga, Pushkarmula;
Oriya	Chittorokudho, Ghanapohora, Kudho;
Tam.	Kottam, Kudaram, Kugaimanjal, Kuiravam;
Tel.	Chengalvakostu.

Distribution Throughout India.

Description Erect, leafy rhizomatous herbs with stout stem and oblong leaves which are silky beneath. Spikes very dense. Bracts ovate, bright red. Calyx cuspidate. Lip white, margins incurved. Capsule globose, red, crowned with persistent calyx. Seeds black with fleshy white appendage.

Flowering & Fruiting July-August.

Habitat Ecology Wastelands, grasslands, moistlands; **Chamyater Khad -** 613 m.

Material Examined EBH-WL-1191; 15.08.2007.

Parts Used Rhizome. Stem.

Folk Uses Stem paste good for healing **wounds**. Rhizomes **cooked** and eaten, 10ml extract of rhizome and stem given empty stomach for 10 days against **acidity** and **jaundice**. Also, rhizome pounded along with leaves of *Swertia chirayita* taken orally thrice a day for **fever** and **urinary troubles** whereas paste of rhizome taken with milk and sugar cures sensation of **internal heat** and **stomach inflammation**. Chutney made from its burnt rhizome, sugar and tamarind taken for **dysentery** and other **digestive troubles**.

Chemical Constituents These are diaosgenin, sitosterol **(stem)**, 8- hydroxy-triacontane-25-one, methyl tritriacontanoate **(roots)**, trigogenin, diogenin **(rhizomes, stems)**, saponins A, B and C, β-sitosterol glucoside, a-amyrin stearate, b-amyrin and lupeol palmitates **(leaves)**.

Biological Activity Antifertility, antispasmodic, antiinflammatory, hypotensive, spasmolytic and CNS depressant activities confirmed.

Uses in Literature Reported earlier as anticholinesterase antiinflammatory, antihelmintic, antidemic, antiexudative, antispasmodic, antitumorous, aphrodisiac, bitter and ophthalmic; and for backache, burning sensation in the eyes, cattarrh, chicken pox, conception, cooling, earache, fever, gastric troubles, giddiness, headache, lactation, piles, rheumatism, skin diseases, snake bite, ulcers and wounds **(Ambasta, 1986; Arora, 1997; Barua et al., 2000; Bhogaonkar & Kanerkar, 2007; Biswas et al., 2010; Das & Mishra, 2000; Gogoi et al., 2003; Hajra & Baishya, 1997; Hosagoudar & Henry, 1996a; Islam, 1996; Jain & Saklani, 1992; Jain & Singh, 1997; Jain et al., 1991; Kaushik & Dhiman, 2000; Khan & Khanum, 2005; Kothari, 2001; Kumar, 2002; Lalramnghinglova, 2003; Mahato & Mahato, 1996; Mandal & Basu, 1996; Masih, 2003; Panda & Das, 2000; Parrotta, 2001; Prajapati et al., 2006; Rajendran & Aswal, 2000; Rana et al., 2003; Rao & Henry, 1995; Retnam & Martin, 2006; Sharma, 2000; Singh et al., 1996a; Siwakoti & Siwakoti, 2000; Sood et al., 2009b; Watt, 1972; Yasodamma et al., 2009).**

Crotalaria alata **Buch.-Ham. (Plate No. 16A)**

Syn.   *C. bialata* Roxb.; *C. sagitticaulis* Wall

Family   Fabaceae

Vern.	Jhunka.
Sanskrit Name	
Sans.	Nagphani.
Distribution	Kumaon–Assam; Khasia, ascending to 1,850m.
Description	A suberect undershrub with stem and leaves clothed with short silky pubescence. Leaves subsessile, stipule forming wing from one node to the next. Racemes 2-3 flowered. Bracts small persistent. Calyx densely silky. Corolla pale. Pods linear-oblong, many seeded.
Flowering & Fruiting	July-November.
Habitat Ecology	Near water channels, grassland; **Karyal Khad** - 615; **Nali Khad** - 600 m.
Material Examined	EBH-WL-1192; 10.11.2008.
Part Used	Whole Plant.
Folk Uses	Plant considered an excellent **green manure**. Plant paste rubbed on body for **curing joints** and **muscular pains**. Decoction of plant prescribed as **tonic** for pregnant woman to facilitate **childbirth**.
Uses in Literature	Known earlier as green manure; and for muscular pains, scorpion bite, snake-bite, terracing and urinary complaints **(Bhogaonkar & Kanerkar, 2007; Guha Bakshi et al., 1999; Jain, 1991; Maheshwari et al., 1996; Uphof, 2001).**

Crotalaria mysorensis Roth (Plate No. 16B)

Syn.   *C. hirsuta* Roxb.; *C. stipulata* Roxb
Family   Fabaceae

Vern.	Sirsan.
Distribution	Tropical region throughout India; ascending to 1,300m in Kumaon.
Description	Copiously branched annuals with stem clothed with long dense silky hair and thin silky, membranous leaves. Racemes stalked. Calyx acuminate, upper lanceolate. Corolla yellow. Pods oblong with 20-30 seeds.
Flowering & Fruiting	July-October.
Habitat Ecology	Moist shady places near water; **Auhr Khad -** 610 m.
Material Examined	EBH-WL-1260; 22.09.2008.
Part Used	Seeds.
Folk Use	2-3 powdered seeds taken orally with water against **stomache pain**.
Uses in Literature	Recorded earlier for fever and stomache pain **(Duthie, 1973)**.

Cryptolepis buchanani Roem. & Schult. (Plate No. 16C)

Syn.   *C. reticulata* Wall.; *Echites reticulata* Roth; *E. cuspidata* Hayne; *Nerium reticulatum* Roxb
Family   Asclepiadaceae

Vern.	Chink booti, Dudhli, Taern.

English, Hindi, Sanskrit and Regional Names

Eng.	Karanta;
Hindi	Dudhi, Karanta;
Sans.	Krishna, Saariva;
Ass.	Gamar;
Mal.	Kalipalvalli;
Oriya	Gopikonioro, Maloti;
Tam.	Umi-thekku;
Tel.	Adavipalatige.

Distribution	Throughout India: W. Kashmir - Assam.
Description	Twining glabrous shrubs having opposite, coriaceous leaves which are shining above. Cymes axillary, many flowered, shortly peduncled, Branches short, divariculate and paniculate. Corolla yellow, lobes lanceolate. Seeds long, oblong-ovate, contracted below the tip.
Flowering & Fruiting	April-November.
Habitat Ecology	Wastelands near water; **Marsand Khad -** 610m; **Sir Khad** - 620m.
Material Examined	EBH-WL-1127; 12.10.2008.
Parts Used	Roots. Latex. Stem Bark. Leaves.
Folk Uses	Paste of its stem bark and that of *Litsea glutinosa* (Lour.) Robins used as a **plaster** cast for the fractured limb of cattle. Latex mixed in hot water applied on knees for **rheumatism**. Root extract given internally to cure **fits** (2-4ml, twice daily, 15 days). 15-25ml decoction of leaves prescribed on empty stomach as **tonic**. 1-3g powdered plant given with honey once daily for atleast 30 days against **rickets**.
Chemical Constituents	Reported to possess a pyridine alkaloid-b achananine, sarmentogenin, sarmentocymarin **(roots)**, sarverogenin-3-β-O-α-L-oleandropyranoside, cryptanoside A and cryptanoside B **(leaves)**.
Uses in Literature	Recorded earlier as fibre yielding, galactogogue; and for abdominal pain, anasarca, ascites, babesiosis, bleeding, body ache, bone fracture, cholera, cuts, dropsy, dysentery, fever, fissures on sole, fits and venereal diseases **(Ambasta, 1986; Hosagoudar & Henry, 1996a; Jain, 1991; Jain & Singh, 1997; Kirtikar & Basu, 1984; Mudgal et al., 1999; Pande et al., 2006; Parrotta, 2001; Rama Rao et al., 2008; Rana et al., 2003; Rao & Henry, 1995; Ramdas et al., 2000; Sahu et al., 2009; Singh & Prakash, 1996; Subramaniam 2000; Sood & Thakur, 2004; Sood et al., 2009b; Watt, 1972).**

Cucumis pubescens **Willd. (Plate No. 16D)**

Syn.   *C. trigonus* C.B. Clarke; *C. pubescence* Willd.; *C. pseudo-colocynthis* Wall

Family   Cucurbitaceae

Vern.	Photnu.
Sanskrit and Regional Names	
Sans.	Bahaphala, Chitra, Chitraphala, Chitavali, Deis, Garakshi vrishka, Godumba, Katpahla, Kumbhari, Laghuchirbhita, Maruya, Mrigadani, Mrigkshi, Mrigeshana, Pathya, Schvetaplishpa, Vichitra, Vishala;
B.	Bankumak, Gomuk;

H.	Bhak wia, Bislambhi, Bislombi, Gorakhhakakadi, Janglindrayan;
Tam.	Adavipuchcha, Chukkangai, Janglindrayan, Thumattikai;
Tel.	Budamkaya.

Distribution	Throughout India, Sri Lanka, Afghanistan, Malaya, N. Australia.
Description	Climbing or trailing annual herbs. Leaves deeply cut into 5-7 lobes. Flowers monoecious, yellow. Fruits fleshy, globose or cylindrical. Seeds smooth, compressed and without margin.
Flowering & Fruiting	May-August.
Habitat Ecology	Grasslands, Corn fields; **Lyond Khad** - 617 m; **Nalti Khad -** 650 m.
Material Examined	EBH-WL-1128; 22.06.2008.
Part Used	Fruits.
Folk Use	Ripe **fruits** eaten.
Chemical Constituents	**Seed oil** contains linoleic (59.1), oleic (17.5), stearic (10.0), arachidic (1.6), behenic (0.6), myristic (0.6) and palmitic (0.6%) acids.
Biological Activity	Alcoholic extract of plant exhibited marked anabolic activity.
Uses in Literature	Described as an astringent; and for bilious disorders, fever and snake bite **(Agarwal, 2003; Ambasta, 1986; Chopra et al., 1956; Das & Agarwal, 1991; Kirtikar & Basu, 1984; Roy et al., 1998; Sharma & Rana, 2005).**

Curcuma longa L. (Plate No. 16E)

Syn.   *C. angustifolia* Roxb.; *C. domestica* (Medik.) *Valeton*

Family   Zingiberaceae

Vern.	Ban Haldi.

English, Hindi, Sanskrit and Regional Names

Eng.	East indian arrowroot, Travancore starch;
Hindi	Haldi, Haladu, Halud, Hardi, Tikhur;
Sans.	Aneshta, Bahula, Haridra, Gauri, Kurkum, Nisa, Rajani, Tavakshira;
Mar.	Tavakhira;
Tam.	Ararutkizhangll;
Tel.	Ararut gaddalu, Palagunda.

Distribution	Throughout India.
Description	Herbs upto 1m tall. Leaves 15-30 cm; petiole as long as blade. Peduncle 18 cm long. Flowers yellowish in compound spikes. Calyx cylin-

dric, toothed. Corolla tube funnel-shaped. Stamen 1. Anther petaloid. Ovary 3-celled, style filiform. Sigma 2-lipped. Capsule ovoid opening by three valves; seeds many, small, oblong.

Flowering & Fruiting	April-June.
Habitat Ecology	Common in moist places; **Ali Khad** - 580 m; **Lyond Khad** - 617 m.
Material Examined	EBH-WL-1193; 22.05.2009.
Part Used	Rhizome.

Folk Uses

Rhizome powder in mustard oil or its paste applied on cuts and wounds as an **antiseptic**. Paste of roots of *Oroxylum indicum* (L.) Vent. and rhizome of *Curcuma longa* or its fresh rhizome grounded with leaves of *Annona squamosa* L. or paste of its leaves and roots of *Aristolochia indica* L. mixed with salt applied to heal **wounds**. Grounded fresh rhizome mixed in half litre of luke warm milk given for healing **fractured bones**. Paste of leaves of *Aristolochia bracteolata* Lamk., rhizome of *Curcuma longa* and fruit of *Piper nigrum* L. applied topically on cattle **wounds**. Paste of rhizome of *Curcuma longa* grounded with stem barks of *Holarrhena antidysentrica* Wall., *Litsea glutinosa* (Lour.) Robbins, seeds of *Sesamum indicum* L. and salt applied as a **plaster** cast for healing fractured limb of cattle.

Chemical Constituents

Important constituents are curcimin **(rhizome)** consisting of sessquiterpenes, zingiberene, D-a-phellandrene, turmerone, dihydroturmerone, γ- and α-alantolactone, curcumene and cineol **(essential oil)**.

Biological activities

Alternative amoebicidic, analgesic, antiulcer, hepatotoxic, hypocholesterolemic, mucogenic, ODC, phagocytotic, TNF, vermifuge activities +ve.

Uses in Literature

Describe so far for abscess, adenopathy, allergy, alzheimer, amoeba, amenorrhoea, anorexia, arthrosis, asthma, atherosclerosis, bacteria, bite, bleeding, boils, bronchosis, bruise, burns, cancer (abdomen, breast, colon), cardiopathy, cataract, catarrah, chestache, childbirth, cholecytosis, cold, coma, congestion, conjunctivitosis, constipation, cramp, cytosis, dyspepsia, dysuria, eczema, edema, elephantiasis, enterosis, epilepsy, epistaxis, fever, fibrosis, fungus, gall stone, gray hair, headache, hematesis, hematuria, high blood pressure, high cholesterol, hysteria, infection, inflammation, itch, jaundice, laryngosis, leprosy, leukaemia, leukoderma, lymphoma, malaria, mannia, morning sickness, mucososis, osteoarthrisis, sore, sore throat, rheumatism, ringworm, scabies, tumour, ulcer, worm and wound **(Biswas et al., 2010; Duke et al., 2002; Islam, 1989, 1996; Hosagoudar & Henry, 1996b; Jain, 1991, 1997; Kapur et al., 1996; Khan, 2003; Kumar, 2002; Lalramnghinglova, 2003; Martin & Retnam, 2005; Maheshwari et al., 1996; Pande et al., 2006; Prajapati et al., 2006; Ramdas et al., 2000;**

Rana et al., 2003; Rao & Henry, 1995; Rout & Panda, 2010; Saren et al., 2000; Singh & Prakash, 1996; Sinha & Sinha, 2001; Thakur et al., 1989; Tiwari & Pande, 2006).

Cyathocline purpurea (Buch.-Ham. ex D. Don) *Kuntze* (Plate No. 16F)

Syn. *C. lyrata* Cass.; *C. stricta* DC.; *C. lawii* Wight; *Tanacetum purpureum* Don; *T. viscosum* Wall.; *Artemisis hirsuta* Rottb

Family Asteraceae

Vern.	Gandh-bhadra.
Distribution	W. Himalaya.
Description	Much-branched glandular hairy, odorous annuals with purple heads in terminal corymbose cymes. Fruits achenes. Pappus none.
Flowering & Fruiting	April-May.
Habitat Ecology	Rocky slopes; **Sir Khad -** 620 m.
Material Examined	EBH-WL-1129; 22.05.2008.
Part Used	Whole Plant.
Folk Use	2 tsp, infusion of plant recommended thrice daily for 15 days for **bronchial affections**.
Chemical Constituents	**Essential oil** from plant contains mainly cyathocol, lyratol and lyratic acid, isoivangustin and a guaionlide.
Biological Activity	Et OH(50%) extract and essential oil from plant anthelmintic, antimicrobial, hypotensive, cardiac depressant, spasmolytic and oxytocic.
Uses in Literature	Known to be used as anthelmintic, antimicrobial, fish poison, spasmolytic, stomachic and for headache and intermittent fever **(Guha Bakshi et al., 1999; Jain, 1991; Maheshwari et al., 1996; Pande** *et al.*, **2006; Parrotta, 2001; Rastogi & Mehrotra, 1995b; Sood & Thakur, 2004).**

Cymbopogon citratus Stapf (Plate No. 17A)

Syn. *Andropogon citratus* DC

Family Poaceae

Vern.	Makoda Ghas.

English, Hindi, Sanskrit and Regional Names

Eng.	Lemon grass;

Hindi	Gandhatrina;
Sans.	Abschlatraka, Atigandha, Badhira, Badhiradhvani, Bodhana, Bhustrina, Bhutica, Bhutina, Chhatra, Goch hakka, Guhhala, Guhyabija, Gundardha, Jambu, Kapriya, Rohisha, Samalambi, Sugandha;
B.	Gandhabena;
Kan.	Majjigehullu;
Mal.	Vasanappulla;
Tam.	Karrapurapillu, Varanappilu;
Tel.	Chippagaddi, Nimmagaddi.
Distribution	Widely distributed over the tropics.
Description	Densely tufted, aromatic perennial grasses having underground, short, whitish or pale-violet stems and linear-amplexicaul, rough-margined leaves in dense clusters which are glaucous green on both sides. Inflorescence solitary with 1 fascicled spike. Rachilla hairy, glumes 2. Stamens 3. Fruit caryopsis.
Flowering & Fruiting	June -August.
Habitat Ecology	Meadows, Near Water channels in grassy grounds; **Lyond Khad -** 617 m; **Sauli Khad** - 765 m; **Sir Khad** - 620 m.
Material Examined	EBH-WL-1214; 22.06.2008.
Parts Used	Whole Plant. Roots. Stem. Leaves.
Folk Uses	15-20 ml decoction of the plant prescribed twice daily for 3-5 days for **abdominal diseases, jaundice** and **fever**. Bath with decoction of leaves used for high **fever** and **influenza**; also its extract (3-5ml) given on empty stomach for 3 days as an **insect repellent**.
Chemical Constituents	These are cymbopogone, cymbopogonol, α- and β- citrals, myrcene, methylleptanone, linalool, linalyl acetate, geranyl acetate, citronellol, nerol, luteolin and –c- glucosides **(whole plant)**.
Biological Activity	Abortifacient, analgesic, antimicrobial, antifungal, antiinflammatory, CNS depressant, glutathionigenic, stimulant, uterotonic and vermifuge activities found to be +ve.
Uses in Literature	Known as an appetizer, aphrodisiac, anthelmintic, diaphoretic, febrifuge, laxative, sedative, stimulant; and for anxiety, arthritis, ascaris, backache, bacteria, bronchitis, bruise, cold and cough, coryza, cholera, dermatosis, diabetes, diarrhoea, digestive problems, dysmenorrhea, dyspepsia, dysuria, flatulence, headache, high blood pressure, high cholesterol, infection, insomnia, influenza, jaundice, leprosy, malaria parasite, pneumonia, pulmonosis, pyorrhoea, pyrexia, rheumatism, ringworm, sprains, stomach ache, toothache, tuberculosis,

water retention, worm, wound, vomiting and yeast   (**Ambasta, 1986; Bhat Gopalakrishna & Nagendran, 2001; Chevallier, 1996; Chopra et al., 1956; Chun-Lin & Jieru, 1995; Duke et al., 2002; Gill et al., 1993, 1997; Harborne & Baxter, 2001; Idu et al., 2008; Jain, 1991; Kamble et al., 2010; Kirtikar & Basu, 1984; Kumar & Narain, 2010; Kurian, 1999; Nunez & Castro, 1995; Parrotta, 2001; Patil, 2009; Prajapati et al., 2006; Rana et al., 2003; Shiddamallayya et al., 2010**).

Cymbopogon martinii (Roxb.) Wats. (Plate No. 17B)

Syn.   *Andropogon martinii* Roxb

Family   Poaceae

Vern.	Palmarosa, Rusha Ghas.

English, Hindi, Sanskrit and Regional Names

Eng.	Palma rose, Rosha grass, Rusha grass;
Hindi	Gandhejas, Rushaghas;
Sans.	Dhyamakah, Rohisa;
Kan.	Harehullu;
Mal.	Sambharapullu;
Tam.	Kavattampillu, Munkilpul, Wraippul;
Tel.	Kamakchi-kassuvu.
Distribution	Cultivated in India.
Description	Erect, glabrous tufted perennials with flat, linear-lanceolate, amplexi-caul leaves and membranous ligules. Inflorescence compound. Spathe 3-6 cm long. Spikelets sessile. Lemma hyaline, palea absent. Anthers 3, yellow brown, Caryopsis oblong.
Flowering & Fruiting	September-February.
Habitat Ecology	Wastelands, Grasslands; **Chamyater Khad -** 613 m.
Material Examined	EBH-WL-1224; 28.08.2007.
Part Used	Whole Plant.
Folk Uses	Plant used as **fodder** grass. Fumigation using the whole plant considered good for **fever**.
Chemical Constituents	Geranoil yields gingergrass oil and palmrosa oil **(aerial parts)**.
Biological Activity	EtOH (50%) extract of plant exhibits CNS depressant property.
Uses in Literature	Known in India as acrid, antibacterial, antifungal, antimicrobial, antiseptic, colic, febrifuge, galactagogue and sudorific; and for baldness

and skin diseases, bronchitis, cardiac debility, cosmetics, dyspepsia, epileptic fits to children, heart diseases, pains, perfumery, leprosy, lumbago, soap, stiff joints and throat affections **(Asolkar et al., 1992; Bor, 1973; Jain, 1991; Maheshwari et al., 1996; Nag et al., 2009; Pande et al., 2006; Prajapati et al., 2006; Roy, 1984; Subramaniam, 2000; Uphof, 2001).**

Cynodon dactylon (L.) Pers. (Plate No. 17C)

Syn.   *C. erectus* Presl; *C. filiformis* Voigt Hort. Suburb.; *C. linearis* Willd.; *C. martimus* H. B. & K. Nov.; *C. repens* Dulac; *C. sarmentosus* S.F.Gray Nat.; *C. stellectus* Willd.; *C. virgatus* Nees; *Digitaria dactylon* Scop.; *D. littoralis* Salisb.; *D. maritima & radiata* Spring

Family   Poaceae

Vern.	Doob.

English, Hindi, Sanskrit and Regional Names

Eng.	Dhub grass, Bermuda grass, Bahama grass, Conch grass;
Hindi	Dhub, Hariyali;
Sans.	Dhruva;
B.	Durba, Dubhu, Dubla;
Kan.	Kudigarikai, Garikaihalu;
Mal.	Karuka, Karukappullu;
Mar.	Haryali, Karala;
P.	Dhub, Khabal, Talla;
Tam.	Arugam-pullu, Harvali;
Tel.	Gericha, Gaddi, Harveli.
Distribution	Distributed and grows almost everywhere.
Description	Creeping herbs which are rooting at the joints and sending smooth upward stems. Roots whitish, tough and creeping. Leaves tapering to a sharp point, ribbed with smooth sheath. Flowers purplish, arranged in spikes. Grains oblong, free with persistent glumes.
Flowering & Fruiting	August-September.
Habitat Ecology	Wild throughout India; **Gobind Sagar Lake** - 490 m; **Karyal Khad** - 615 m; **Sir Khad** - 620 m.
Material Examined	EBH-WL-1171; 18.08.2007.
Parts Used	Whole Plant. Leaves.

Folk Uses	Fresh juice of the herb as such or its extract with garlic and warm mustard oil or the leaf paste applied locally for **body pains** and **rheumatism**. Paste of the plant mixed with cow's or goat's milk given to check bleeding from **piles**. Also, 3-4 of its fresh leaves with 1-2 black pepper and a little sugar given to **strengthen memory** in children. Plant also well known as a pasture and **lawn grass**.
Chemical Constituents	Herb contains β-sitosterol, β-carotene, vitamin C, palmitic acid, triterpenoids, alkaloids-ergonavine, ergonavivine; ferulic, syringic, vanillin, p-coumaric acid; furfural, alcohol, glucose, fructose etc.
Biological Activity	Basal stems increase potency.
Uses in Literature	Reported earlier as antidiabetic, antiviral, emetic; and for abortion, amoebiasis, bleeding from nose, cough, diarrhoea, dog bite, dysentery, dysuria, fever, haemorrhoides, head ruling, leucorrhoea, restoring vigour, stomach ache and wounds **(Ambasta, 1986; Banerjee, 2000; Bhat Gopalakrishna & Nagendran, 2001; Bhogaonkar & Kanerkar, 2007; Chaudhary & Neogi, 2000; Dash & Misra, 2000; Hajra & Baishya, 1997; Jain, 1991; Kapur & Singh, 1996; Khan & Khanum, 2005; Khare & Khare, 2000; Kumar & Narain, 2010; Pande et al., 2006; Patil et al., 2007; Prajapati et al., 2006; Prakasha et al., 2010; Rana et al., 2003; Ranjan et al., 2000; Retnam & Martin, 2006; Rosakutty et al., 2000; Rout & Panda, 2010; Saini, 1996a; Saren et al., 2000; Satapathy & Brahmam, 2000; Semwal et al., 2010; Sharma, 2003; Sharma & Rana, 2000; Singh & Prakash, 1996; Singh & Pandey, 1996; Singh et al., 1996a; Siwakoti & Siwakoti, 2000; Siwakoti & Varma, 1996).**

Cyperus compressus L. (Plate No. 17D)

Family Cyperaceae

Vern.	Tuli Ghas.
Distribution	Throughout India.
Description	Tufted annuals with stems upto 50 cm high and obtuse leaves which are shorter than stem. Umbels simple. Spikelets 4-9, subdigitate, linear-oblong. Glumes 20-30, long, ovate. Stamens 3. Achenes obovate, triquetrous, dark-brown.
Flowering & Fruiting	August-December.
Habitat Ecology	Moist soils and water-logged fields; **Sauli Khad -** 765 m.
Material Examined	EBH-WL-1272; 12.11.2008.
Part Used	Leaves.
Folk Use	Used as a **fodder** grass.

Use in Literature	Described in India as fodder grass (**Bhat Gopalakrishna & Nagendran, 2001**).

Cyperus distans L.f. (Plate No. 17E)

Syn.   *C. elatus* Rottb.; *C. nutans* Presl; *C. jacquini* Schrad.; *C. graminicola* Staudt

Family   Cyperaceae

Vern.	Ganeechi.
Distribution	Himalaya upto 1,000 m, Sri Lanka and Singapore. All warm regions.
Description	Glabrous herbs with stolons clothed by dark brown elliptic acute scales. Stems upto 90 cm. Leaves often as long as stem, broad. Umbels copiously compound. Bracts leaf-like. Spikelets nodding; mature ones spreading at right angles, red, caducous. Stamens 3; anthers oblong. Achenes trigonous, brown.
Flowering & Fruiting	August-December.
Habitat Ecology	Damp shady places; **Matwana Khad -** 610 m; **Sir Khad -** 620 m.
Material Examined	EBH-WL-1131; 23.10.2007.
Part Used	Leaf-buds.
Folk Use	Leaf-buds used for **flavouring** curries.
Uses in Literature	Oil obtained used for various purposes (**Bhat Gopalakrishna & Nagendran, 2001; Biswas & Calder, 1984; Bor, 1973; Guha Bakshi et al., 1999; Jain, 1991**).

Cyperus flabelliformis Rottb. (Plate No. 17F)

Syn.   *C. involucratus* Rottb

Family   Cyperaceae

Vern.	Sola.
Distribution	Throughout India.
Description	Perennials with stout, horizontal rhizome and stems upto 90 cm tall. Leaves reduced to sheaths. Umbels compound, subcorymbose. Spikelets digitate, 10-20 flowered. Flowers greenish. Glumes imbricate, broadly ovate. Stamens 3. Achenes long, brown.
Flowering & Fruiting	August-December.

Habitat Ecology	Common in moist shady places; **Sir Khad** - 620 m.
Material Examined	EBH-WL-1132; 14.09.2007.
Part Used	Whole Plant.
Folk Use	Used as a good **fodder** grass.
Uses in Literature	Used earlier for fever, digestive disorders, dysmenorrhea, indigestion, vomiting; and as diuretic, emenagogue and ornamental plant (**Bhat Gopalakrishna & Nagendran, 2001**).

Cyperus iria L. (Plate No. 18A)

Syn.   *C. songaricus* Kard.; *C. semirudus* Moritzi Verz.; *C. diaphaniria* and *microiria* Steud.; *C. microlepis* Baker

Family   Cyperaceae

Vern.	Doob.
Hindi and Regional Names	
Hindi	Burachucha, Nagarmotha;
B.	Burachucha;
Mal.	Rumpul.
Distribution	Kashmir–Sri Lanka, Andamans.
Description	A glabrous short-lived weed with caespitose stems upto 60 cm high and grass-like leaves. Umbels compound. Spikes loosely spicate; rachis hairy. Glumes 3-5 nerved. Stamens 2 or 3. Anthers oblong, muticous. Achenes obovate-elliptic, triquetrous, surface black.
Flowering & Fruiting	July-October.
Habitat Ecology	Common weed of fields, wastelands and marshy places; **Auhr Khad** - 610 m; **Ghambar Khad** - 528 m.
Material Examined	EBH-WL-1146; 14.10 2007.
Parts Used	Whole Plant. Tubers.
Folk Uses	Plant used as **fodder**. 15-20ml decoction of the plant mixed with honey given twice daily for 3-5 days against **diarrhoea**. Decoction of its tuber and that of *C. rotundus* (1:1 ratio) checks **fever** (20 ml, thrice daily till cure).
Chemical Constituents	These are sesquiterpenoids, including alpha and beta-rotunol, mustakone, copane, sugeonol, sugetrin, cyperol, iso-kobusone, (+)-epoxy guaiene, (-rotundone, 4-alpha, 5-alpha- oxidoeudosm-11-en-3-alpha-ol, cyperenone and cyperone triterpenoid glycoside (**essential oils**).

Uses in Literature	Reported to be used as an astringent, stimulant, stomachic and tonic (**Asolkar et al., 1992; Bhat Gopalakrishnan & Nagendran, 2001; Caius, 2003; Clarke, 1973; Das et al., 1996; Jain, 1991; Kirtikar & Basu, 1984; Kumar & Narain, 2010; Parrotta, 2001; Rana et al., 2003; Samwatsar & Diwanji, 1999; Shukla & Verma, 1996; Tarafder et al., 1997; Uphof, 2001).**

Cyperus rotundus L. (Plate No. 18B)

Syn. *C. hexastachyus* Rottb.; *C. leptostachyus* Griff.; *C. retzii* Nees non Poir.; *C. tenuiflorus* Royle; *C. albidus* Herb.; *C. tuberosus* Sensu Cl. non Roxb.

Family   Cyperaceae

Vern.	Motha.

English, Hindi, Sanskrit and Regional Names

Eng.	Coco grass, Nut grass;
Hindi	Mustha, Motha, Mutha;
Sans.	Abda, Arnoda, Bhadrakshi, Bhadramusta, Gangeya, Granthi, Gundra, Hima, Kachola, Kakshottra, Kasheru, Krodeshtha, Kuru, Kurubilva, Kutannata, Muetaka, Murta,Musta, Mustaka, Sugandhigranthila, Valya, Varahi, Varida, Vindakhya;
B.	Motha, Mutha;
Kan.	Tungegadde;
Mar.	Motha;
Tam.	Korai;
Tel.	Tungamuste.
Distribution	A pestiferous weed. All warm regions.
Description	Glabrous weeds with subsolitary stems, slender stolons, hardening into wiry roots and long leaves often overlapping stem. Umbels frequently compressed, large. Glumes imbricating, ovate. Nuts oblong, trigonous, apiculate.
Flowering & Fruiting	July-October.
Habitat Ecology	Weed of moist soils, irrigated lands, marshy areas; **Gobind Sagar Lake -** 490 m.
Material Examined	EBH-WL-1173; 12.08.2008.
Parts Used	Roots. Rhizome. Tubers.
Folk Uses	Leaves used for **thatching** huts; its decoction given for healing **wounds**; and the juice taken orally to cure **fever** (10-20 ml, thrice daily

for 3-5 days). Root paste applied on wounds to remove **foetid smell** and **pus**. Tuberous rhizomes eaten **fresh**.

Chemical Constituents   A new saponin-oleanolic acid-3-oneohesperiloside (I), rotundone, flavonoid glycoside, cyponene 1& 2, patchoulenone, cyperol and isocyperol isolated from **tubers**.

Biological Activity   Alcoholic extract of tuber exhibits anti-inflammatory, antipyretic, analgesic, hypotensive activity in cats.

Uses in Literature   Known so far as an anthelmintic, diaphoretic, diuretic, emmenagogue, galactagogue; and for abdominal pains, dermatitis, diarrhoea, fever, hyper-acidity, dysmenorrhea, indigestion, intermittent fevers, jaundice, joint pains, lactation, malarial fever, nausea, white leucorrhoea, wounds, sores, stomach pains, traumatic injuries, ulcers and vomiting **(Ansarali & Sivadasan, 2009; Bhat Gopalakrishna & Nagendran, 2001; Cauis, 2003; Das et al., 1996; Goud et al., 2000; Islam, 1996; Jain, 1991; Jain et al., 1991; Joshi, 2009; Kapur & Singh, 1996; Khan & Khanum, 2005; Kirtikar & Basu, 1984; Kumar & Narain, 2010; Martin & Retnam, 2005; Mishra, 2008; Parabia & Pathak, 2007; Panda & Das, 2000; Pande et al., 2006; Parrotta, 2001; Prajapati et al., 2006; Punjani, 2002; Retnam & Martin, 2006; Rout & Panda, 2010; Saini, 1996b; Sarkar et al., 2000; Sen & Behera, 2009; Shukla & Verma, 1996; Singh, 2000, 2003c; Singh & Rao, 2003; Siwakoti & Varma, 1996; Sudhakar & Vedavathy, 2000; Thomas & Britto, 2000).**

Datura stramonium L. (Plate No. 18C)

Syn. *D. ferox* Nees; *D. inermis* Jacq.; *D. stramonium* var. *tatula* L.; *D. tatula* L.; *D. inermis* Jacq.; *D. wallichii* Dunal; *Stramonium vulgatum* Gaertn

Family Solanaceae

Vern.	Dhatura.

English, Hindi, Sanskrit and Regional Names

Eng.	Apple of peru, Devil's apple, Devil's trumphet, Dewtry, Jamestown weed, Jimson weed, Mad apple, Stink weed, Thorn apple;
Hindi	Dhatura;
Sans.	Devika, Dhattura, Dhatura, Dhurtakrit, Dhustura, Dhultura, Ghanta-pushpa, Kalama, Kanaka, Kanakaohaya, Kantaphala, Madanaka, Mahamohi, Mahashatha, Matulaka, Savisha, Shivapriya, Shivasekhara, Shyama, Turi, Unmatta, Unmattaka;
B.	Sadadhutura;
G.	Dholo dhaturo;
Kan.	Ummatta;
Mal.	Matulam, Ummam, Ummata;
Oriya	Dhtura, Sukladhutura;
P.	Dattura, Tattur;
Tam.	Emanamam, Simalyumattai, Turutturam, Umattai, Vellumattai;
Tel.	Dutturamu, Tellavummetta, Ummetta.
Distribution	Temperate Himalaya, (Kashmir-Sikkim) : 200-2,200 m.
Description	Coarse annuals with ovate, toothed or sinuate leaves and ovate-lanceolate calyx. Flowers axillary. Corolla white. Capsule equally spinous on all sides. Seeds reniform.
Flowering & Fruiting	June-October.

Habitat Ecology	Waste grounds, cultivated areas; **Drobad Khad -** 712 m; **Tatoh Khad -** 710 m.
Material Examined	EBH-WL-1195; 19.06.2008.
Parts Used	Roots. Leaves. Flowers. Fruits. Seeds.
Folk Uses	Flowers considered sacred for offering to **appease** Lord Shiva. An amulet of its root worn on hand to cure **fever**. Inhalation of smoke of burning leaves believed to cure **asthma**. Fruits **intoxicating**. 1 g powdered seeds prescribed with cow's milk for 3 days to check **fever**.
Chemical Constituents	Active properties due to hyoscamine, hyoscine, scopolamine, fluorodaturative, homofluoro-daturatine **(seeds)**, a new vitanolide-vitastra-monolide along with daturalactone and its ketoderivative characterized as 5a, 12a, 27-trihydroxy-1-oxo-6a, 7a- epoxy-22R-vita2, 24-dienolide **(leaves)**, hyoscine, 3a, 6b-apoatropine, tigcoidine, tropine, meteliodine **(plant)**, 0.3 to 0.5% of alkaloids chiefly as hypslyamine with smaller amount of hyoscine and atropine **(dried leaves, flowering tops)**.
Biological activities	Analgesic, anesthetic, antiinflammatory, antiseptic, aphrodisiac, CNS-stimulant, fungicidal, hallucinogenic, hypnotic, sedative activities found to be +ve.
Uses in Literature	Recorded in India as anasarca, anodyne, antirheumatic, antispasmodic, mydriatic, narcotic in treatment of bronchitis and psychoactive plant; and for abscess, acidity, adenopathy, alopecia, asthma, boils, breast inflammation due to excessive formation of milk, burn, cancer, carcinoma, cardiopathy, catalepsy, catarrh, childbirth, cholera, cigarettes (leaves), controlling salvation and exparasite, dental caries, diarrhoea, dislocation of joints, dysmenorrhoea, enlarged testicles, epilepsy, haemorrhoids, headache, inflammatory pains, itch, jaundice, lactation (vet), milching (vet), nervous disorder (vet), nymphomania, painful diseases of rectum, paralysis, post encephalitic parkinsonianism, purpurea, radiculosis, respirosis, rheumatism (vet), sciatica, skin diseases, stomach complaints, swelling of neck (vet), thirst, tremor, trismus, tonsilitis, toothache, travel sickness, tuberculosis, tumour, typhus, ulcers, uterosis vicious indulgence, wart, whitlow and wound **(Chopra et al., 2005; Duke et al., 2002; Henry, 1999; Jain, 1968, 1991; Jain et al., 1994; Jha et al., 1996; Kaushik & Dhiman, 2000; Lindley, 1981; Manandhar, 1996a; Pande et al., 2006; Prajapati et al., 2006; Rana et al., 2003; Rastogi & Mehrotra, 1995b; Satapathy, 2008; Singh, 1991; Sood et al., 2009b; Upadhye et al., 1994; Watt, 1972).**

Debregeasia hypoleuca **Wedd. (Plate No. 18D)**

Syn.   *D. bicolor* Wedd.; *Urtica bicolor* Roxb.; *Boehmeria salicifolia* Don; *B. hypoleuca* Hochst.; *Missiessya hypoleuca* Wedd.; *Morocarpus salicifolius* Blume

Family   Urticaceae

Vern.	Shyaru.
Hindi and Regional Names	
Hindi	Sansaran, Siran;
P.	Sansaru, Pincho;
U.P.	Sansaru, Siaru, Tusarro, Tushiari.
Distribution	W. Temperate Himalaya : Kashmir - Kumaon.
Description	A large sturdy shrub having alternate leaves and branches and oblong-lanceolate leaves which are clothed with snow-white wool. Flowers yellow, unisexual. Heads conglobate sessile. Fruits yellow.
Flowering & Fruiting	January-April.
Habitat Ecology	Frequent on rocky slopes, forests, shrubberies; **Tatoh Khad -** 710 m.
Material Examined	EBH-WL-1121; 06.03.2009.
Parts Used	Roots. Bark. Leaves. Fruits. Wood.
Folk Uses	Leaves considered as good cattle **fodder**. 10-20ml decoction of leaves good against **hernia** and **epilepsy** (once daily for 15-20 days). Raw fruits eaten to dissolve **kidney stones**. Poultice of paste of roots applied for killing maggots in **wounds** of cattle and goats. Bark paste bandaged for treating **bone fracture**. Wood used as **fuel**.
Uses in Literature	Used in India for cordage, edible purposes, fishing lines, stomach pain, ulcers and killing maggots in wounds of cattle and goats **(Ambasta, 1986; Kapur & Nanda, 1996b; Rana et al., 2003; Raju, 2000; Rama Rao & Henry, 1996; Roy et al., 1998; Sood et al., 2009b; Uniyal, 2003; Watt, 1972).**

Dichanthium annulatum (Forssk.) Stapf. (Plate No. 18E)

Syn.   *Andropogon annulatus* Forsk

Family   Poaceae

Vern.	Kirdu.
Distribution	Throughout India.
Description	A variable perennial grass with dense tufts and pubescent nodes. Inflorescence subdigitate. Racemes 7 cm long. Spikes digitate, sub-sessile, joints many. Pedicels linear-filiform. Lower pairs of spiklets homogamous male or neuter. Sessile spiklets dorsally compressed. Pedicelled spikelets awnless.
Flowering & Fruiting	September-December.
Habitat Ecology	Grows along roadsides and near water channels; **Karyal Khad** - 615 m; **Lyond Khad -** 617 m.
Material Examined	EBH-WL-1120; 12.10.2007.
Part Used	Whole Plant.
Folk Use	Regarded a good **fodder plant** by the locals.
Uses in Literature	Reported as diuretic and for kindey complaints **(Asolkar et al., 1992; Bhat Gopalakrishna & Nagendran, 2001; Bor, 1973; Roy, 1984, Uniyal et al., 2002).**

Dicliptera roxburghiana Nees (Plate No. 18F)

Syn.   *D. bupleuroides* Nees

Family   Acanthaceae

Vern.	Saundi.
Regional Name	
P.	Kirach, Somni.
Distribution	Throughout India.
Description	Erect hairy herbs, which are 30-90 cm tall with grooved stem. Flowers purple, in compact terminal and axillary clusters. Capsules 4-seeded. Seeds distinctly verrucose.
Flowering & Fruiting	October-December.

Habitat Ecology	Frequent on slopes, rock crevices; **Ali Khad -** 680m; **Sauli Khad -** 765 m.
Material Examined	EBH-WL-1153; 03.10.2007.
Part Used	Whole Plant.
Folk Use	Used as a **tonic**.
Chemical Constituents	3,5-pyridinedicarboxamide, lupeol and allantoin, stigmasterol and β-sitosterol with their glucosides isolated from **aerial parts**.
Uses in Literature	Known earlier for checking bleeding, cough, gastroenteritis and wounds **(Kapur & Nanda, 1996a; Kapur & Singh, 1996; Sood & Thakur, 2004; Pande et al., 2006; Rastogi & Mehrotra, 1995b).**

Digitaria griffithii (Hook.f.) Henn. (Plate No. 19A)

Syn.   *D. sanguinalis* Scop.; *Panicum corymbosum* Thw.; *P. griffithii* Arn

Family   Poaceae

Vern.	Ghuni.
Distribution	Throughout India.
Description	Prostrate annuals having glabrous leafsheath, linear- flat, acuminate leaf-blade and drooping inflorescence of 5-10 racemes. Glumes 4. Lodicules 2, small. Racemes with interrupted loose groups of spikelets; pedicels of upper spikelets up to 0.5 cm long, scabrid. Lower glumes sometimes missing in the upper spikelets. Fruit a caryopsis with adherent pericarp.
Flowering & Fruiting	October-December.
Habitat Ecology	Found growing in shady places; **Ali Khad -** 580 m.
Material Examined	EBH-WL-1215; 06.11.2007.
Part Used	Whole Plant.
Folk Use	Reported to be used as good **fodder** grass by the locals.
Uses in Literature	Known earlier in India as fodder grass **(Bhat Gopalakrishna & Nagendran, 2001; Bor, 1973; Mudaliyar, 1921).**

Dioscorea belophylla (Prain) Voigt. ex Haines. (Plate No. 19B)

Syn.   *D. cliffortiana* Lam.; *D. decemanglelaris* Herb.; *D. glabra* Auct.; *D. hetero-phylla* Roxb.; *D. pulchella* Roxb.; *D. thunga* Herb.; *D. versicola* Herb.; *Helmia bulbifera* Kunth

Family   Dioscoreaceae

Vern.	Tardi.
Distribution	Sub-Tropical Himalaya.
Description	Climbers possessing opposite, ovate-lanceolate leaves and flowers in spikes. Capsules membranous. Seeds winged all round.
Flowering & Fruiting	September-November.
Habitat Ecology	Climbers near waterbodies in forests; **Papral Khad** - 740 m.
Material Examined	EBH-WL-1154; 11.09.2007.
Parts Used	Tubers. Leaves. Seeds.
Folk Uses	Tubers relished as **vegetables**. 5 mg powdered leaves given thrice daily for 10 days to cure **leucorrhoea**. Powder of seeds commonly used by magicians for performing water related **magics**.
Chemical Constituents	**Tubers** rich in diosgenin.
Biological Activity	Antiinflammatory and antifertility activities (+)ve.
Uses in Literature	Useful earlier as an abortifacient, anthelmintic, antifertility, aphrodisiac, blood purifier, edible tubers, stomachic, tonic; and for abdominal pains, asthma, biliousness, bronchitis, dysentery, dyspepsia, flatulence, headache, leucoderma, piles, sores and tumours **(Ambasta, 1986; Jain, 1991; Lal et al., 1996; Lalramnghinglova, 1996; Pande et al., 2006; Rana et al., 2003; Singh, 1996; Sinha, 1996; Sood et al., 2009b).**

Dioscorea bulbifera L. (Plate No. 19C)

Syn. *D. alata* L.; *D. acutangula* Wall.; *D. anguliflora* Wall.; *D. japonica* Wall.; *D. odoratissima* Wall.; *D. pulchella* Hohen.; *D. pulchelia* Roxb.; *Helmia bulbifera* Kunth

Family Dioscoreaceae

Vern.	Dragal.

English, Hindi, Sanskrit and Regional Names

Eng.	Aerial yam; Black bearing yam, Bulp-bearing yam, Potato yam, Air potato;
Hindi	Pitaalu, Ratalu, Sinalu;
Sans.	Badar kachna, Vishvaksenapriya;
Ass.	Kathalu, Mati-alu;
B.	Banalu, Gaichalu, Kukuralu;
Bo.	Hadukaranda;
G.	Salvinavelya, Suariya;
Kan.	Heggenasu, Onthalaigasu, Tung-genasu;
Mal.	Kattu-kachil;
Mar.	Manakund, Gathalu, Karukarinda;
Oriya	Pita-alu;
P.	Zaminkand;
Tel.	Chedupaddu-dumpa, Malakakayapendalamu, Pannukelangu.
Distribution	Tropical India, Himalaya upto 2,100 m.
Description	Glabrous climbers with very large roots and stout acutely angled or winged stem. Leaves opposite, ovate and stout petiole. Flowers yellowish-green but fading later to dull red. Capsules with seeds winged at the base.
Flowering & Fruiting	September-February.
Habitat Ecology	Deciduous moist forests, near water channels, Wastelands, Roadsides; **Panyala Khad -** 810 m.
Material Examined	EBH-WL-1194; 12.08.2007.
Parts Used	Roots. Tubers. Leaves.
Folk Uses	Leaves used in the preparations of local **tea.** Tubers eaten as a **vegetable.** Poultice of its tuber paste used effectively as an **antidote** against

scorpion sting or insect bites or given orally to **alleviate pain** in hernia. Also, tubers often employed for **sacrificial** rites by Tantriks.

Chemical Constituents   **Tubers** contain diosbulbin, three new furanoids nortiterpenes- diosbulbins A, diosbulbin B and disobulbin C.

Biological Activity   Antiinflammatory and antifertility activities (+)ve.

Uses in Literature   Described so far as acrid, contraceptive, febrifuge; and for abdominal pain, boils, bone fracture, cough, cold, cuts, dysentery, epitaxis, goitre, gout, haemoptysis, indigestion, insect bite, itching, jaundice, lactation, leucoderma, orchitis, pharyngitis, piles, pyogenic infections, rheumatism, scrofula, skin infections, snake bite, sprains and injuries, starch extraction after elimination of calcium oxalate, stomach pain, syphilis, ulcers and wounds **(Ambasta, 1986; Arora, 1997; Borthakur, 1996; Brahmam et al., 1996; Chowdhery, 1996; Dam & Hajra, 1997; Deokota & Chhetri, 2009; Girach et al., 1999; Hosagoudar & Henry, 1996a; Jain, 1991; Jain et al., 1994; Lal et al., 1996; Lalramnghinglova, 1996; Maheshwari et al., 1996; Mandal & Basu, 1996; Mishra, 2008; Nag et al., 2009; Pande et al., 2006; Parrotta, 2001; Patil, 2009; Prajapati et al., 2006; Rana et al., 2003; Rao & Henry, 1995; Retnam & Martin, 2006; Saini, 1996a, b; Singh & Prakash, 1996; Singh & Zaheer, 1996; Sinha, 1996; Siwakoti & Varma, 1996; Uniyal et al., 2002; Watt, 1972).**

Echinochloa frumentacea Link (Plate No. 19D)

Syn.   *Panicum frumentaceum* Roxb

Family   Poaceae

Vern.	Madir.

Hindi, Sanskrit and Regional Names

Hindi	Janlisamak, Samak, Sannwa, Sawa, Shama, Shamula;
Sans.	Aripriya, Rajadhanya, Shyama, Shyamaka, Sukamara, Tribija;
B.	Samrashama, Sanwa, Saon, Shamula, Syamadhan;
G.	Samo, Samoghas;
Kash.	Karin, Soak;
P.	Chandra, Sama, Samuka, Sanwak, Sanwank, Soak;
Tam.	Kudraivallipillu, Railpillu;
Tel.	Bontachamala, Sanwa, Saon, Shamula, Syamadhan.
Distribution	Cultivated over the greater parts of India; ascending to Himalaya upto 2,100 m.
Description	Tall robust herbs with erect stems which are 60-120 cm high and nodding panicles. Spikes incurved, crowded; spikelets, unequally pedicelled, glabrous, cuspidate. Anthers 3, yellow. Caryopsis long, elliptic.
Flowering & Fruiting	July-February.
Habitat Ecology	Hilly areas near water channels; **Koshriyan Khad -** 520 m.
Material Examined	EBH-WL-1196; 02.09.2007.
Parts Used	Whole Plant. Seed.
Folk Uses	Plant **eaten** raw for its sweet taste. Also, considered good **fodder** for livestock.

Uses in Literature	Useful in biliousness, constipation, flatulence, sterlity; and as acrid and oleaginous coolant **(Bews, 1979; Bor, 1973; Cauis, 2003; Kirtikar & Basu, 1984; Pande et al., 2006; Rana et al., 2003; Rao & Henry, 1995; Roy, 1984; Uphof, 2001).**

Eleusine indica Gaertn. (Plate No. 19E)

Syn.   *E. distachya* Trin.; *E. distans* Moench; *E. domingensis* Sieber; *E. gouini,inaequalis, rigidifolia & scabra* Fourn.; *E. gracilis* Salisb.; *E. marginata* Lindl.; *E. marginata* Lindl.; *E. tristachya* Lamk.; *Leptochloea pectinata* Kunth; *Panicum compressum* Forsk.; *Paspalum dissectum* Kniphof.; *Triticum germinatum* Spreng

Family   Poaceae

Vern.	Rajputana.

English, Hindi and Regional Names

Eng.	Crowfoot, Dog's tail grass, Crabgrass, Goose grass, Wire grass;
Hindi	Malankuri, Mandla;
G.	Abdaunagli;
Oriya	Nandia;
Tam.	Thipparagi;
Tel.	Karuchodi, Kuror, Karsodi.
Distribution	Throughout India (ascending 1,750in the Himalaya) - Penang and Sri Lanka.
Description	Erect annuals with stout, soft stem and distichous flaccid leaves. Sheaths flattened, ciliate. Ligules obsolete. Spikes in terminal whorls; spikelets glabrous, ovate, subacute. Rugose glumes membranous, unequal, lower oblong-ovate; upper longer than the lower. Caryopsis oblong, obtusely trigonous. Seeds oblong.
Flowering & Fruiting	August-November.
Habitat Ecology	Very common in damp places; **Ali Khad** - 580m; **Sir Khad -** 620 m.
Material Examined	EBH-WL-1197; 16.09.2007.
Part Used	Whole Plant.
Folk Uses	Plant regarded as **soil binder** and good **fodder** grass. 5-10 ml decoction of the plant given to children thrice daily till cure against **convulsions**. Stems used for making **mats**. Paste of its rhizome crushed with tuber of *Stephania glabra* applied locally to cure **itches**. 1-3 g powdered roots given in the morning with milk for 3-5 days to check **fever**.

Chemical Constituent	The **plant** contains nitrates.
Uses in Literature	So far used as antipyretic, depurative, diaphoretic, diuretic, febrifuge, food for livestock, laxative, stomachic, sudorific; and for influenza, hypertension, liver complaints, oliguria and retention of urine **(Bews, 1979; Bhat Gopalakrishna & Nagendran, 2001; Bor, 1973; Bora, 2000; Cauis, 2003; Duthie, 1888; Jain, 1991; Kirtikar & Basu, 1984; Kumar & Narain, 2010; Mehra, 1982; Mudaliyar, 1921; Prajapati et al., 2006; Rana et al., 2003; Retnam & Martin, 2006; Saxena et al., 1997; Singh et al., 1996a; Uphof, 2001).**

Emilia sonchifolia (L.) DC. DC. (Plate No. 19F)

Syn.   *Cacalia sonchifolia* (L.) DC

Family   Asteraceae

Vern.	Pooshathala.

English, Hindi, Sanskrit and Regional Names

Eng.	Chikweed;
Hindi	Hirankhuri;
Sans.	Susarulish;
Ass.	Moralia;
B.	Sadimodi, Sudhimudi;
Mal.	Mulshevi; Muyaluevi;
Mar.	Sadamandi;
Tam.	Muyalccevi.

Distribution	Common throughout India from Punjab – Sri Lanka, ascending to 1,350m in the hills.
Description	Scabrid or puberulous, glabrous herbs with purple florets in the capitulum. Achenes 5-ribbed, scabrid; style-arms cylindric, tip conic.
Flowering & Fruiting	December-April.
Habitat Ecology	Moist places, paddy fields, disturbed grounds, wild; **Barthin Khad -** 585 m.
Material Examined	EBH-WL-1152; 28.03.2009.
Parts Used	Roots. Leaves.
Folk Uses	Plant paste applied externally on the **painful** area. Decoction of the plant mixed with sugar given for **bowel complaints**. Tender leaves and stalks eaten raw or boiled as a **vegetable**. 20 g of its leaves. boiled

	with 50 ml Gingelly oil and white of an egg consumed for 3 days to check **body pain**. A garland of its roots tied around neck of babies before sunrise considered good against **fever** due to evil eyes.
Chemical Constituents	These are flavonoids including rutin, n-hexacosanol, triacontane, ursolic acid and vitamin E **(aerial parts)** and toxic nitrate concentrates **(herb)**.
Biological Activity	Leaves eaten as **potherb**.
Uses in Literature	Plant reported earlier as antiasthmatic, antimicrobial, antipyretic, astringent, febrifuge, ophthalmic, sudorific, thermogenic, vulnerary; and for diarrhoea, eye diseases, gastropathy, night blindness, otalgia **(Ambasta, 1986; Bhalla et al., 1996; Guha Bakshi et al., 1999; Jain, 1991; Kaushik & Dhiman, 2000; Kirtikar & Basu, 1984; Lindley, 1981; Manandhar, 1996a; Pande et al., 2006; Parrotta, 2001; Patil, 2009; Patil et al., 2007; Prajapati et al., 2006; Rastogi & Mehrotra, 1995a; Siwakoti & Varma, 1996; Tiwari & Tiwari, 1996; Uphof, 2001; Viswanathal et al., 2006)**.

Equisetum arvense L. (Plate No. 20A)
Family   Equisetaceae

Vern.	Brahmgund, Rugosika, Sehet bund.
English Name	
Eng.	Field horsetail.
Distribution	Himalayan regions.
Description	Perennial herb with a yellowish fruiting stem growing upto 30 cm. Leaves needle-shaped. Spores in clusters. Cones at tip of fertile stems.
Flowering & Fruiting	July-October.
Habitat Ecology	Common on moist grounds; **Sir Khad -** 620 m; **Karyal Khad -** 615 m.
Material Examined	EBH-WL-1198; 24.09.2007.
Part Used	Whole Plant.
Folk Uses	Two tsp decoction of fresh plant taken thrice a day for ten days against burning **sensation** in urine and as **diuretic**. Its decoction also used as drops for nasal **polypus** and as wash for **tumours** and **cancerous lesions** of bones. Poultice of plant good against **cancerous ulcers**.
Chemical Constituents	These are triacontanedioic, octacosanedioic acids polyphenolic acids **(plant)** and Me esters of protocatechuic and caffeic acids **(stem)**.
Biological activities	Antibacterial, antiedemic, antispasmodic, aquaretic, bitter, carminative, diaphoretic, diuretic, tonic, cooling, vulnerary activities exhibited.

Uses in Literature	Considered useful for alopecia, arthrosis, bacteria, bladder stone, bleeding, brittle nails, burns, cancer (breast, gastro-urinary system intestines, kidneys, liver, stomach, tongue,) choleocytosis, chronic swelling of the legs, conjunctivosis, constipation, dentition, emphysema, fever, dropsy, gout, gravel and renal affecions, nose bleeding, rheumatic and arthritic problems **(Ambasta, 1986; Asolkar et al., 1992; Duke et al., 2002; Kaul, 1997; Manandhar, 1996b; Pande et al., 2006; Prajapati et al., 2006; Rastogi & Mehrotra, 1993; Saini, 2008; Singh & Viswanathan, 1996; Vasudeva, 1999; Viswanathan, 2000; Uphof, 2001).**

Equisetum debile Roxb. (Plate No. 20B)

Syn.   *E. ramossisimum* Desf. ssp. *debile* Hauke

Family   Equisetaceae

Vern.	Fox tailed asparagus.
English, Hindi and Regional Names	
Eng.	Bandukei, Horse tail, Nari, Trotak;
Hindi	Jortor;
P.	Trotak, Nari, Bandukeri, Buki, Malli.
Distribution	Along shady streams all over India. Temperate regions of Uttrakhand and upto 2,100 m.
Description	Rigid aquatic pteridophytic perennial herbs having long subulate-acuminate, black leaf teeth. Cones or spikes 0.8-1.8 cm long, sessile. Peltate sporophylls orbicular or oblong. Sporangia oblong, yellow. Spores 5 cm long, oval shaped.
Flowering & Fruiting	August-December.
Habitat Ecology	Bogs, margins of ponds, moist shady corners; **Ali Khad -** 580 m; **Marol Khad -** 610 m.
Material Examined	EBH-WL-1225; 25.11.2007.
Parts Used	Whole Herb. Leaves.
Folk Uses	Poultice of the plant applied for curing **bone fracture**. Decoction of the plant (15-20ml) given thrice daily for 15 days against **gonorrhoea**. 20-30 ml juice of leaves given thrice a day till cure for **checking fever**.
Chemical Constituents	Kaempferol-3-sophroside -7- glucoside and kaempferol -3- scophoroside isolated from the **plant**.
Uses in Literature	Recorded in India as coolant, diuretic; and for bone fracture, diarrhoea, enteritis, eye inflammation, fever, influenza, hepatitis and

gonorrhoea (**Ambasta, 1986; Bhattacharjee, 2001; Chopra et al., 1956; Joshi, 2008; Pande et al., 2006; Rastogi & Mehrotra, 1993; Sharma, 2003; Singh & Viswanathan, 1996; Uphof, 2001**).

Eriophorum comosum **Wall. (Plate No. 20C)**

Syn.   *E. arundinaceum* Wall.; *Scirpus comosus* Wall.; *S. elongatus* Wall.; *Trichophorum comosum* Strachey.; *T. arundinaceum* Strachey

Family   Cyperaceae

Vern.	Bagad.
English Name	
Eng.	False bhabbar.
Distribution	Occuring commonly in Himalaya.
Description	Glabrous robust stems with leaves only near base and overtopping stem. Umbels compound. Spikelets mostly solitary, narrowly ellipsoid; bracts long. Glumes acute, green. Nuts smooth brown-black.
Flowering & Fruiting	July-September.
Habitat Ecology	Moist shady places; **Karyal Khad** - 615 m; **Panyala Khad** - 650 m.
Material Examined	EBH-WL-1156; 24.09.2007.
Parts Used	Stem.
Folk Uses	**Brush** made from its stem used for giving a fresh coat of soil mixed with coal and dung to 'chullha' (earthern furnace) and **walls**. **Ropes** made of its stem fibres used for various purposes.
Use in Literature	Known for making ropes (**Clarke, 1973; Rana et al., 2003**).

Eupatorium adenophorum **Spreng. (Plate No. 20D)**

Syn.   *Ageratina adenophora* (Spreng.) King & Roxb

Family: Asteraceae

Vern.	Kalibasuti.
Distribution	Native of Mexico; found in Himalaya upto 1,800m.
Description	Branched herbs with homogamous heads and glandular hairy peduncles. Achenes 5-ribbed. Pappus 1-seriate.
Flowering & Fruiting	February-May.

Habitat Ecology	Common on moist slopes; **Sauli Khad -** 765 m.
Material Examined	EBH-WL-1216; 24.03.2009.
Part Used	Leaves.
Folk Uses	Paste of crushed leaves applied on **cuts** and **injuries** to check **bleeding**; taken against severe **stomach aches**. Also, juice of leaves used as drops for **eye diseases**.
Chemical Constituents	Found to contain isohexacosane, n-hexacosanoic acid, β-amyrin, stigmasterol, lupeol, taraxasterol, salvigenin, epifriedelinol **(aerial parts)**, n-dotriacontane, β-sitosterol, stigmasterol, tarasteryl palmitate and taraxasteryl acetate **(leaves)**.
Biological Activity	Plant spasmolytic.
Uses in Literature	Considered useful for body swellings, cancerous growth, cuts, injuries, malaria, snake-bite, stomach ache, toothache, tumour, wounds; and as shampoo **(Abraham, 1997; Biswas et al., 2010; Deokota & Chhetri, 2009; Gangwar & Ramkrishnan, 1990; Hajra & Baishya, 1997; Hosagoudar & Henry, 1996b; Jamir et al., 2008; Joshi, 2008; Kharkongor & Joseph, 1997; Kumar, 2002; Mandal & Basu, 1996; Panda, 1996; Pande et al., 2006; Rao & Jamir, 1982; Semwal et al., 2010).**

Euphorbia geniculata Ort. (Plate No. 20E)

Syn.   *E. hetrophylla* L.; *E. prunifolia* Jacq.; *Poinsettia hetrophylla* (L.) Klotz. & Garcke *ex* Klotz

Family   Euphorbiaceae

Vern.	Khabad-dudhia.
Distribution	Found apparently wild and as an escape in the Sutlej valley.
Description	Hispid small plants possessing leafy prostrate stems, perennial root, opposite, coriaceous, obovate oblong leaves and minute stipules. Involucres axillary, gland usually without a limb. Capsules small with quadrangular seeds which are bluntly pointed.
Flowering & Fruiting	August-November.
Habitat Ecology	Screes, Moist slopes; **Kothi Khad -** 810 m.
Material Examined	EBH-WL-1217; 04.09.2008.
Parts Used	Aerial Plant Parts. Leaves. Seeds.
Folk Uses	2 tsp decoction of the plant prescribed once every morning for 5 days against **constipation**. Paste of seeds and leaves useful against **skin eruptions**.

Chemical Constituents	Alanine, cysteine, serine, aspartic acid, methionine, proline, glutamic acid, glucose, rhamnose, galactose **(leaf extract)**, kaempferol, its 3-rutinoside, quesactin, its 3- rhamnoside, quercitrin, β-amyrin acetate, β- sitosterol, campesterol, stigmasterol, cholesterol, geniculatoside, steroidal galactoside, stigma 16-en-3α-o-(beta-D-galactopyranoside), geniculatoside F **(aerial parts)** have been detected.
Biological Activity	Plant spasmogenic.
Uses in Literature	Used earlier against asthma, blisters, boils, bronchiatis, scorpion sting **(Asolkar et al., 1992; Jain, 1991).**

Euphorbia helioscopia L. (Plate No. 20F)

Syn. *Euphorbion helioscopium* (L.) St.-Lag.; *Galarhoeus helioscopius* (L.) Haw.; *Tithymalus helioscopius* (L.) Hill

Family Euphorbiaceae

Vern.	Bada-dudhia.
English, Hindi and Regional Names	
Eng.	Cat's milk, Churn-staff, Sun spurge;
Hindi	Hirruseeah, Mahab;
P.	Chatriwal, Dudal, Gandabute, Kulfadpdak.
Distribution	Punjab and W. Himalaya; introduced elsewhere.
Description	Erect annuals which are dichotomously branched above. Stems stout and copiously branched. Leaves long and membranous. Involucres turbinate; lobes oblong. Glands fimbriate. Capsules smooth globose. Seeds reticulately pitted, subglobose.
Flowering & Fruiting	March-August.
Habitat Ecology	Moist places, irrigated channels, nallahs; **Sir Khad -** 620 m.
Material Examined	EBH-WL-1176; 08.03 2009.
Parts Used	Leaves.
Folk Uses	Paste of leaf applied externally (twice daily) against **scorpion sting**.
Chemical Constituents	Twelve euphoscopins A-L, four epieuphoscopine A,B,D and F, eleven euphorins A-K, two euphohelioscopins A ans B, euphorin and euphohelinone isolated from **leaves**.
Biological Activity	Aerial parts diuretic.

Uses in Literature	Known to be used against cholera, eruptions; and as anthelmintic, cathartic and laxative **(Ambasta, 1986; Guha Bakshi et al., 2001; Kapur & Singh, 1996; Kaul, 1997; Kirtikar & Basu, 1984; Kumar & Nagiyan, 2006; Kumar & Naqshi, 1990; Rastogi & Mehrotra, 1991; Sood & Thakur, 2004; Swami & Gupta, 1996; Watt, 1972).**

Euphorbia hirta L. (Plate No. 21A)

Syn. *E. pilulifera* L.; *E. capitata* Wall

Family Euphorbiaceae

Vern.	Bada-dudhia.

English, Hindi, Sanskrit and Regional Names

Eng.	Asthma plant, Cat's milk, Churn-staff, Pill-bearing Spunge, Sun spurge;
Hindi	Dudhi, Hirruseech, Mahabi;
Sans.	Puritoa;
B.	Barokhervie;
G.	Dudheli, Dudhi;
Mal.	Nelapalai;
Mar.	Dudnoli, Dudhi, Govardhan, Mothidudhi;
P.	Chatriwal, Dudal, Gandabute, Kulfadodak;
Tam.	Amampatchai arisi;
Tel.	Bidarie, Nanabala, Nanabiyan, Reddimanabrolu.
Distribution	Punjab - W. Himalaya; introduced elsewhere.
Description	Erect, ascending hispid annuals. Leaves opposite elliptic-oblong, acute toothed. Stem often very stout and umbellately or dichotomously branched. Involucres numerous in axillary terminals. Seeds pale brown, acutely angled.
Flowering & Fruiting	March-August.
Habitat Ecology	Damp places, irrigated channels, roadsides, gardens and nallahs; **Karyal Khad -** 615 m; **Gobind Sagar Lake -** 490 m.
Material Examined	EBH-WL-1177; 24.06.2008.
Parts Used	Whole Plant. Leaves. Seeds.
Folk Uses	Poultice of warmed infusion of leaves considered a good remedy for **renal stones**. Latex of the plant (twice daily till cure) applied to remove **warts**. 15-20 ml decoction of seeds prescribed for **cholera**.

Chemical Constituents	Properties exhibited due to quercetin -3β- glucoside, quercetin -3β-galactoside-2'' –gallate isolated, wax esters of neutral lipids lauric, palmitic, stearic, oleic, linoleic, arachidic and behenic acids **(leaves)**, octa cosyl alcohol, β- sitosterol, β-dihydrofucosterol, helioscopiol and a triterpenoid, flavone, tithymalin **(plant)** and esters of 12-deoxyphorbol **(aerial parts)**.
Biological Activity	Antiprotozoal, anticancerous, CVS, diuretic hypoglycaemic and spasmolytic activities confirmed.
Uses in Literature	Described earlier as anticancerous, colic, hypoglycaemic, spasmolytic; and for asthma, body pain, carbuncle, conjunctivitis, cuts, cough, dysentery, excessive urination, fertility, fever, gonorrhoea, headache, itches, leprosy, leucorrhoea, skin diseases, ulcers and fissures in mouth, worms and wounds **(Ambasta, 1986; Anonymous, 1948-76a; Balasubramanian & Prasad, 1996; Beg et al., 2006; Bhattacharjee, 2001; Biswas et al., 2010; Brahmam, 2000; Chopra et al., 1956; Farooq, 2005; Gogoi & Das, 2003; Guha Bakshi et al., 2001; Hosagoudar & Henry, 1996a; Idu et al., 2008; Jain, 1991; Jain & Sharma, 2000; Jain et al., 2010; Jha et al., 1996; Kaushik & Dhiman, 2000; Khan & Khanum, 2005; Kirtikar & Basu, 1984; Mishra & Sahu, 1984; Noumi, 2010; Pande et al., 2006; Patil et al., 2007; Prajapati et al., 2006; Rana et al., 2003; Ranjan, 2003; Rao & Henry, 1995; Retnam & Martin, 2006; Rout & Panda, 2010; Samant et al., 2001a; Silori et al., 2009; Sinha et al., 1996; Singh & Srivastava, 2000; Singh et al., 1996a; Siwakoti & Siwakoti, 2000; Siwakoti & Varma, 1996; Solanki et al., 2007; Srivastava et al., 2003; Swami & Gupta, 1996; Thakor, 2009; Viswanathan, 2000).**

Euphorbia parviflora L. (Plate No. 21B)

Syn.   *E. cassioides* Presl; *E. decumbens* Willd.; *E. androsaemoides* Dennst.; *E. papilliogera* Boiss.; *E. bracteolaris* L.; *E. hypericifolia* L

Family   Euphorbiaceae

Vern.	Lal Dudhi.
Hindi, Sanskrit and Regional Names	
Hindi	Dudhi, Dudhikalave, Hakshardana;
Sans.	Dugdhika;
G.	Dadheli;
P.	Hazardana.
Distribution	Hotter parts throughout India, from Punjab ascending to 1,350m in the Himalaya.

Description	Glabrous erect pubescent annuals with opposite, oblong, obtuse serrulate leaves having distinct nerves. Involucres axillary terminal. Capsules subglobose; cocci pubescent. Seeds bluish, obsolete.
Flowering & Fruiting	May-June.
Habitat Ecology	Common at moist places; **Landy Khad -** 615 m; **Lyond Khad** - 617 m.
Material Examined	EBH-WL-1178; 26.06.2008.
Part Used	Whole Plant.
Folk Uses	1-3 g powdered plant given with milk to children in **colic**. 5-10ml decoction of the plant given twice a day till for curing **measles**. Crushed plant also given to cattle to expel without any **internal injury** iron nails and pointed objects eaten along with meals.
Uses in Literature	Known earlier for dysentery, diarrhoea, menorrhagia, leucorrhoea and as astringent and feeble narcotic **(Kirtikar & Basu, 1984; Cauis, 2003; Rana et al., 2003; Siwakoti & Varma, 1996; Solanki et al., 2007).**

Ficus hispida L.f. (Plate No. 21C)

Syn.   *F. oppostitifolia* Willd.; *F. daemomum* Koen.; *F. prominens* Wall.; *F. mollis* Willd.; *F. scabra* Jacq.; *Convellia daemomum* Miq.; *C. oppositifolia* Gasp.; *C. setulosa* Miq.; *Sycomorpha roxburghi* Miq

Family   Moraceae

Vern.	Dabru.

English, Hindi, Sanskrit and Regional Names

Eng.	Citron-leaved ficus;
Hindi	Gobla, Degar, Katgularia, Kathumaea, Totmila;
Sans.	Ajali, Ajakashi, Asuma, Bhadrodumbarika, Chitrabhashaja;
Ass.	Khoskadumar;
B.	Dumamoor, Jogdumus, Kakdumar, Kakodumar;
Bo.	Dhedhu, Mira;
G.	Dhedaumaro, Dhedumeru, Jangliangir;
Kan.	Adhviatti, Kadatti;
Mal.	Kabut;
Mar.	Bhokada, Boksia, Dherumbar, Kalaumbar, Kaharawat;
Oriya	Bhaidimiri, Korotosamo;
P.	Daduri, Degar, Rumbal;
Tam.	Ottonnlam, Peyatti, Sonatti;
Tel.	Bhummari;
U.P.	Ban-gular.
Distribution	Throughout India; H.P.- Bhutan: 450-1,100 m.
Description	Shrubby trees with opposite leaves which are hispid beneath. Petioles also hispid. Stipules pubescent often in whorls on yellowish recep-

tacle with leafless branches. Bracts scattered on sides of receptacles. Figures globose, pale yellow when ripe.

Flowering & Fruiting	March-May.
Habitat Ecology	Common in valley; **Kothi Khad -** 810 m.
Material Examined	EBH-WL-1226; 12.05.2009.
Parts Used	Leaves. Stem Latex. Fruits. Bark. Wood.

Folk Uses — Stem latex applied in dental caries as a relief from **toothache**. 3-4 g powdered bark in combination with 'soa' (*Anthum sowa* Roxb.) given with cow's milk for **spermatorrhoea**. 5-10ml juice of leaves with milk given thrice a day to check **dysentery**. Ripe fruits can be eaten raw **pickled** or made into **jam** where as unripe ones and tender leaves cooked as **vegetable** for checking **flatulence, dysentery** and **diabetes**. Wood used as **fuel.**

Chemical Constituents — Active properties associated to bergapten, psoralin, β-amyrin, β-sitosterol **(fresh leaves),** n-triacontanyl, β-amyrin and gluconol acetates **(bark)**.

Biological Activity — Stem bark antibacterial, antiprotozoal, anthelmintic and antiviral.

Uses in Literature — Recorded to be used earlier as an antiperiodic, astringent, cattle fodder, edible, emetic, fibre, galactagogue (vet), purgative, tonic, vegetable (leaves sometimes cooked with pork); and for blood dysentery, bubo (vet), colic pains, curries, cuts, delivery (vet), diarrhoea (vet), earache, easy shedding of placenta (vet), giddiness, gonorrhoea, jams, headache, mouth and stomach ulcers, preserving foetus in the womb, removal of placenta (vet), warts and wounds **(Ambasta, 1986; Geetha et al., 1996; Girach et al., 1997, 1999; Jain, 1991; Kirtikar & Basu, 1984; Lalramnghinglova, 2003; Mohanty & Padhy, 1996; Pande et al., 2006; Rama Rao & Henry, 1996; Rana et al., 2003; Rastogi & Mehrotra, 1993; Retnam & Martin, 2006; Saini, 1996a; Sharma et al., 1979; Singh et al., 1996a; Sinha, 1986, 1987; Siwakoti & Verma, 1996; Solanki et al., 2007; Sood & Thakur, 2004; Sood et al., 2009b; Watt, 1972).**

Ficus roxburghii Wall. (Plate No. 21D)

Syn.   *F. macrophylla* Roxb.; *F. scleroptera* Griff.; *F. regia* Miq.; *Covellia macrophylla* Miq.

Family   Moraceae

Vern.	Trayambalu.
Hindi and Regional Names	
Hindi	Timal, Timbal;

B.	Demur, Doomur;
P.	Burh, Timbal.

Distribution	Outer Himalaya; Kashmir–Bhutan, ascending to 1,600 m.
Description	Low spreading trees with brown bark and leaves having cordate base. Stipules ovate-lanceolate, pubescent. Receptacles reddish-brown or purplish and spotted. Achenes granulate, viscid.
Flowering & Fruiting	April-August.
Habitat Ecology	Grasslands, wastelands, rocky slopes; **Kothi Khad -** 810 m.
Material Examined	EBH-WL-1111; 12.05.2008.
Parts Used	Leaves. Fruits. Bark. Wood.
Folk Uses	Leaves lopped for **fodder**. Fruits **edible**, cooked as well as made into **jams**; and its powder (250 g) along with 'shilajeet' (25 g) and 'saunf' 25 g (*Foeniculum vulgare*) beneficial in **spermatorrhoea** (twice daily for 5-7 days). Paste of its bark applied on boils for quick discharge of **pus**; also **ropes** are made from it. Wood used as **fuel**.
Chemical Constituents	β-sitosterol, friedelin epifriedelanol and epifriedelanol isolated **(plant)**.
Uses in Literature	Used in India as edible, fodder, galactagogue (vet), vegetable; and for breast ulcers, delivery (vet), leucoderma, meal plates for feast in villages, ringworm and ropes (bark) **(Jain, 1991; Parmar & Kaushal, 1982; Rama Rao & Henry, 1996; Rana et al., 2003; Sood & Thakur, 2004; Sood et al., 2009b; Watt, 1972).**

Fragaria indica Andr. (Plate No. 21E)

Syn.   *Duchesnea indica* Focke

Family   Rosaceae

Vern.	Puin Aakha.

English, Hindi and Regional Names

Eng.	Indian or mock strawberry;
Hindi	Pahari rashbhari;
P.	Banphal, Ingrach, Kanzar, Palzar, Tawai.

Distribution	Himalaya upto 3,400 m.
Description	Stoloniferous herbs with hairy stem and 3, digitate, coarsely serrate-dentate or incised leaflets. Flowers axillary, solitary. Bracteoles appressed hairy. Petals yellow. Achenes long, red, smooth.

Flowering & Fruiting	June-August.
Habitat Ecology	Forests. Shady banks, Edges of cultivation; **Matwana Khad -** 610 m.
Material Examined	EBH-WL-1218; 28.07.2008.
Parts Used	Whole Plant. Fruits.
Folk Uses	Ripe fruits **edible**. Poultice of plant paste applied as demulcent over **blisters**.
Chemical Constituents	Methoxydehydrocholesterol isolated.
Biological activities	Circulostimulant.
Uses in Literature	Used so far for edible fruits, abscess, bug bites, burns, cachexia, cancer, cough, dermatosis, eczema, laryngosis, pulmonosis, rheumatism, ringworm, snake bite, sting, swelling, tonsilosis, trauma and tuberculosis **(Ambasta, 1986; Arora & Pandey, 1996; Barua et al., 2000; Duke et al., 2002; Hajra & Baishya, 1997; Jain, 1991; Kapur & Nanda, 1996a; Rana et al., 2003).**

Fragaria nubicola (Hook. f.) Lindl. ex Lacaita (Plate No. 21F)

Syn.   *F. vesca* L

Family   Rosaceae

Vern.	Bhee-kaphal.
English Name	
Eng.	Wild strawberry.
Distribution	Temperate regions of Asia.
Description	Herbs with creeping stolons. Leaves mostly radical, tufted, long-stalked. Stipules narrow, entire. Flowers white. Bracteoles small, entire.
Flowering & Fruiting	May-October.
Habitat Ecology	Common in grassy areas, near water sources; **Sauli Khad -** 765 m.
Material Examined	EBH-WL-1219; 24.10.2008.
Parts Used	Leaf. Fruits.
Folk Uses	Decoction of leaves used as a gargle for **sore throats**. Fruits considered to have **cooling, diuretic properties** and **edible**.
Chemical Constituents	Properties due to flavonoids, tannins and a volatile oil **(leaves)**, a volatile oil, methyl slicylcate and borneol **(fruits)**.

Biological activities	Plant exhibits alternative, antinitrosamicnic, antioxidant, antipyretic, antiseptic, catabolic, depurative, discutient and nervine properties.
Uses in Literature	Useful earlier for anaemia, arthritis, blennorrhagia, bronchosis, calculus, cancer, cerebrosis, diabetes, diarrhoea, dysentery, earache, gout, headache, impotence, inflammation, jaundice, odontosis, rheumatism, tuberculosis; and as astringent and diuretic **(Duke et al., 2002; Kapur & Srivastava, 1996; Panda, 1996; Pande et al., 2006; Prajapati et al., 2006; Rana et al., 2003).**

Fumaria indica (Hausskn.) Pugsley. (Plate No. 22A)

Syn.   *F. parviflora* Lamk.; *F. vaillantii* Loisel. *var. indica* Hausskn.; *F. vaillantii* Loisel

Family   Fumariaceae

Vern.	Pitpapda.

English, Hindi, Sanskrit and Regional Names

Eng.	Fine-leaved fumitory;
Hindi	Pitpapda;
Sans.	Araka, Charaka, Katupatra, Pansu;
B.	Bansulpha;
G.	Khasirdlio, Pitpapda;
Mar.	Pitpapda;
Tam.	Tura;
Tel.	Chata-rashi.

Distribution	Indo-Gangetic plains, a weed of cultivation.
Description	Diffuse annual branched herbs. Leaves much divided, segments very narrow. Flowers whitish or rose coloured. Racemes lax-flowered. Pedicels exceeding the bracts. Fruits globose-rugose when dry, rounded at the top with 2 pits.
Flowering & Fruiting	February-March.
Habitat Ecology	Moist slopes, cultivation fields; **Sir Khad -** 620 m; **Karyal Khad -** 615 m; **Chamyater Khad -** 613 m.
Material Examined	EBH-WL-1220; 05.03.2009.
Part Used	Whole Plant.
Folk Uses	50-100ml decoction of the plant given once every morning for 3-5 days against **fever, skin diseases** and as **blood purifier.**

Chemical Constituents These are 19-methyloctacosan-1-ol, 3-methyloctacosan-1,3-diol, C27-29 n-alkanes, β-sitosterol, stigmastrol, campesterol; (-) tetrahydrocoptisine, protopine, (+) -, (+-) bicuculline and (+-) tetrahydrocoptisine, phthalide **(seeds)** alkaloid –(+) papraine-, benzylisoquinoline alkaloid-fumarizine **(whole plant)**.

Biological Activity Dried plant-anthelmintic, diuretic, diaphoretic and aperient.

Uses in Literature Recorded earlier as aperient, diaphoretic, diuretic, laxative; and for blood diseases, fever, influenza, urinary troubles and wounds **(Ambasta, 1986; Chandra, 1997; Jain, 1991; Kaul; 1997; Kaushik & Dhiman, 2000; Pande et al., 2006; Rana et al., 2003; Ranjan, 2000; Rastogi & Mehrotra, 1993; Saini, 1996b; Siwakoti & Varma, 1996, Sood & Thakur, 2004)**.

Galium aparine L. (Plate No. 22B)

Syn. *G. agreste* var. *echinospermon* L.; *G. aparine* Auct

Family   Rubiaceae

Vern.	Dhanpatri.
English Names	
Eng.	Cleavers, Goosegrass.
Distribution	Temperate Himalaya (Kashmir – Sikkim), N. and W. and Central Asia.
Description	Rambling or climbing annuals having mucronate leaves with scabrid midrib and margins. Peduncles axillary and in terminal leafy panicles. Fruits clothed with spreading hooked bristles.
Flowering & Fruiting	July-August.
Habitat Ecology	Grassy slopes, wastelands, screes; **Sauli Khad** – 765 m; **Sir Khad -** 620 m.
Material Examined	EBH-WL-1227; 05.07.2007.
Part Used	Whole Plant.
Folk Uses	Pulverized dried herb given with honey and black pepper against **cough** and **urinary disorders**; 1-3 g twice a day for 5-7 days.
Chemical Constituents	Known to possess iridoid glycoside, asperuloside **(aerial parts)** and luteolin **(plant)**.
Biological Activity	Extract of the plant lowers arterial pressure upto 50%; and also found to exhibit alterative, antibacterial, antipyretic, antispasmodic, haemostatic and larvicidal properties.
Uses in Literature	Used in India as antiscorbutic, aperient, diuretic, refrigerant sinusitis, a substitute for coffee; and for adenopathy, amenorrhoea, ascites, bacteria, bleeding, bites, burn, calculus, cancer, cholecytosis, dropsy, eczema, hysteria, ischuria, itch, jaundice, kidney stone, leprosy, leukaemia, migraine, otosis, psoriasis and seborrhea **(Ambasta, 1986;**

Duke et al., 2002, Kapur & Nanda, 1996b; Pande et al., 2006; Prajapati et al., 2006; Rana et al., 2003; Rastogi & Mehrotra, 1991; Sood & Thakur, 2004; Wangchuk et al., 2008).

Geranium nepalense Sweet (Plate No. 22C)

Syn.   *G. radicans* DC.; *G. pallidum* Royle; *G. patens* Royle; *G. affine* W. & A.; *G. arnottianum* Steud.

Family   Geraniaceae

Vern.	Bakra.

English, Hindi and Regional Names

Eng.	Nepal geranium, Napalese cranolesbill;
Hindi	Bhanda;
Kash.	Roel;
P.	Bhanda.
Distribution	Throughout Temperate Himalaya: 1,700-3,000 m.
Description	Much branched, diffuse, hairy or villous slender herbs with 5-gonal deeply lobed leaves and slender peduncles. Flowers purple. Sepals shortly-awned. Petals entire. Seeds shining smooth.
Flowering & Fruiting	June- September.
Habitat Ecology	Rock crevices, moist areas, wastelands, roadsides; **Gobind Sagar Lake -** 490 m; **Sir Khad** - 620 m.
Material Examined	EBH-WL-1201; 12.07.2008.
Part Used	Whole Plant.
Folk Use	Roots used for **tanning**.
Uses in Literature	Used in India as an astringent, diuretic and dye; and for cuts, jaundice, toothache, ulcer, wounds, renal diseases, itching, eczema and stomach ache complaints **(Ambasta, 1986; Jain, 1991; Krishna & Singh, 1987; Kumar, 2002; Pande et al., 2006; Rana et al., 2003; Rastogi & Mehrotra, 1991; Sood & Thakur, 2004; Uniyal & Chauhan, 1973; Uphof, 1968; Watt, 1972).**

Girardinia heterophylla Decne. (Plate No. 22D)

Syn.   *G. diversifolia* Friis; *Urtica heterophylla* Vahl; *U. diversifolia* Link; *U.palmata* Forssk.; *U. horrida* Link

Family   Urticaceae

Vern.	Badi-kogsi.
Hindi Names	
Hindi	Awa, Alla, Bichua, Chikri.
Distribution	Temperate and subtropical Himalaya.
Description	Tall stout tufted herbs with perennial roots, furrowed pubescent hispid or hirsute stem and branches and broad leaves. Male cymes loosely paniculate. Flowers subsessile hispid; fruiting cymes elongate. Achenes broadly ovate or subcordate, punctate, black; style persistent.
Flowering & Fruiting	September-October.
Habitat Ecology	Common in depressions; **Tatoh Khad -** 710 m.
Material Examined	EBH-WL-1113; 02.10.2008.
Parts Used	Whole Plant. Leaves.
Folk Uses	100-125 g powdered plant mixed with wheat flour given twice a day for 10 days to cattle for overcoming general **weakness**. Paste of leaves applied to hasten **suppurations, rheumatic aches** and **pains**. Leaves also used for making **'chutney'**.
Chemical Constituents	5-OH-tryptamine, histamine, acholine **(leaves)** and β-sitosterol **(plant)** have been isolated.
Uses in Literature	Used in India to cure fever, headache and for making ropes and twines **(Deokota & Chhetri, 2009; Parrotta, 2001; Rana et al., 2003; Watt, 1972).**

Gloriosa superba L. (Plate No. 22E)

Family   Liliaceae

Vern.	Langhi.
English, Hindi, Sanskrit and Regional Names	
Eng.	Super lily, Tiger lily;
Hindi	Karihari, Languli;

Sans.	Agnimukhi, Langli;
Kan.	Agnisikhe;
Mal.	Mettonni;
Mar.	Karianag;
Tam.	Kanvali poo, Kaandal, Kalappai kizhangu;
Tel.	Kannu noppi mokka, Nabhi.

Distribution   Throughout India, ascending upto 1,650m in the Himalaya.

Description   Climbing annual herbs with alternate, sub-sessile, linear-lanceolate leaves. Flowers showy, solitary, axillary, crimson red; petals with wavy margins. Capsule with numerous, black seeds.

Flowering & Fruiting   July-September.

Habitat Ecology   Near water sources, moist areas; **Kothi Khad -** 810m.

Material Examined   EBH-WL-1228; 12.08.2008.

Part Used   Tuberous Roots.

Folk Uses   Root tuber paste applied externally against **skin diseases, swellings** of the joints and rheumatic pains. Paste of a bit of tuberous root along with black pepper given for inducing **abortion** upto three months of pregnancy in women. Also, 1-3g shade dried, powdered tuberous roots soaked in cow's urine given to treat **gonorrhoea**.

Chemical Constituents β-sitosterol, its glucoside and 2-hydroxy-6-methoxybenzoic acid isolated from **rhizomes**.

Biological Activity   Plant exhibits antimicrobial activity.

Uses in Literature   Recorded to be used earlier as anthelmintic, colic, muscle relaxant, stomachic, tonic; and for baldness, bleeding piles, child birth, chronic ulcers, gout, gonorrhoea, infection, inflammations, intermittent fever, leprosy, painful delivery, piles, rheumatic pain,   scrofula, skin diseases, scorpion sting, snakebite, smallpox, sores, syphilis and tumour **(Ansarali & Sivadasan, 2009; Biswas et al., 2010; Jain, 1991; Jain & Singh, 1997; Pande et al., 2006; Parrotta, 2001; Prajapati et al., 2006; Rama Rao et al., 2008; Rao & Henry, 1995; Retnam & Martin, 2006; Rastogi & Mehrotra, 1995b; Saxena et al., 1997; Singh & Pandey, 1996; Singh & Rao, 2003; Solanki et al., 2007; Tarafder et al., 1997; Yasodamma et al., 2009).**

Gnaphalium pensylvanicum **Willd. (Plate No. 22F)**

Syn.   *G. peregrinum* Fernald; *G. purpureum* Auct

Family   Asteraceae

Vern.	Sukaru.
Distribution	Throughout India.
Description	Erect, cottony, branched annual herbs upto 50 cm tall having stem coated with grey cottony tomentum. Leaves spathulate-obovate, apiculate, glaberate. Heads in terminal clusters. Involucral bracts 2-3 seriate, ovate-lanceolate, acuminate. Corolla five- lobed. Achenes brown, oblong. Pappus white.
Flowering & Fruiting	December-May.
Habitat Ecology	Common in fields near water sources; **Sir Khad - 620 m.**
Material Examined	EBH-WL-1115; 12.04.2009.
Part Used	Leaves.
Folk Use	10-15 ml decoction of leaves given twice a day to check **headache**.
Uses in Literature	Earlier known for body pains and fever **(Duthie, 1973; Hooker, 1872-97).**

Gomphrena celosioides **Mart. (Plate No. 23A)**

Family   maranthaceae

Vern.	Kangu.
Distribution	Throughout India.
Description	Herbs. Leaves opposite. Flowers capitate, white. Sepals 5, lanceolate. Stamens 5, filaments linear. Ovary subglobose. Seeds lenticular.
Flowering & Fruiting	January-April.
Habitat Ecology	Along roadsides; **Sir Khad -** 620 m.
Material Examined	EBH-WL- 1111; 12.02. 2008.
Part Used	Whole Plant.
Folk Uses	Considered a good **fodder** grass and usually given to increase the **quantity of milk**.

Hedera helix Clarke (Plate No. 23B)

Syn.   *H. nepalensis* K. Koch; *H. himalaica* Tobler.; *H. rhombea* Sieb. & Zucc

Family   Araliaceae

Vern.	Dakari.
English, Hindi and Regional Names	
Eng.	Barren ivy, Beene tree, Bent wood, Black ivy, Creeping ivy, Nepal ivy, Wood bind;
Hindi	Lab-Lab;
Ass.	Mej-peosree;
Mal.	Maravala;
P.	Banbakri, Banda, Kadloli, Karbaru;
Tam.	Maravalai.
Distribution	N. W. Himalaya: 1,800-3,000 m.
Description	A woody climber climbing by aerial rootlets. Leaves thick, 3-5 lobed, shining ovate. Flowers yellow-green. Fruits black.
Flowering & Fruiting	June-November.
Habitat Ecology	Near water sources; **Papral Khad -** 740 m.
Material Examined	EBH-WL-1229; 27.10.2008.
Parts Used	Stem. Leaves. Fruits. Bark. Young Branches.
Folk Uses	Berries **edible**. Branches and leaves lopped for **fodder**. Leaves and bark constituent of local tea. One tsp of powdered climber prescribed with cow's milk, twice a day for overcoming **gastric disorders**. 10ml decoction of bark prescribed twice daily for 3-5 days as **stomachic**. Also, massage of paste of its stem along with seeds of *Pinus roxburghii* and 'alsi oil' (*Linum usitatissimum* L.) effective against **arthritis**.
Chemical Constituents	Constituents isolated from different parts are rutin, kaempfecol-3-glu- coside, coumric acid, ferulic acid, chlorogenic acid, scopoline **(bark)**,

aphedrine, hederasaponin C, a-hederin, two saponin complexes-CS60 and CS90 **(leaves)**, eleven triterpene glycosides- kizuta saponins K2, K4, K5, K7, K7A, K7B, K7C, K8, K9, K11, and K13 **(stem bark)**.

Biological Activity Hedaragenin exhibits highly significant anti-inflammatory activity against carrageenin-induced oedema and formaldehyde induced arthritis in cat.

Uses in Literature Recorded in India as cathartic, diaphoretic, fodder, purgative, stimulant; and for amenorrhoea, arthrosis, bacteria, bronchitis, burn, cancer, catarrh, constipation, cough, dysentery, dysmenorrhea, fever, gout, headache, hepatosis, high blood pressure, induration, infection, insomnia, intoxication, pertusis, polyp, mumps and sores **(Ambasta, 1986; Chhetri, 2005; Duke et al., 2002; Rana et al., 2003; Retnam & Martin, 2006; Sood et al., 2009b; Watt, 1972).**

Hedychium spicatum Buch.-Ham. ex Sm. (Plate No. 23C)

Syn. *H. acuminatum* Rose

Family Zingiberaceae

Vern. Ban haldi.

English, Hindi, Sanskrit and Regional Names

Eng.	Lesser gangal, Spiked ginger lily;
Hindi	Gandhapalashi, Kapurakachari, Sit-ruti;
Sans.	Amlaharida, Durva, Kachhora, Karpura, Kapurakachali, Palashasathi, Samudra, Shathi, Savata, Tuni;
B.	Ada, Arna;
Bo.	Sir, Sutti;
G.	Kapurkachari;
Mar.	Kapurakachari, Sanatakka;
P.	Banhaldi, Khor, Saki;
Tam.	Simaikkichilikkilhangu.

Distribution Subtropical Himalaya.

Description Scapigerous stout leafy herbs with oblong-lanceolate leaves, and flowers in terminal spikes. Capsule globose.

Flowering & Fruiting July-September.

Habitat Ecology Frequent on grassy slopes, nallahs, moist areas; **Papral Khad** - 740 m.

Material Examined EBH-WL-1116; 02.09.2008.

Part Used	Rhizome.
Folk Uses	Powdered rootstock prescribed for **liver disorders** (1 tsp twice daily for 5-7 days). Poultice of its paste also applied for relief from **painful joints**. Small pieces of rhizome mixed with green fodder fed to livestock to **enhance lactation**.
Chemical Constituents	**Rhizome** shows presence of ethyl ester of p-methoxy cinnamic acid a new diterpene-6-oxo-lobada-7, 11, 13-trien-16-oic acid lactone(1).
Biological Activity	EtOH (50%) extract of rhizome hypoglycaemic and antiinflammatory; plant vasodilator, mild hypotensive and antiseptic.
Uses in Literature	Used in India as an aromatic, brain tonic, carminative, emmenagogue, expectorant, stimulant; and for bronchial asthma, cuts, diarrhoea, food poisoning, headache, inflammation, internal injuries, liver complaints, nausea, pain, snake-bite, stomach ache and vomiting **(Bhat Gopalakrishna & Nagendran, 2001; Chopra et al., 1956; Dey, 1973; Gupta, 1981; Jain, 1991; Kumar, 2002; Lalramnghinglova, 1996; Pande et al., 2006; Prajapati et al., 2006; Semwal et al., 2010; Sharma & Sood, 1997).**

Heteropogon contortus (L.) Beauv. ex Roem. & Schult. (Plate No. 23D)

Syn. *H. allinii* Roem.; *H. firmus* Presl; *H. glaber* Pers.; *H. hirtus* Pers.; *H. hirsutus* Beauv.; *H. hispidissimus* Steud.; *H. hohenackeri* Hochust. *ex* Miq.; *H. messanensis* Guss.; *H. polystachyus* Nees; *H. roxburghii* Arn.; *Andropogon allionii* DC.; *A. messannensis* Biv.; *A. bellard* Burb.; *A. besukiensis* Steud

Family Poaceae

Vern.	Bandarpuncha.

English, Hindi and Regional Names

Eng.	Spear grass, Wild oats;
Hindi	Sarol, Shurighas, Shurval;
B.	Kher;
G.	Dabhjuliyan;
Oriya	Dauria, Sinkola, Sinkolo;
P.	Suryala;
Tam.	Karusipullu, Panipullu, Usipullu;
Tel.	Dubbagasarigaddi, Eddigaddi, Kaserigaddi, Yedda, Yeddi, Yerragoyi.
Distribution	Throughout India, Myanmar and Sri Lanka to the straits of Malacca, ascending Himalaya to 1,700 m.

Description	Perennial grasses with tufted culms upto 100 cm which are erect or decumbent below. Leaves linear, flat, rigid. Ligules short, truncate, ciliate. Racemes 4-7.5 cm long. Spikelets imbricating. Anthers 3. Caryopsis cylindric, hairy.
Flowering & Fruiting	September-December.
Habitat Ecology	Hill slopes, waste places; **Lyond Khad -** 617 m.
Material Examined	EBH-WL-1117; 04.11.2007.
Part Used	Whole Plant.
Folk Uses	Plant used for making coarse **mats, broom ('bonkri')** for sweepers and **thatching.** Locals also regard it as good **soil binder** and **fodder** for livestock.
Chemical Constituents	These are oil **(awns)**, myo-inositol, galactinol and raffinose **(grass)**.
Uses in Literature	Described so far as diuretic, stimulant; and for asthma, atrophy, cachexy, dysentery, emaciation, fever, haemal, muscular pain, scorpion sting, toothache and rheumatism **(Asolkar et al., 1992; Bews, 1979; Bor, 1973; Cauis, 2003; Duthie, 1888; Guha Bakshi et al., 1999; Jain, 1991; Jain et al., 1991; Maheshwari et al., 1996; Mudaliyar, 1921; Rana et al., 2003; Roy, 1984; Uphof, 2001).**

Hydrilla verticillata (L.f.) Royle (Plate No. 23E)

Syn. *H. angustifolia* Blume; *H. dentata* Casp.; *Leptanthes verticillata* Herb.; *Serpicula verticillata* L.f

Family Hydrocharitaceae

Vern. **Hindi and Regional Names**	Jala, Thangi.
Hindi	Jhangi, Kureli;
B.	Jhangi, Kureli, Saola;
P.	Jala;
Tel.	Punachu, Pachu, Nachu.
Distribution	Still and slowly running waters throughout India and Sri Lanka.
Description	Submerged leafy dioecious herbs with 4-8 leaves in a whorl which are linear and serrulate. Flowers long. Fruits 2-3 seeded; smooth, squarrose muricate.
Flowering & Fruiting	January-April.

Habitat Ecology	Common in fresh water lakes, mostly floating in masses; **Ali Khad -** 580 m.
Material Examined	EBH-WL-1118; 05.03 2009.
Part Used	Whole Plant.
Folk Uses	Poultice of the plant paste applied against **skin diseases**. Plant used as **green manure, fodder** for animals and **food** for aquatic animals.
Uses in Literature	Known so far as green manure and food for fish **(Ambasta, 1986; Arber, 2003; Biswas & Calder, 1984; Das et al., 1996; Jain, 1991; Joshi, 2009; Kumar & Narain, 2010; Maheshwari et al., 1996).**

Impatiens balsamina L. (Plate No. 23F)

Syn.   *I. balsamina* var. *vulgaris* L.

Family   Balsaminaceae

Vern.	Teurya.

English, Hindi, Sanskrit and Regional Names

Eng.	Garden balsam;
Hindi	Gulmenhdi;
Sans.	Dushpatrijati;
B.	Dupati;
G.	Gulmendi, Pahtanbal;
Mal.	Muchingam;
Mar.	Terada;
Oriya	Haragaura;
P.	Bantil, Trual, Halu, Tatura, Tilphar, Juk;
Tam.	Kasittumbai.

Distribution	Throughout tropical and subtropical India.
Description	Pubescent or glaberate herbs possessing deeply serrated leaves and glandular petiole. Flowers rose coloured, spurred. Sepals minute. Standard orbicular retuse, wings very broad and lateral lobe much rounded. Capsules with globose seeds having black testa.
Flowering & Fruiting	July-September.
Habitat Ecology	Along water courses, damp grounds, fields, wastelands; **Panyalya Khad -** 810 m.
Material Examined	EBH-WL-1030; 12.08.2008.
Parts Used	Roots. Leaves. Flowers.

Folk Uses	Flowers and leaves used as a **substitute** for henna. Leaf juice applied against **skin diseases**. Roots regarded as **anticarcinogenic** (3-5 g powder, twice a day for 15-30 days).
Chemical Constituents	Active properties due to cyanidin, leucocyanidin, myricetin, quercetin, kaempferol, 2-methoxy-1,4-naphthaquinone **(flowers)**, caffeic, p-coumaric, ferulic, gentisic, p-hydrobenzoic, protocatechuic, salicylic, sinapic, syringic and vanillic acids, 2-hydorxy-1,4-naphthaquinone and naphtholglycosidase **(plant)**.
Biological Activity	50% EtOH extract of flower exhibits marked antibiotic activity against *Sclerotinia fructicola* and *Colletotrichum lindemuthianum*.
Uses in Literature	Used in India for lumbago and intercostal neuralgia and as an illuminant **(Ambasta, 1986; Anonymous, 1948-76a; Chatterjee & Pakrashi, 1997; Chopra et al., 1956; Girach et al., 2000; Jain, 1991; Kapur & Nanda, 1996b; Kirtikar & Basu, 1984; Parrotta, 2001; Prajapati et al., 2006; Rana et al., 2003; Rao & Henry, 1995; Rastogi & Mehrotra, 1991; Siwakoti & Varma, 1996).**

Ipomoea cairica (L.) Sweet (Plate No. 24A)

Syn.   *Convolvulus cairicus* L.; *I. palmata* Forsk.; *I. senegalensis* Lamk.; *I. stipulaceae* Jacq.; *I. pendulla* Br.

Family   Convolvulaceae

Vern.	Railway creeper.
Distribution	Tropical India, Malaya, Sri Lanka Asia and Africa.
Description	Perennial twining herbs with a tuberous root. Stems tuberculate or smooth. Petiole 8 cm. Leaves palmate, large entire, apex acute. Sepals glabrous. Corolla purple, Stamens unequal. Ovary glabrous. Capsule globose. Seeds black, hairy.
Flowering & Fruiting	May-July.
Habitat Ecology	Wastelands, dry slopes, ditches; **Ali Khad -** 580m; **Tatoh Khad -** 710 m.
Material Examined	EBH-WL-1159; 16.04.2009.
Parts Used	Roots. Stem.
Folk Use	Stem and roots **consumed** occasionally or during times of scarcity.
Chemical Constituents	**Stem** possesses ergosnine, ergocornine and ergocistrine.
Uses in Literature	Used as food, purgative; and for cordage, cough and pain **(Ambasta, 1986; Hooker, 1872-97; Jain, 1991; Pande et al., 2006; Parrotta, 2001; Sood et al., 2009b; Uphof, 2001).**

Ipomoea carnea **Jacq. (Plate No. 24B)**

Syn. *I. cassicaulis* (Benth.) Robyns; *I. fistulosa* Mart. *ex* Choisy
Family   Convolvulaceae

Vern.	Naktibuti, Jablota.
Distribution	W. Himalaya and Nepal: 1,400 m.
Description	Perennial, erect or straggling shrubs having ovate-oblong acuminate leaves with cordate base. Flowers pink in dichotomous axillary and terminal cymes. Capsules globose. Seeds hairy.
Flowering & Fruiting	Throughout the year (flowers) July- September (fruiting).
Habitat Ecology	Wastelands, dry slopes, ditches, hedges; **Ali Khad -** 580 m.
Material Examined	EBH-WL-1055; 10.09.2008.
Parts Used	Whole Plant. Leaves. Latex.
Folk Uses	Commonly planted as a hedge plant to **check soil erosion**. Smashed leaves applied externally in **snake bite**. A gargle of leaf extract beneficial in healing **mouth sores** and **pyorrhea**; also its juice used as an effective remedy for **toothache**. Latex as such applied for healing **wounds** or used after mixing with 'ragi' (*Elusine coracana* Gaertn.) flour against **pains** due to swollen body parts.
Chemical Constituents	Known to contain arerin, β-sitosterol, triacontane, ipomose an anthracene, β-sitosterol, glucoside, gum, alapin and saponins **(leaves) (Anonymous, 1976b)**.
Biological Activity	Aerial parts effective on CNS.
Uses in Literature	Used earlier as poison (vet.), purgative, vegetable; and for arthritis, cuts, sprain, ulcer and wounds **(Ambasta, 1986; Banerjee, 2000; Dash & Misra, 1999a, 2000; Girach et al., 2000; Jain, 1991; Kaushik & Dhiman, 2000; Kumar & Narain, 2010; Pande et al., 2006; Singh, 2000; Singh & Pandey, 1996; Singh & Prakash, 1996, Singh & Srivastava, 2000; Sood et al., 2009b)**.

Ipomoea muricata (L.) Jacq. (Plate No. 24C)

Syn.   *Convolvulus muricatus* L.; *Calyonyction muricatuma* G. Don; *C. bona-nox* var. *muricata* Chois

Family   Convolvulaceae

Vern.	Ghaudan.

English, Hindi and Regional Names

Eng.	Bharmardi, Gario;
Hindi	Michai;
B.	Michai;
Bo.	Gariya;
G.	Garayo;
Mar.	Bhonvari;
Tam.	Kattutali.

Distribution	Himalaya, Kangra–Sikkim; very common.
Description	Twiners possessing muricate stem and entire, cordate leaves. Corolla tube linear. Capsule globose, apiculate.
Flowering & Fruiting	June-August.
Habitat Ecology	Damp places; **Ali Khad -** 580 m.
Material Examined	EBH-WL-1157; 10.07.2008.
Parts Used	Leaves. Seeds.
Folk Uses	Seed oil considered **sacred** for offering during worshipping; also its paste applied to check **falling hairs**. 1g powdered seeds given with hot water once a day in the morning to cure **constipation**. Paste of leaves and seeds (2:1) applied locally in case of **dry skin** to restore its normal colour and texture.
Chemical Constituents	**Seeds** yield behenic acid (3.78%), five resin glycosides- Mb-1, Mb-2, Mb-3, Mb-4, and Mb-5.
Uses in Literature	Known to be used in India as cathartic, purgative; and for abdominal disease, biliousness, edible pedicels, inflammation, joint pains, leucoderma and scabies **(Ambasta, 1986; Chandra, 1997; Pande et al., 2006; Ranjan, 1999; Rastogi & Mehrotra, 1995b; Saini, 1996b; Sood & Thakur, 2004; Sood et al., 2009b; Trivedi, 2004).**

Ipomoea nil (L.) Roth (Plate No. 24D)

Syn.   *I. hederacea* Jacq.; *I. coerulea* Koen.; *I. Dillenii* Roem. & Sch.; *I. punctata* Pers.; *Convolvulus coeruleus* Spreng.; *C. nil* L.; *Convolvuloides triloba* Moench; *Pharbitis hederacea* Chois.; *P. diversifolia* Lindl.; *P. punctata* G. Don; *P. purshii* G. Don; *P. variifolia* Decne.; *P. barbata* G. Don

Family   Convolvulaceae

Vern.	Ghaudan.

English, Hindi, Sanskrit and Regional Names

Eng.	Morning glory;
Hindi	Ghota, Kaladana;
Sans.	Krishnabji;
B.	Ghota;
G.	Ghota, Kaladana;
Kan.	Ganribija;
Mal.	Taliyari;
Mar.	Nilpushpi;
Oriya	Kanikhonda;
Tam.	Kodi kakkatam, Sirikki;
Tel.	Jiriki.
Distribution	Throughout India.
Description	Twinner with tri-lobed leaves, tubular, funnel-shaped corolla and cymose inflorescence. Capsule subglobose. Seeds black.
Flowering & Fruiting	June-September.
Habitat Ecology	Moist-shady places; **Gobind Sagar Lake -** 490 m.
Material Examined	EBH-WL-1156; 07.08.2008.
Parts Used	Whole Plant. Seeds.
Folk Uses	1-3 g powdered seeds taken with a glass full of luke warm water in the morning as a **laxative**; also its paste given to check **headache, fever** and **stomach** disorders. Application of extract of the whole plant mixed with hot mustard oil promotes **hair growth.**
Chemical Constituents	**Seeds** possess chanoclavine, lysergol, penniclavine, iso-penniclavine and olymoclavine.
Biological Activity	Seeds hypotensive, psychotropic, analgesic, analeptic and uterus and intestinal stimulant.

Uses in Literature	Known to be used as an anthelmintic, cathartic, galactagogue, purgative, tonic and vegetative; and for abdominal diseases, constipation, diseases of liver, fever, headache, joint pains, leucoderma, scabies, skin diseases, spleen and verminosis **(Hooker, 1872-97; Jain, 1991; Kaushik & Dhiman, 2000; Kirtikar & Basu, 1984; Lindley, 1981; Pande et al., 2006; Patil, 2009; Prajapati et al., 2006; Rana et al., 2003; Retnam & Martin, 2006; Sebastian, 1984; Sharma et al., 1979; Singh & Pandey 1980; Siwakoti & Varma, 1996).**

Ipomoea pestigridis L. (Plate No. 24E)

Syn. *Convolvulus pestigridis* Spreng.; *C. bryoniacfolius* Salisb.

Family Convolvulaceae

Vern.	Photial wagpadi.
English and Regional Names	
Eng.	Tiger's foot, Bindweed;
B.	Langulilata;
Mal.	Pulichuvata;
Oriya	Bilaipado;
Tam.	Pulichavadi, Punaikkirai;
Tel.	Chikunvvu, Mekamaduga, Puritikada.
Distribution	Drier low hills throughout India from Punjab - Malacca and Sri Lanka..
Description	Twiners with elliptic, acuminate leaves. Inflorescence capitate, few flowered, peduncle 4-11 cm; Sepals lanceolate, acute. Corolla sparsely hairy. Capsule ovoid, glabrous. Seeds ellipsoid, grey, tomentose.
Flowering & Fruiting	September-December.
Habitat Ecology	Common in waste places, along bushes; **Sir Khad -** 620 m.
Material Examined	EBH-WL-1158; 16.11.2007.
Parts Used	Roots. Leaves. Seeds.
Folk Uses	Leaf paste applied on **cuts**. Seeds eaten by the children for **strength**. Poultice of roots used against **dog-bites**.
Biological Activity	Plant spasmolytic.
Uses in Literature	Roots used as purgative; and for boils, carbuncles and dog bites **(Biswas et al., 2010; Chandra, 1997; Hooker, 1872-97; Jain, 1991; Kirtikar & Basu, 1984; Pande et al., 2006; Parrotta, 2001; Rana et al., 2003; Singh & Srivastava, 2000; Trivedi, 2004).**

Justicia simplex Don (Plate No. 24F)

Syn.   *J. mollissima* Wall.; *J. procumbens* Wall.; *Rostellularia rotundifolia* Nees
Family   Acanthaceae.

Vern.	Juffa, Pitpapda.
Distribution	Abundant in W. India; 700-1,700 m, Kashmir, Scinde Hills.
Description	Hairy herbs with long branches, petioled ovate hairy leaves and cylindric spikes. Bracts elliptic. Sepals lanceolate. Capsules oblong, pubescent. Seeds long, tuberculate.
Flowering & Fruiting	June-August.
Habitat Ecology	Rock-crevices, moist slopes; **Gobind Sagar Lake -** 490 m.
Material Examined	EBH-WL-1060; 02.07.2008.
Parts Used	Whole Plant. Leaves.
Folk Uses	20 ml decoction of whole plant along with cinnamom leaves given thrice a day to **cure cough, cold** and **chest congestion**; aq extract of plant (1/2 tsp) together with sugar given to children twice a day for **digestive disorders**. Also, paste of this plant with that of *Achyranthes aspera* applied on face for curing **pimples** and **blisters** and to check **itching**. Juice of leaves used as drops against **ophthalmic ailments**.
Chemical Constituents	These are simlexolin, seramin, sesamolin, helioxanthin, neoxanthin B, justician C, justicidin E, justisolin, simplexoside, β-sitosterol, lupeol, justasaponin **(plant)** and peonidin-3-glucoside **(flowers)**.
Uses in Literature	Plant known to be used earlier as expectorant, laxative, refrigerant; and for biliousness, burning of the body, indigestion, fever, intoxication, tired feeling and wounds **(Chatterjee & Pakrashi, 1997; Hooker, 1872-97; Kirtikar & Basu, 1984; Rana et al., 2003; Rastogi & Mehrotra, 1993; Sood & Thakur, 2004; Trivedi, 2004).**

Lactuca dissecta Don (Plate No. 25A)

Syn. *L. arvensis* Edgew.; *L. stocklsii* Boiss.; *Chondrilla auriculata* Wall

Family Asteraceae

Vern.	Khal.
Distribution	Temperate Himalaya (Kashmir – Bhutan): 1,350-2,700 m.
Description	Glabrous or sparsely pubescent annuals. Stem dichotomously branched with slender leafy branches. Leaves entire, petiole cauline. Heads narrow cylindric. Achenes oblanceolate.
Flowering & Fruiting	May-September.
Habitat Ecology	Wastelands near water sources; **Sir Khad -** 620 m.
Material Examined	EBH-WL-1032; 12.05.2008.
Part Used	Leaves.
Folk Uses	Leaf juice used as **eye drops** (2-3 times a day till cure) against redness and burning sensation.
Uses in Literature	Known for relieving burning sensation and redness **(Jain, 1991; Kapur & Srivastava, 1996).**

Lannea coromandelica (Houtt.) Merr. (Plate No. 25B)

Syn. *Dialium coromandelicum* Houtt.; *Odina wodier* Roxb

Family Anacardiaceae

Vern.	Gujar.

English, Hindi, Sanskrit and Regional Names

Eng.	Wodier jhingam;
Hindi	Jhingan, Jingan, Kaimil, Mohi, Mohin, Mohini, Moyen, Thingan;
Sans.	Jingini;

B.	Jiol;
Kan.	Manjishtha;
Mal.	Karasu;
Tam.	Oti, udi;
Tel.	Oddimanu.
Distribution	Sub-Himalayan tract extending to the Indus; ascending upto 1,219 m in the Outer Hills.
Description	Deciduous trees. Branches softly pubescent. Leaves imparipinnate. Leaflets 7-9 in terminal panicles. Flowers small, unisexual. Fruits reniform, 1-seeded. Drupes fleshy.
Flowering & Fruiting	March-June.
Habitat Ecology	Occasionally in village outskirts near water channels; **Karyal Khad - 615 m.**
Material Examined	EBH-WL-1158; 05.05.2008.
Parts Used	Leaves. Stem. Bark. Wood.
Folk Uses	100ml of its stem juice and that of *Erythrina variegata* L. taken orally to stop **dysentery**. Paste of bark applied on **skin diseases**. Leaves and twigs chewed in **toothache**. Timber used for **house construction** and **agricultural** implements.
Chemical Constituents	Known to possess rutin, leucocyanidin, β-sitosterol, leucodelphinidin, quercetin, 3-arabinoside, de- epicatechin, phlobatanins, physcion, anthranol B, a neutral polysaccharide, galactose, arabinose **(leaves)**, rhammnose, uronic acid **(bark)**, lanosterol and cluytl ferulate **(heartwood)**.
Uses in Literature	Described to be used as astringent, galactagogue; and for asthma, body-ache, cholera, cuts, diarrhoea, dysentery, impetigenous eruptions, local swellings, ulcer, sprains, sores, stomach ache and wounds **(Ambasta, 1986; Banerjee, 2000; Biswas et al., 2010; Chatterjee & Pakrashi, 1997; Chopra et al., 1956; Dager & Dager, 1996; Dash & Misra, 1999a; Jain, 1991, 2010; Kaushik & Dhiman, 2000; Lalramnghinglova, 1996, 2003; Maheshwari et al., 1996; Pande et al., 2006; Parrotta, 2001; Prajapati et al., 2006; Rao & Henry, 1995; Singh et al., 2001; Siwakoti & Varma, 1996; Vartak & Suryanarayana, 1995).**

Lantana camara L. (Plate No. 25C)

Syn.   *Camara aculeata* (L.) Kuntze; *L. aculeata* L.; *L. camara* L.; *L.mixta* L.; *L. spinosa* L. ex* Lecointe

Family   Verbenaceae

Vern.	Ujadu.

English, Hindi, Sanskrit and Regional Names

Eng.	Lantana, Wild sage;
Hindi	Lantana;
Sans.	Caturangi;
G.	Ghanidalia;
Kan.	Hasike, Kakke, Nata hu gida;
Mal.	Arippu;
Mar.	Chadrung, Chanderi;
Oriya	Naga-airi;
P.	Desilantana;
Tam.	Arippu.
Tel.	Pulikampa.
Distribution	Native of America; wild in India, ascending to 1,500 m, in Himalaya.
Description	Rambling, densely fulvous-hairy shrubs with long branches. Leaves opposite or ternate. Peduncles numerous, axillary. Calyx small, membranous. Corolla-tube slender. Drupes cylindric, fleshy or nearly dry, black and shining.
Flowering & Fruiting	April-June.
Habitat Ecology	Rocky wet slopes, waste places, semi-naturalised and common in cultivated areas; **Sir Khad -** 620 m.
Material Examined	EBH-WL-1062; 12.04.2009.
Parts Used	Stem. Leaves. Fruit.
Folk Uses	Ripe fruits **edible** and its stem used as **toothbrush**. Poultice of leaves considered good for **swellings** due to internal injuries; also dried leaves mixed with grains used in storage containers as an **insect repellent**. Plant grown as **hedges**.
Chemical Constituents	Found to possess α-amyrin, β-sitosterol, a triterpene acid, lantadene A, lantadene B, lantalonic acid, lantic acid, ursonic acid **(leaves)**,

geroniol, linalool, cedrene, caryophyllene, farnesol, cis-nerolidol and trans-nerolidol **(essential oil)**.

Biological Activity	Lowering of blood pressure and acclerated respiration exhibited.
Uses in Literature	Recorded earlier as antibacterial, antispasmodic, carminative, diaphoretic, edible (fruit), fish bait, mosquito repellent, poison (vet), stimulant, toothbrush; and for anaemic patients, burns, chest diseases, chronic inflammation of skin, cold, cough, cuts, diabetes, diarrhoea, dysentery, dyspepsia, fever, fistulae, fits, giddiness, infection of respiratory tract, itching, jaundice, kidney and liver lesions, mouth ulcers, photosensitization, rheumatism, tuberculosis and wounds **(Ambasta, 1986; Aminuddin & Girach, 1991; Ansarali & Sivadasan, 2009; Balasubramanian & Prasad, 1996; Bhattacharyya, 1996; Bhogaonkar & Kanerkar, 2007; Biswas et al., 2010; Chatterjee & Pakrashi, 1997; Chopra et al., 1956; Dwarakan & Ansari, 1996; Girach & Amminuddin, 1995; Goud & Pullaiah, 1996; GuhaBakshi et al., 1999; Henry et al., 1996; Hosagoudar & Henry, 1996a, b; Jadhav, 2009; Jain 1991, 1996; Khan & Khanum, 2005; Kurian, 1999; Lal et al., 1996; Lalramnghinglova, 1996, 2003; Maheshwari et al., 1996; Masih, 2003; Mishra et al., 1996; Noumi, 2010; Pande et al., 2006; Parrotta, 2001; Prajapati et al., 2006; Rana et al., 2003; Retnam & Martin, 2006; Saini 1996b; Solanki et al., 2007; Sood & Thakur, 2004; Sood et al., 2009b; Watt, 1972; Yoganarasimhan, 1996).**

Lathyrus aphaca L. (Plate No. 25D)
Family   Fabaceae

Vern.	Sudu.

English, Hindi and Regional Names

Eng.	Yellow-flowered pea. Yellow vetching;
Hindi	Jangli-matar, Masur-chuna;
B.	Janglimatar, Masur-chuna;
P.	Gagla, Rawan, Rawari.
Distribution	Spread through the northern provinces, ascending from the plains of Bengal to the temperate zone in Hazara, Kashmir and Kumaon.
Description	Much branched, slender, wingless annuals. Stipules entire, truncate, hastate, leaf-like and in pairs. Peduncle 2-3 times the stipule. Calyx teeth equal, lanceolate, exceeding the tube. Corolla yellow, twice the calyx. Pods linear-oblong, 4-6 seeded.
Flowering & Fruiting	April-May.

Habitat Ecology	Moist places; **Sauli Khad -** 765 m.
Material Examined	EBH-WL-1031; 04.04.2008.
Parts Used	Whole Plant. Seeds.
Folk Uses	Plant used as **fodder**. Powdered seeds (1/2 tsp) given once every day for 5 days to **cure cough** and **diarrhoea**.
Uses in Literature	Reported as astringent, fodder, narcotic, psychoactive and resolvent (**Ambasta, 1986; Kirtikar & Basu, 1984; Jain et al., 1994; Lindley, 1981; Manandhar, 1987; Saini, 1996a; Samant et al., 2001a; Sharma & Rana, 2005; Sood & Thakur, 2004**).

Leea crispa L. (Plate No. 25E)

Syn.   *L. asiatica* (L.) Ridl.; *L. aspera* Edgew.; *L. edgeworthi* Santin; *L. pinnata* Andr

Family   Leeaceae

Vern.	Ghangolae.
Regional Names	
B.	Banchalita;
Mal.	Nalugu, Nellu;
Oriya	Hatikanopotro.
Distribution	Tropical Himalaya.
Description	Erect spreading shrubs, upto 3 m tall with woody, jointed stem and flowers in small terminal cymes. Berries globular, succulent, black.
Flowering & Fruiting	August-September.
Habitat Ecology	Open slopes, screes, near water; **Lyond Khad -** 617 m; **Tatoh Khad -** 710 m.
Material Examined	EBH-WL-1262; 17.07.2008.
Parts Used	Leaves. Fruits.
Folk Uses	Fruits **edible**. Leaves lopped for **fodder**.
Uses in Literature	Considered useful for cuts, wounds, rheumatism, snake–bite, urinary complaints; and as anthelmintic and edible (**Jain, 1991; Sood & Thakur, 2004; Watt, 1972**).

Lepidagathis cuspidata Nees (Plate No. 25F)

Syn.   *Ruellia cuspidata* Wall

Family   Acanthaceae

Vern.	Puthkanda.
Distribution	Tropical India extending from W. Himalaya.
Description	Erect or diffuse shrubs with leaves which are multicose and attenuate at both ends. Spikes dense, flowers in opposite pairs. Calyx 5-partite. Corolla scarcely whitish with purple spots. Bracts ovate. Capsules 4-seeded.
Flowering & Fruiting	December-June.
Habitat Ecology	Rock-crevices, wastelands, roadsides, frequent; **Sir Khad -** 620 m.
Material Examined	EBH-WL-1033; 02.03.2009.
Part Used	Leaves.
Folk Uses	Plant poultice good for **itchy affections** of the skin; also its decoction (15-25 ml, twice daily, 5-7 days) used as a tonic for **fever**. 2-4 g powdered leaves given with cold water once every morning to check mouth **ulcers**. Leaves used as **fodder**.
Use in Literature	Used against fever **(Pande et al., 2006; Parrotta, 2001; Rana et al., 2003; Sood & Thakur, 2004).**

Lespedeza sericea Miq. (Plate No. 26A)

Syn.   *L. cuneata* G. Don; *L. juncea* Wall.; *L. argyraea* Sieb.; *Hedysarum sericeum* Thunb.; *Anthylis cuneata* Dum

Family   Fabaceae

Vern.	Binni.
Distribution	Himalaya (Hazara and Kashmir–Assam): 1,000-2,100 m.
Description	Erect undershrubs having crowded leaves and long coriaceous, truncate leaflets. Flowers pedicellate. Bracteoles linear, minute. Calyx covered with greyish hairs. Corolla white. Pods silky.
Flowering & Fruiting	November-February.
Habitat Ecology	Moist places; **Gobind Sagar Lake -** 490 m.
Material Examined	EBH-WL-1034; 22.01.2009.

Part Used	Whole Plant.
Folk Use	Regarded as an important **forage** plant.
Chemical Constituents	Succinic acid isovitexin, isoorientin, vicenin-2, lucenin-2 **(leaves)**, potassium lespedezate and potassium isolespedzate **(plant)** have been characterized.
Biological Activity	Potassium lespedezate and potassium isolespedzate at 0.8 μM quite effective in opening leaf of *Cassia mimosoides.*
Uses in Literature	Recommended for forage, soil conservation and erosion control **(Hooker, 1872-97; Kapur & Srivastava, 1996; Rastogi & Mehrotra, 1993; Uphof, 2001).**

Lindernia ciliata (Colsm.) Pennell (Plate No. 26B)

Syn.   *Gratiola ciliata* Colsm.; *G. serrata* Roxb.; *Bonnaya brachiata* Link & Otto; *Vanellia brachiata* (Link & Otto) Haines

Family   Scrophulariaceae

Vern.	Gingu.
Distribution	Throughout India, Indo-Malaya region.
Description	Erect, branched, glabrous annuals which are hairy at nodes. Leaves oblong, serrated, sessile. Flowers in terminal lax racemes; pedicels slender. Bracts linear, subulate. Calyx linear, ciliate, aristate. Corolla white. Stamens 2. Capsules oblong with ellipsoidal seeds.
Flowering & Fruiting	August-February.
Habitat Ecology	Common in moist shady places; **Ali Khad -** 580 m; **Sir Khad -** 620 m.
Material Examined	EBH-WL-1266; 12.02.2009.
Part Used	Whole Plant.
Folk Use	Plant used as a good **fodder** for livestock.
Use in Literature	Used as a remedy for gonorrhoea **(Cook, 1996; Kumar & Narain, 2010; Mukerjee, 1984).**

Macrotyloma unifloruma (Lam.) Verdc. (Plate No. 26C)

Syn.   *Dolichos uniflorum* Lamk.; *D. biflorus* Baker

Family   Fabaceae

Vern.	Kulthi.

English, Hindi, Sanskrit and Regional Names

Eng.	Horse gram;
Hindi	Kultthi;
Sans.	Kulatthah;
Kan.	Huruli;
Mal.	Mutira;
Tam.	Kollu;
Tel.	Ullavalu.
Distribution	Throughout India.
Description	Suberect or trailing much branched annuals with trifoliate leaves. Leaflets entire, membranous. Flowers yellow, 1-3 in the axils of leaves. Pods sword-shaped, 5-6 seeded, which are reniform and reddish brown.
Flowering & Fruiting	July-September.
Habitat Ecology	Moist places, near water sources; **Karyal Khad -** 615 m.
Material Examined	EBH-WL-1299; 02.09.2008.
Part Used	Seed.
Folk Uses	Soup of seeds drunk for treating **menstrual disorders** and expelling **kidney stones** (once everyday for atleast 5 days).
Chemical Constituents	**Seeds** contain several phytosterols including 24-methylene-25-methylcholesterol and 24α-ethyl-5α-cholest-9-en-3β-ol.

Uses in Literature	Recorded as abortifacient, acrid, anthelmintic, antifertility, astringent, diaphoretic, diuretic, emmenagogue, expectorant, febrifuge, ophthalmic and tonic; and for body warm up, burns, dysentery, dysuria, ear diseases, haemorrhoids, kidney stones, leprosy, leucorrhoea, measles, nephrolithiasis, renal calculi, thermogenic and toothache **(Jain, 1991; Pande et al., 2006; Prajapati et al., 2006; Malla & Chhetri, 2009; Semwal et al., 2010).**

Marchantia palmata **Nees (Plate No. 26D)**

Family　Marchantiaceae

Vern.	Matakain.
Distribution	Kumaon Himalaya and Outer Himalaya upto 2,700m; fairly common in plains.
Description	Green, dichotomous, dioecious thallus with broad, flat, emarginate apex. Dorsal surface with a dark line in the middle. Ventral surface brown. Scales coarsely irregularly toothed. Male receptacles circular, 3-9 lobed. Female receptacles 7-11 rayed. Perianth purple. Capsules spherical, dark brown with yellowish brown spores and elaters.
Flowering & Fruiting	January-April.
Habitat Ecology	Moist rocks and water banks; **Sir Khad -** 620 m.
Material Examined	EBH- WL-1283; 16.01.2009.
Part Used	Whole Plant.
Folk Uses	Fleshy leaf paste applied as **poultice** against acute burns due to fire and hot water. Plant used as a substitute of **'bhang'** (*Cannabis sativa*).
Use in Literature	Earlier known for body pains, fever and snakebite **(Ganglee, 1985; Kashyap, 1993; Kumar, 1995; Teg et al., 2007).**

Marsilea minuta **L. (Plate No. 26E)**

Family　Marsileaceae

Vern.	Tripatre.

English, Hindi, Sanskrit and Regional Names

Eng.	Water clover fern;
Hindi	Chaupatira, Godhi, Paflu;
Sans.	Sunisannah, Sunisannaka;
B.	Sunsnishak;

Kan.	Chitigina soppu;
Kash.	Paflu;
P.	Godhi, Tripattra;
Tam.	Araikeerai;
Tel.	Chick-lintakura; Mudugo-tamara.

Distribution	W. Himalaya and throughout the plains of India.
Description	Hydrophytic ferns with long, creeping, thin glabrous rhizome having hairy apex. Stipes 2-5 cm distant on rhizome. Pinnae 1-2 cm long, obcuneate, veins many, anastomosing obliquely. Sporocarp in pairs, pedicellate, bean-shaped. Megaspores ellipsoidal; microspores yellowish, oval 48-51 μm.
Flowering & Fruiting	November-April.
Habitat Ecology	Ditches, small ponds, rice-fields, bank of lakes: plains - 1,800 m; **Ali Khad -** 580 m.
Material Examined	EBH-WL-1055; 04.04.2009.
Parts Used	Leaves. Sporocarp.
Folk Uses	Leaves and sprouts sold in the market for cooking as **pot herb**. In times of scarcity, stipes too are **eaten**. Sporocarps considered rich source of starch and also **cooked** for consumption.
Uses in Literature	So far used for alexitric, antibacterial, anticonvulsant, antifungal, diuretic, edible (stalk of leaves), refrigerant, resolvent; and for abscesses, backache, boils, cough, diarrhoea, dyslactation fracture, insomnia, menorrhagia, myalgia, sedation, snakebite, sore, spastic condition of leg muscles and trauma **(Ambasta, 1986; Banerjee & Ghora, 1996; Biswas & Calder, 1984; Chatterjee & Pakrashi, 1997, Chopra et al., 1956; Das et al., 1996; Dixit & Vohra, 1984; Jain, 1991; Joshi, 2009; Khullar, 1994; Lalramnghinglova, 1996; Rastogi & Mehrotra, 1993; Saini, 2008; Seth & Kumar, 2007; Singh & Viswanathan, 1996; Singh et al., 2007).**

Martynia annua L. (Plate No. 26F)

Syn.   *M. diandra* Gloxin

Family   Martyniaceae

Vern.	Kanv.

English, Hindi, Sanskrit and Regional Names

Eng.	Devil's claw, Snake's head, Tiger's claw;

Hindi	Bichu, Hathajori;
Sans.	Kakanasa;
B.	Baghnoki;
Mar.	Vinchu;
P.	Bichu, Hathajori;
Tel.	Garudamukku, Telukonddichhettu.

Distribution — Naturalised in India.

Description — Stout clammy-pubescent herbs possessing large, opposite, cordate, lobed, dentate leaves. Racemes long terminal, erect with drooping foxglove-shaped flowers which are pink and ill-smelling. Corolla glandular-hairy, lobes unequal; upper lip reflexed. Pods green and fleshy and turn black woody on drying. Seeds brown to black, 2 to each pod.

Flowering & Fruiting — August-November.

Habitat Ecology — Common in moist and shady places; **Tatoh Khad -** 710 m.

Material Examined — EBH-WL-1232; 28.11.2008.

Parts Used — Leaves. Fruit.

Folk Uses — Leaf paste applied externally on **wounds** of cattle to kill worms, and on neck to heal **tuberculous glands** also its juice used as a gargle for **sore throat**. 15-20 ml decoction of leaves given twice daily for 20 days to cure **epilepsy**. External application of seed oil considered good for **eczema** and **suppuration of boils**.

Chemical Constituents — Pelargonidin-3,5-diglucoside and cyanidin-3-galactoside isolated from **flowers**.

Biological Activity — Seeds- CVS active.

Uses in Literature — Known so far as blood purifier and alexeteric (fruit); and for tubercular glands of the neck, scorpion sting, epilepsy, wounds, bronchial disorders, cold, inflammation and painful urination **(Ambasta, 1986; Bhogaonkar & Ahmed, 2007; Girach et al., 2000; Jain, 1991; Jain et al., 1991; Khan & Khanum, 2005; Kirtikar & Basu, 1984; Maheshwari et al., 1996; Pande et al., 2006; Patil, 2009; Patil et al., 2007; Rana et al., 2003; Rao & Henry, 1995; Retnam & Martin, 2006; Singh, 2000; Singh & Srivastava, 2000; Siwakoti & Varma, 1996).**

Medicago denticulata **Willd. (Plate No. 27A)**

Syn.   *M. canescens* Grah.; *M. polymorpha* Roxb

Family   Fabaceae

Vern.	Khokani.
Regional Names	
B.	Maina;
P.	Maina.
Distribution	Tropical zone of the N.W. India.
Description	Annuals with subglabrous stems. Stipules lanciniated, leaflets long and obovate-cuneate. Peduncles short. Pods spirals, muricated.
Flowering & Fruiting	February-April.
Habitat Ecology	Barren fields, edges of cultivated areas, moist places; **Koshriyan Khad** - 520 m.
Material Examined	EBH-WL-1157; 28.03.2009.
Part Used	Whole Plant.
Folk Use	Plants used as **fodder**.
Chemical Constituents	Biochanin A (0.0098) and genistein (0.016%) characterized in **plant**.
Biological Activity	Saponin (I) (40.0 µg/ml) exhibited antimycotic activity against *Sclerotium rolfsii* and *Fusarium oxysporum*.
Uses in Literature	Known earlier as cattle fodder, green manure vegetables; and for preventing land erosion **(Ambasta, 1986; Arora & Pandey, 1996; Jain, 1991; Singh & Singh, 1981; Sood & Thakur, 2004; Watt, 1972).**

Melothria heterophylla **Pittier (Plate No. 27B)**

Syn.   *Solena amplexicaulis* (Lamk.) Gandhi; *S. heterophylla* (Lour.) Cogn.; *Zeehneria umbellate* Thw.

Family   Cucurbitaceae

Vern.	Bhains.
Hindi and Regional Names	
Hindi	Amantmul;
B.	Kudari;

Bo.	Gametta;
Mal.	Njerinjanpuli;
Oriya	Karakai, Makira, Matka;
P.	Bankakra;
Tam.	Pulivanji;
Tel.	Thiyyadonda.
Distribution	Throughout India.
Description	Perennials having branched, glabrous stems and simple tendrils. Leaves polymorphous. Petioles pubescent. Calyx glabrous. Corolla small, yellowish white. Fruits cylindric, bright red when ripe. Seeds obovoid, smooth, white.
Flowering & Fruiting	May-October.
Habitat Ecology	Common on moist rocky slopes; **Barthin Khad** - 585 m.
Material Examined	EBH-WL-1179; 08.09.2007.
Parts Used	Fruits. Roots.
Folk Uses	Fruits used as **vegetable**. 5-10 ml root juice with cumin and sugar given with cold milk once daily in the morning for 15-20 days for **spermatorrhoea**. Root paste applied against **mouth ulcers**; its mixture with turmeric powder (2:1) cures **gastric** problems and discomfort in stomach (3-5 g, thrice daily, 5 days).
Chemical Constituents	Known to contain lignoceric, tricosanoic and behinic acids **(plant)**, columbin **(roots)**, linolaic, linolenic and oleic acids **(seeds)**.
Uses in Literature	Useful earlier as coolant, edible (fruits), invigurating purgative, stimulant, tonic; and for antifertility, cuts, diabetes, dysuria, earache, fever, inflammation, jaundice, labour, paralysis, snake bite, sore eyes, sores of small pox, spermatorrhoea, syphilis and unconsciouness **(Ambasta, 1986; Arora & Pandey, 1996; Bennet et al., 1991; Chopra et al., 1956; Jain, 1991; Kapur & Srivastava, 1996; Kirtikar & Basu, 1984; Negi & Gaur, 1991; Pande et al., 2006; Pandey et al., 1996; Rawat & Chawdhury, 1998; Roy et al., 1998; Singh & Kumar, 2000b; Singh & Pandey, 1998; Sinha, 1996; Sood & Prakash, 2007; Sood & Thakur, 2004; Sood et al., 2009b; Tiwari & Tiwari, 1996).**

Mentha longifolia L. (Plate No. 27C)

Syn.   *M. spicata* L. var. *longifolia* L.; *M. sylvestris* L.; *M. royleana* Wall. *ex* Benth.; *M. longifolia* var. *royleana* Rech.f.; *M. longifolia* var. *royleana* (Benth.) Raim. & Sachs

Family   Lamiaceae

Vern.	Jungli-pudina, Pudina.

English, Hindi, Sanskrit and Regional Names

Eng.	Horse mint;
Hindi	Podina;
Sans.	Ajirnahara, Pudina, Rochani, Ruchishya, Shakashobana, Sugandhipatra, Vatihara, Vyanjana;
Bo.	Pudina, Vartalau;
P.	Baburi, Belanne, Koshu, Vien, Yura.
Distribution	Temperate W. Himalaya from Kashmir - Garhwal.
Description	Strong-seeded perennial aromatic herbs possessing creeping rootstock and robust or slender, hairy-tomentose stem. Leaves rounded or cordate. Bracts lanceolate, pedicels hairy. Nutlets usually pale, smooth, brown and reticulate.
Flowering & Fruiting	November-February.
Habitat Ecology	Common along water sources, nallahs, waste places; **Karyal Khad - 615 m.**
Material Examined	EBH-WL-1029; 02.02.2009.
Parts Used	Whole Plant. Leaves.
Folk Uses	Infusion of leaves soaked in water drunk as a **cooling medicine** and for **vomiting**; also its grounded leaves mixed with seeds of *Punica granatum* or pulp of fruit of *Tamarindus indica* and a pinch of salt used for making **chutney** which is considered good against indigestion, vomiting and loss of appetite. Leaves also added for **flavouring soups, tea and dishes**.
Chemical Constituents	Carvone (76.9%), trans-carveol (18.8). cis-carveol (2.2), trans-dihydrocarveol, cis-dihydrocarveol, limonene, 1,8-cineole, humulene and caryophyllene characterized from **leaf oil**.
Biological Activity	Plant exhibits antihemolytic and antioxidant activities.

Uses in Literature	Useful as antiseptic, carminative, cooling medicine, diaphoretic, digestive, diuretic, stimulant; and for chest, deafness, diseases of the blood, dropsy, dyspepsia, headache, lessening burning sensations, leucoderma, liver, mental diseases, skin erruptions, spleen, strengthening kidney, throat troubles, vomiting and wounds to kill maggots **(Chopra et al., 1956; Dash & Misra, 2000; Duke et al., 2002; Gaur et al., 1983; Jain, 1991; Jamir et al., 2008; Joshi, 2009; Kapur & Singh, 1996; Kaul, 1997; Kirtikar & Basu, 1984; Koelz, 1979; Lindley, 1981; Pande et al., 2006; Sood & Thakur, 2004; Rana et al., 2003; Rastogi & Mehrotra, 1993; Uphof, 1968).**

Mentha piperita L. (Plate No. 27D)

Syn. *M. viridis* L.; *M. sativa* L.; *M. aquatica* L.

Family Lamiaceae

Vern.	Pipermint.
English, Hindi and Regional Names	
Eng.	Brandy mint, Pepermint, Marshmint;
Hindi	Gamathi phudina, Peppermint;
P.	Vilayeti podina.
Distribution	Occurs in Indian gardens and as an escape.
Description	Strong-scented perennial herbs with creeping rootstock, small flowers, coarsely serrated petioled leaves. Whorls many flowered in axillary and terminal spikes. Nutlets smooth or reticulate.
Flowering & Fruiting	July-September.
Habitat Ecology	Commonly near water channels, moist places; **Ali Khad** - 580 m; **Matwana Khad -** 610 m.
Material Examined	EBH-WL-1043; 05.08.2007.
Part Used	Whole Plant. Leaves.
Folk Uses	Poultice of bruised fresh leaves relieves **local pains** and **headache**. Hot infusion of the plants taken as tea for checking **stomach ache** and **allaying nausea** and relief against **menstural colic**.
Chemical Constituents	**Essential oil** from leaves possesses methofuran, menthone, isomenthone, menthol, isomenthol, neomenthol, neoisomenthol, pulegone, piperitone, α-pinene, β-pinene, cineale and carvone.
Biological Activity	Plant exhibits antibacterial and antioxidant activities.

Uses in Literature	Considered useful as astringent, dye, emetic, carminative, digestive, essential oil confectionary, infantile colic, infants cordial, stimulant, stomachic and toothpastes; and for allaying nausea, disorders of stomach and gall bladder, dry cough, flatulence, headache, herpes zoster, hoarseness influenza, liquors, throat sore, voice weakness and vomiting **(Ambasta, 1986; Baburaj et al., 2000; Duke et al., 2002; Jain, 1991; Khan & Khanum, 2005; Kirtikar & Basu, 1984; Lindley, 1981; Pande et al., 2006; Prajapati et al., 2006; Rajan, 2000; Rastogi & Mehrotra, 1991).**

Micromeria biflora (Buch.-Ham.) Benth. (Plate No. 27E)

Syn.   *M. ovata* Beck; *Thymus biflorus* Ham

Family   Lamiaceae

Vern.	Pushanbanda.
English Name	
Eng.	Indian wild thyme.
Distribution	Tropical and Temperate Himalaya (Kashmir – Bhotan): 300-2,350m.
Description	Dwarf plants with woody rootstock and excessively numerous stems or branches. Leaf margins thick. Flowers small in terminal spikes, pedicelled. Calyx hirsute, teeth subulate. Corolla white, 2-lipped, hairy; upper lip straight, margin entire whereas lower ones spreading and 3-lobed. Stamens 4. Style subulate. Nutlets ovoid, smooth.
Flowering & Fruiting	January-July.
Habitat Ecology	In crevices of rocks; **Sir Khad -** 620m.
Material Examined	EBH-WL-1044; 02.07.2007.
Part Used	Whole Plant.
Folk Uses	Decoction of the plant mixed with a pinch of black pepper used as an effective remedy for **urine blockage;** ½-1 cup thrice daily for 3 days.
Uses in Literature	Considered useful for eczema, cold, gastroenteritis, infested wounds of cattle, postnatal care and as carminative **(Abraham, 1997; Ambasta, 1986; Jain, 1991; Pande et al., 2006; Rana et al., 2003; Singh et al., 1980).**

Mirabilis jalapa L. (Plate No. 27F)
Family Nyctaginaceae

Vern.	Shivkali.

English, Hindi, Sanskrit and Regional Names

Eng.	Four o'clock flower, Marvel of peru;
Hindi	Gulabbas;
Sans.	Krishnakeli, Sandhyakali;
B.	Gulabas, Krishnakeli;
Mal.	Antimalari;
P.	Abari, Gulabbas;
Tam.	Antinarulu, Pattarachi;
Tel.	Batharachi, Chanrakanta.

Distribution	Throughout India.
Description	Erect leafy herbs upto 70 cm high. Leaves cordate. Flowers long tubular, yellow, purple or magenta, fragrant and many flowered in terminal spikes. Corolla white, 2-lipped, hairy, upper lip straight, margin entire where as lower ones spreading and 3-lobed. Stamens 4. Style subulate. Fruits elliptical, leathery black. Nutlets ovoid, smooth.
Flowering & Fruiting	June-September.
Habitat Ecology	Moist localities, Wastelands near habitations; **Tatoh Khad -** 710 m; **Papral Khad -** 740 m; **Sir Khad -** 620 m.
Material Examined	EBH-WL-1045; 05.08.2008.
Parts Used	Roots. Leaves. Flowers. Seeds.
Folk Uses	Flowers used as an **offering** to various deities. Leaves largly used as a **vegetable**; and its poultice effective against **suppuration of boils**. 30-50ml decoction of seeds taken on empty stomach to check **constipation**.
Chemical Constituents	**Flowers** yield miraxanthins I,II,III and IV, indicaxanthin and vulgaxanthin.
Biological Activity	Plant exhibits alternative, antiabortive, antiseptic, candidicide and spasmolytic activities.
Uses in Literature	Plant known to be used as aphrodisiac, antipyretic, diuretic, edible, ornamental, purgative, sacred, spasmolytic, tonic and vegetable; and for blisters, boil, bruise, cancer, candida, cold at child birth, conjuctivi-

tis, constipation, diabetes, diarrhoea, dermatosis, dropsy, dysentery, earache, edema, enlarged lymph nodes, enterosis, hepatosis, herpes, hypochondria, inflammation, itch, leucorrhoea, mycosis, otosis, piles, phlegmones, scorpion bite, skin freckles, stomach ache, sun stroke and urticaria **(Ambasta, 1986; Ansarali & Sivadasan, 2009; Bhogaonkar & Kanerkar, 2007; Biswas et al., 2010; Dastur, 1970; Dwarakan & Ansari, 1996; Duke et al., 2002; Gogoi & Borthakur, 1991; Henry et al., 1996; Hosagoudar & Henry, 1996b; Jain et al., 1991; Islam, 1990, 1996; Jain, 1991; Kamble et al., 2010; Kirtikar & Basu, 1984; Kurian, 1999; Mahato & Choudhary, 2005, Lalramnghinglova, 1996; Lindley, 1981; Molla & Roy, 1996; Pande et al., 2006; Rana et al., 2003; Rao & Henry, 1995; Retnam & Martin, 2006; Saklani & Jain, 1994; Samant et al., 2001a, b; Sanyal, 1994; Saxena et al., 1997; Sharma, 2000; Singh & Kumar, 2000b; Singh & Srivastava, 2000; Sood & Prakash, 2007; Sood & Thakur, 2004; Upadhye et al., 1994).**

Momordica dioica Roxb. ex Willd. (Plate No. 28A)

Syn.   *M. humilis* Wall.; *M. muricata* DC.; *M. senegalensis* Lamk.; *Cucumis africanus* Bot

Family   Cucurbitaceae

Vern.	Meetha karela.

English, Hindi, Sanskrit and Regional Names

Eng.	Bitter gourd, Carilla fruit;
Hindi	Karela, Kareli;
Sans.	Sushari;
B.	Karela;
Kan.	Hagal;
Mal.	Kaippa, Kaippavalli;
Mar.	Karle;
Tam.	Pakal, Pavakka.
Distribution	Throughout India.
Description	Climbing herbs with simple tendrils, orbicular petioled, glabrous pubescent leaves. Male flowers 5-6 cm, yellow; bracts large, sessile. Female flowers yellow, ebracteate. Calyx linear-lanceolate, villous. Corolla ovoid. Fruits rostrate. Seeds compressed, corrugate.
Flowering & Fruiting	July-September.
Habitat Ecology	Moist Places; **Chamyater Khad -** 613 m.

Material Examined	EBH-WL-1046; 24.07.2008.
Parts Used	Roots. Fruit.
Folk Uses	Young fruits cooked as a **vegetable;** 3-5g powdered dried fruit taken on empty stomach with luke warm water till cure against **diabetes.** 15-20 ml decoction of roots prescribed twice daily for 8-10 days against **leucorrhoea.**
Chemical Constituents	β-sitosterol, β-D-glucoside and stearic acid, octacosane, 1-triacontradiene, 7-stigmasten-3β-ol, 7,25-stigmastadien-3β-ol, 5,25-stigmastadien-3β-ol glucosides, phytosphingosine A and B, zeatin, zeatin riboside **(seeds)** calcium, carotene, riboflavin, ascorbic acid **(stem),** lectins, proteins, triterpenes, vitamins, ascobic acid, iodine, alkaloid, flavonoids, glycosides, amino acids, 6-methyl-tritriacont-50on-28-of, 8-methyl hextracont-3-ene-sterol pleuchiol **(fruits),** α-spinasterol octadecanonate (1), α-spinasterol-3-o-β-D-glucopyranoside (11), 3-o-β-D-glucuronopyranosyle gypsogenin (IV), 3-o-β-D-glucopyransyl hederagenin (V) **(roots)** have been isolated.
Uses in Literature	Useful earlier as anthelmintic, carminative, condiment, refrigerant, stomachic, tonic; and for antifertility, diptheria, diseases of liver, gout, rheumatism, spleen and stomach disorders **(Ambasta, 1986; Jain, 1991; Jain et al., 1991; Khanna et al., 1996; Painuli & Maheshwari, 1996; Pande et al., 2006; Rao & Henry, 1995; Rout & Panda, 2010; Singh & Rao, 2003).**

Mucuna pruriens (L.) DC. (Plate No. 28B)

Syn.   *Carpopogone pruriens* Roxb.; *Dolichos pruriens* L.; *M. prurita* Hook.; *M. utilis* Wall

Family   Fabaceae

Vern.	Dryagal.

English, Hindi, Sanskrit and Regional Names

Eng.	Common cowwitch, Cowhage;
Hindi	Kiwach, Kaunch, Goncha;
Sans.	Altamagupba;
B.	Alkushi, Bichchoti;
G.	Kivanch, Kavatch;
Kan.	Klasukunni, Haraguni;
Mal.	Kavancha, Kuhli, Kanchkuri;
Mar.	Kavancha, Kuhili, Kanchkuri;

Oriya	Kaincho;
P.	Kawach, Kaunch, Goncha;
Tam.	Poonaipidukkan, Poonikalei;
Tel.	Dulagondi, Pillaidugu.
Distribution	Cosmopolitan in the tropics.
Description	Annual glabrescents having grey silky, ovate–rhomboidal leaves and short peduncled drooping racemes. Corolla purplish, broad. Pods long, turgid, ribbed. Seeds dark brown.
Flowering & Fruiting	July-September.
Habitat Ecology	A climber on trees near water sources, moist places; **Papral Khad - 740 m; Auhr Khad -** 610 m.
Material Examined	EBH-WL-1048; 24.08.2008.
Parts Used	Root. Leaves. Fruit. Seed.
Folk Uses	3-5 ml root extract with honey given against **cholera, urine** and **kidney troubles** (twice a day till cure). 5-10 g powdered seeds taken with cow's milk before bed as an **aphrodisiac** and for checking **piles**. Paste of mixture of its 10 g powdered seeds and 10g sandal powder prepared in 50 ml water, applied twice a day for 12 days on **cracked soles, leucoderma** and **dry skin**.
Chemical Constituents	Reported to contain alkaloids, arachidic acid, alkylamines, betacarboline, behenic acid, beta-sitosterol, cystine, dopamine, bufotenine, flavones, fatty acids, gallic acid, prurienine, serine, riboflavin, saponins and tripsin **(leaves, fruits, seeds)**.
Biological Activity	Hypnosis, hypothermia, catatonia and enhanced amphestamine toxicity exhibited.
Uses in Literature	Known earlier as anthelmintic, aphrodisiac, CNS active, hypoglycaemic, laxative, spasmolytic, nervine tonic; and for depression, fever, filariasis, fractures, headache, impotency, intestinal worms, irregular menstruation, paralysis, parkinsonism, pshychological disorders, rotten teeth, scorpion sting and weakness **(Ambasta, 1986; Bhogaonkar & Kanerkar, 2007; Biswas et al., 2010; Chandra, 1997; Duke et al., 2002; Henry, 1999; Idu et al., 2008; Jain, 1991, 2010; Kaushik & Dhiman, 2000; Khan & Khanum, 2005; Lalramnghinglova, 2003; Lindley, 1981; Pande et al., 2006; Parrotta, 2001; Parabia & Pathak, 2007; Rana et al., 2003; Rao & Henry, 1995; Satapathy, 2008; Singh & Pandey, 1996; Prajapati et al., 2006; Siwakoti & Varma, 1996; Sood et al., 2009b; Thakor, 2009; Watt, 1972)**.

Najas graminea **Dd. (Plate No. 28C)**
Family   Naiadaceae

Vern.	Jalchida.
Distribution	Temperate and Tropical regions.
Description	Submerged, grass-like annual plumose herbs. Stem slender without spines. Leaves narrowly linear, denticulate. Flowers solitary. Achenes ellipsoid-oblong. Areoles minute, polyhedral.
Flowering & Fruiting	August-October.
Habitat Ecology	In ponds and water-logged areas; **Sir Khad -** 620 m.
Material Examined	EBH-WL-1274; 12.07.2008.
Parts Used	Whole Plant. Leaf.
Folk Uses	Poultice of leaf paste applied on **goitre** and **suppuration of boils.** Plant used as **food** for fish.
Use in Literature	Used as biofertilizer **(Das et al., 1996; Hooker, 1872-97; Kumar & Narain, 2010).**

Najas indica **(Willd.) Cham. (Plate No. 28D)**
Family   Naiadaceae

Vern.	Chu.
Distribution	C. & S. Europe.
Description	Floating or submerged aquatic annual herbs which are rooting below in shallow water. Stem without spines. Leaves linear, flat, dentate. Achenes ellipsoid. Seeds tetragonal.
Flowering & Fruiting	August-September.
Habitat Ecology	In ponds and water-logged areas; **Sir Khad -** 620 m.

Material Examined	EBH-WL-1277; 22.08.2008.
Part Used	Whole Plant.
Folk Use	Plant used as **fish food.**

Nasturtium officinale **R. Br. (Plate No. 28E)**

Syn.   *Roripa nasturtium-aquaticum* (L.) Hayek; *N. fontanum* Asch.; *Sisymbrium nasturtium-aquaticum* L.

Family   Brassicaceae

Vern.	Chuch.

English and Regional Names

Eng.	Water cress;
P.	Piriya halim.
Distribution	Throughout Asia and Temperate Europe.
Description	Branched aquatic herbs with entire, lobed or pinnatifid leaves. Flowers small, white. Pods linear, long, cylindric. Seeds small, turgid.
Flowering & Fruiting	April-September.
Habitat Ecology	Bogs, irrigated channels, streamsides, damp places; **Sir Khad -** 620 m; **Matwana Khad -** 610 m.
Material Examined	EBH-WL-1036; 24.08.2008.
Parts Used	Plant. Aerial Plant Parts. Leaves. Seeds.
Folk Uses	Decoction of the plant given for **expelling worms** from the body (1 tsp once every morning for 3-5 days). Aerial plant parts used as a **vegetable.** Leaves consumed as **salad** and its decoction (15-20 ml thrice a day for 3-5 days) considered an effective remedy against common **cold.** Powder of grounded seed with a glass of luke warm water given to cure **stomach ache** (twice daily for 2-3 days).
Chemical Constituents	These contain 2-phenylethyl isothiocyanate (72.9%), pulegene (8.0%), heptyl isothiocyanate (4.9%), 4-phenylbutyl **(leaves)**, 2-phenylethyl isothiocyanate (83.5%), 4-phenylbutyl isohtiocyanate (6.9%), pulegone (2.2%) and sec-butyl isothiocyanate **(stem)**, 3-phenylpropionate, phenethylisothiocyanate, 8-methyliooctanonitrile and 9-methyl-thiononanonitrile, gluconasturtin and a non-drying fatty oil (24%) **(seeds).**
Biological Activity	Antibacterial, anticancer, detoxicant, stimulant, tonic, vermifuge, vulnerary activities confirmed.

Uses in Literature	Recorded earlier as an appetizer, antiscorbutic, blood purifier, detoxifying herb, diuretic, febrifuge, gravel, pickle, pot herb, salad, soups, stimulant, valuable source of vitamins; and for anaemia, asthma, blemishes, boils, bronchosis, cardiopathy, catarrah, chest affections, chronic bronchitis, cold, dry throat, eye sight, fever, goitre, gout, herpes, infection, illness, kidney trouble, polypus of nose, sore throat, splenosis, swelling, toothache, tuberculosis, tumour and wart **(Ambasta, 1986; Arora, 1997; Arora & Pandey, 1996; Bennet et al., 1991; Bhargava, 1959; Chandrasekar & Srivastava, 2003; Chevallier, 1996; Chhetri, 2005; Chopra et al., 1956; Dash et al., 2007; Devi, 2003; Duke et al., 2002; Henry, 1999; Kapoor et al., 2010; Panda, 1996; Prajapati et al., 2006; Ranjan, 2000; Samant et al., 2001b; Sood & Thakur, 2004; Uphof, 1968; Watt, 1972).**

Nerium indicum Mill. (Plate No. 28F)

Syn.   *N. odorum* Sole; *N. odoratum* Lamk.; *N. latofolium* Mill

Family   Apocynaceae

Vern.	Kaner, Ghanira.

English, Hindi, Sanskrit and Regional Names

Eng.	Indian oleander, Sweet-scented oleander;
Hindi	Kaner, Karber, Kuruvira;
Sans.	Karavira;
Ass.	Karabi, Karbira, Karvir;
B.	Karabi;
G.	Kagaer;
Kan.	Dhavekaeri, Paddale, Kanagalu;
Kash.	Gandeela, Gandula;
Mal.	Areli;
Mar.	Kanher, Kaneri;
Oriya	Kenero, Korobiro;
P.	Ganira, Kanhira;
Tam.	Arale, Asuvabari, Irattaichegappayalari, Vellalari;
Tel.	Ganneru, Kastoori pattelu.
Distribution	Found in Himalaya.

Description Evergreen, large-sized shrubs with milky juice and leaves mostly in whorls of 3. Flowers red or white in colour, fragrant and in terminal cymes. Capsules long narrow, 23 cm long. Seeds downy.

Flowering & Fruiting April-June.

Habitat Ecology Near water channels; **Sauli Khad -** 765 m.

Material Examined EBH-WL-1041; 14.05.2008.

Parts Used Leaves. Stem bark. Root bark. Fruit.

Folk Uses A pinch of dried leaves taken as snuff for checking **headache**. Powdered root-bark (2 g) given twice daily with honey for 30 days in **paralysis**. Crushed stem-bark in combination with fennel (*Foeniculum vulgare*) and sugar (1:1 ratio) prescribed as **blood purifier**. Oil prepared from root bark good against **skin eruptions**. Fruits offered to Lord Shiva during **worship**.

Chemical Constituents Cardioactive glycosides as neriodorin and karabin **(root, bark, seeds)**, scopoletin and scopoline **(bark)**, oleandrin, nariodin, ursolic acid **(leaves)** etc have been isolated.

Biological Activity Alcoholic elixir of root, bark, leaf and flower exhibits antibacterial activity. Also, oleanderin and oleandrigenin b-D- diginopyranoside showed cardiopkinetic and diuretic activities.

Uses in Literature Reported to be used as an abortifacient, anthelmintic, attenuant, cardiac, carminative diaphoretic, febrifuge, fodder, gunpowder, insect repellent, insecticide, ornamental garlands, powerful heart poison, rat poison, resolvent, symbolic (sacred), vermic (vet); and for asthma, blisters, boils, bronchitis, copious, ophthalmic, lachrymation, dysentery, epilepsy, eye diseases, gum troubles, hooka tuber, leprosy, menorrhagia, pthisis, purifying air, ringworm, skin eruptions like herpes, snake bite, swellings, ulcers, worms and wounds **(Ambasta, 1986; Chauhan, 1999; Dastur, 1970; Girach et al., 1996; Jain, 1991; Kapur & Singh, 1996; Kaushik & Dhiman, 2000; Khanna et al., 1996; Kurian, 1999; Pande et al., 2006; Parrotta, 2001; Prajapati et al., 2006; Ranjan, 2000; Rastogi & Mehrotra, 1993; Retnam & Martin, 2006; Saini, 1996a; Sood et al., 2009b; Watt, 1972).**

Ocimum basilicum L. (Plate No. 29A)

Syn.   *O. pilosum* Willd.; *O. album* L.; *O. hispidum* Lamk.; *O. menthaefolium* Benth.; *O. caryophyllatum* Roxb.; *Plectranthes barrelieri* Spreng

Family   Lamiaceae

Vern.	Bhabhri.

English, Hindi, Sanskrit and Regional Names

Eng.	Basil, Common basil, Sweet basil;
Hindi	Bahari, Kalitulsi;
Sans.	Ajagendhika, Asurasa, Barbara, Karahi, Surabhi, Talasidvesha, Tungi, Varvara;
B.	Babui-tulsi, Debunsha;
G.	Damaro, Nasabo;
H.	Bahari, Kalitulsi;
Kan.	Kam-kasturi, Sajjebiya;
Mal.	Pachaha;
Mar.	Marva, Sabja;
Oriya	Dhala, Tulasi;
P.	Baburi, Rehan;
Tam.	Tirnutpatchie, Tirunitru;
Tel.	Rudrajada, Vipudi-patri.
Distribution	Throughout India, often planted.
Description	Erect, glabrous or pubescent strongly scented herbs with ovate toothed leaves. Bracts petiolate. Inflorescence in terminal clusters of whorled flowers (verticillasters). Flowers small. Fruiting calyx shortly pedicelled, two lower teeth awned, longer than the rounded upper. Corolla white, 1 cm long, bilabiate. Stamens long. Nutlets long, ellipsoid, black.

Flowering & Fruiting	June-September.
Habitat Ecology	Escape in moist slopes, common in wastelands; **Matwana Khad** - 610 m.
Material Examined	EBH-WL-1042; 06.09.2008.
Part Used	Leaves.

Folk Uses Poultice of leaves used for **sores** and **sinusitis**; its decoction useful as an expectorant to **relieve** mucous secretions from the bronchial tube. Also, the extract of leaves considered good against **stomach** and **intestinal disorders** (5-10 ml, twice daily, 5 days), **toothaches, earaches, headaches**; the juice on mixing with camphor used as drops to check **nasal blockage**.

Chemical Constituents **Oil** from **leaves** contains methylehavicol (70-80%), estragole, linalool, excalyptol, ocimene, linalool acetate, exgenol, 1-epibicyclosesquiphellandrene, menthol, menthone, cyclohexane, cyclohexanone, myrcenol and nerol.

Biological Activity Myrcenol and nerol exhibited antiasthmatic activity.

Uses in Literature Earlier known as anthelmintic, antipyretic, aphrodisiac, appetizer, carminative, condiment, demulcent, diaphoretic, emmenagogue, expectorant, nasal douche, psychoactive, stimulant and stomachic; and for alcoholic intoxication, asthma, blindness, bowel complaints, cholera, chronic dysentery, cough, convulsion, cramp, diseases of heart, diuretic, dropsy, dysentery, edible **(seeds, leaves, stem)**, earache, enlarged spleen, epilepsy, fever, fits, gonorrhoea, headache, joint pains, labour complaints, piles, respiratory troubles, ring worm, sinuses, sores, snake-bite and wounds **(Agarwal, 2003; Ambasta, 1986; Bhatt et al., 1999; Billore et al., 1998; Biswas et al., 2010; Chaudhury & Neogi, 2000; Chevallier, 1996; Chopra et al., 1956; Chun-Lin & Jieru, 1995; Das & Agarwal, 1991; Dash & Misra, 1999b, 2000; Duke et al., 2002; Girach, 1992; Hosagoudar & Henry, 1996b; Idu et al., 2008; Jadhav, 2009; Jain, 1991; Jain et al., 1994; Jamir, 1995; Katewa et al., 2000; Khanna et al., 1996; Khare, 2004; Kirtikar & Basu, 1984; Kumar, 2002; Kurian, 1999; Lal et al., 1996; Mandal & Basu, 1996; Masih, 2003; Molla & Roy, 1996; Noumi, 2010; Pande et al., 2006; Parrotta, 2001; Parabia & Pathak, 2007; Prajapati et al., 2006; Rana et al., 2003; Reddy et al., 2005; Rosakutty et al., 2000; Roy et al., 1998; Saini, 1996a; Samant et al., 2001b; Saxena et al., 1997; Seetharam et al., 1999; Sharma, 2003; Singh, 2000; Singh & Kumar, 2000b; Singh et al., 2001; Saklani & Jain, 1994; Sood & Thakur, 2004).**

Oroxylum indicum Vent. (Plate No. 29B)

Syn.   *B. pentandra* Lour.; *B. indica* L.; *Calosanthes indica* Blume; *Spathodea indica* Pers

Family   Bignoniaceae

Vern.	Arlu.

English, Hindi, Sanskrit and Regional Names

Eng.	Indian treatment tree;
Hindi	Arlu, Saona, Sonapatha, Ullu;
Sans.	Shyonaka;
Ass.	Toguna, Bhatgnila, Dingari;
B.	Sona, Nasona, Sonpatti;
G.	Aralu, Tentu;
Kan.	Tigolu, Bunepale, Sonepattu;
Mal.	Palagapaiyani;
Mar.	Tetu;
Oriya	Phapni, Phonphonia;
P.	Mulin, Tatmorang;
Tam.	Achi, Permpini;
Tel.	Dundilum, Permpini.

Distribution	Throughout India; mostly in the Terai west to the Chenab.
Description	Glabrous trees attaining 13 m. Bark thick. Leaves pinnate, leaflets ovate. Raceme terminal, long. Corolla fleshy. Disc large, fleshy. Capsule thick. Seeds winged all round.
Flowering & Fruiting	June-August.
Habitat Ecology	In forests near water channels; **Karyal Khad -** 615 m.
Material Examined	EBH-WL-1047; 15.08.2008.
Parts Used	Roots. Stem. Pods. Seeds.
Folk Uses	Tree lopped for **fodder**. Extract of its roots crushed with aerial roots of *Ficus benghalensis* and the tail of lizard (*Varanus bengalensis*) prescribed for **giddiness** – 2 tsp twice a day. In veterinary, pods fed orally along with 'rai' (*Brassica juncea* Hook.) and 'ajwain' (*Trachyspermum amoni* L.) for discomforts due to **stomach ache** and **spleen affections**. A garland of its coloured winged seeds used for temple **decoration**. Pills formed by grinding a piece of its 4 cm long stem, 3 barley grains and

one black pepper given with cow'milk on empty stomach (1 tablet, once daily, 30 days) for **bone fracture** in children. Poultice of leaves applied on painful **rheumatic joints**. Powdered pods (2-4 g) given to check **dysentery, stomach ache** and **fever** (twice a day till cure).

Chemical Constituents   Properties associated due to baicalein, its 6-glucuronide, seretellarein, its 7-gluconnonide **(leaves, stem bark)**, oroxylin A, chryrin and scutellarein-7-rutinoside **(stem bark)**, prunetin and β-sitosterol **(heart wood)**, glucuronide- oroxidin, biacatein-β-glucronide and 7-glucononide **(seeds)**.

Biological Activity   Tender fruits known for spasmolytic activity.

Uses in Literature   Useful earlier for bone fracture, bronchitis, diarrhoea, dropsy, dysentery, dyspepsia, flatulence, fractures, gastroenteritis, jaundice, joint pain, lactation, leucoderma, muscular pain, paralysis, piles, rheumatism, snake bite, wounds; and as acrid, anodyne, anthelmintic, antiarthritic, anti-inflammatory, aphrodisiac, appetizing, astringent, bitter, carminative, colic, constipating, diaphoretic, digestive, diuretic, expectorant, febrifuge, refrigerant and tonic **(Ambasta, 1986; Awasthi & Goel, 2000; Barua et al., 2000; Bhattacharyya, 2000; Bhogaonkar & Kanerkar, 2007; Biswas et al., 2010; Bora, 2000; Deokota & Chhetri, 2009; Gupta, 1997; Jain et al., 1991, 1997, 2010; Jamir et al., 2008; Kaushik & Dhiman, 2000; Khan & Khanum, 2005; Kumar, 2002; Lalramnghinglova, 1996, 2003; Panda & Das, 2000; Pandey et al., 1996; Parrotta, 2001; Prajapati et al., 2006; Rama Rao et al., 2008; Rao & Henry, 1995;   Rao et al., 2000; Retnam & Martin, 2006; Samwatsar & Diwanji, 2000; Saxena et al., 1997; Sharma, 2000; Siwakoti & Varma, 1996; Sood et al., 2009b; Thakor, 2009; Yasodamma et al., 2009).**

Oxalis corniculata L. (Plate No. 29C)

Syn.   *O. foliosa* Blatt.; *O. pusilla* Salisb.; *O. repens* Thunb.; *O. villosa* Bieb

Family   Oxalidaceae

Vern.   Malora.

English, Hindi, Sanskrit and Regional Names

Eng.	Indian sorrel, Yellow oxalis, Wood sorrel;
Hindi	Amrul, Amrulsak, Anboti, Chalmoir, Chukatripati, Seh;
Sans.	Ambashta, Amalalonika, Amlika, Amletaja, Changeri, Chukrita;
Ass.	Chengeritenga;
B.	Amrul, Amrulsak, Chalmori, Chukatripati, Omloti;
Kan.	Hulichikkai;

Mal.	Poliyarala;
Mar.	Ambuti, Bhinsarpati;
P.	Amlika, Amrul, Chukha, Khattamitha, Surechi, Travuke;
Tam.	Paliakiri, Puliyarai;
Tel.	Ambotikura, Paflachinta, Pulichinta, Pullachanchali.

Distribution   Throughout warmer parts of India.

Description   Hairy annual acid herbs with long-petioled, compound leaves and obcordate leaflets. Flowers subumbellate, axillary. Capsule tomentose, many. Seeds transversely ribbed.

Flowering & Fruiting   April-November.

Habitat Ecology   Moist places; **Sir Khad -** 620m; **Karyal Khad -** 615 m.

Material Examined   EBH-WL-1049; 25.10.2008.

Parts Used   Whole Plant. Leaves.

Folk Uses   3-5 g paste of plants with cumin seeds taken orally with water thrice a day against **diarrhoea** and **dysentery;** its poultice applied on forehead to relieve **headache** whereas its mixture with rhizome of *Drynaria quercifolia* applied to set **bone fracture.** Seeds **consumed** in times of scarcity. Leaves **refreshing** and eaten raw or as **'chutney'** as a source of **vitamin-c**; its juice applied externally for **suppuration of boils** and on backbone of the infants for **curing rickets.** Also, **fruits** chewed in order to get fresh taste in the mouth especially during pregnancy period. Leaf juice considered good for **countering** 'dhatura' poisoning, good appetiser; removing piles and opacities of cornea.

Chemical Constituents   Constituents isolated are malic acid **(stem)**, tartaric and citric acids **(stem, leaves)**.

Biological Activity   Alcoholic extract of leaves showed complete inhibition of growth of *Staphylococcus typhi, S. aureus, S. albus* and *S. citrarus.*

Uses in Literature   Known in literature for body sores, burns, chronic conjunctivitis, cough, diarrhoea, dizziness, dysentery, fractures, headache, indigestion, insomnia, skin diseases, snake bites, sores, stomach troubles, swelling beneath tongue, wounds and as tonic for nursing mothers **(Abraham, 1997; Alagesboopathi et al., 2000; Ambasta, 1986; Bajpayee & Dixit, 1996; Balasubramanian & Prasad, 1996; Banerjee, 2000; Banerjee & Ghora, 1996; Biswas et al., 2010; Das, 2000; Gogoi et al., 2003; Hajra & Baishya, 1997; Hosagoudar & Henry, 1996a, b; Jain, 1991; Jha et al., 1996; Joshi, 2009; Kapur & Singh, 1996; Kapur & Srivastava, 1996; Kaul, 1997; Khan & Khanum, 2005; Kharkongor & Joseph, 1997; Kumar, 2002; Kumar & Narain, 2010; Lalramnghinglova, 1996; Malla & Chhetri, 2009; Pande et al., 2006; Patil, 2009;**

Parabia & Pathak, 2007; Prajapati et al., 2006; Rana et al., 2003; Ranjan, 2000; Rao & Henry, 1995; Retnam & Martin, 2006; Rout & Panda, 2010; Saini, 1996a; Sasidharan & Swarupanandan, 2000; Semwal et al., 2010; Singh, 2000; Singh et al., 1996a; Siwakoti & Varma, 1996; Siwakoti & Siwakoti, 2000; Swami & Gupta, 1996).

P

Parthenium hysterophorus L. (Plate No. 29D)
Family Asteraceae

Vern.	Chikadu.
English Names	
Eng.	Bitter-broom, Escoba, Amarga.
Distribution	Common weed throughout India.
Description	Profusely branched herbs with angular hairy stem, dissected leaves and numerous white heads peduncled in corymb-like cymes. Achenes compressed.
Flowering & Fruiting	August-September.
Habitat Ecology	Plentiful in wastelands; **Sir Khad -** 620 m.
Material Examined	EBH-WL-1109; 16.09.2007.
Part Used	Whole Plant. Root.
Folk Uses	5-10 ml decoction of plant given in high **fever** (twice daily for 3-5 days till cure). Decoction of roots used as a **tonic**. Poultice of root paste in ghee recommended for **slipped navel**.
Chemical Constituents	Constituents isolated are bornylacetate, coronopilin, dihydroiso-parthenin, hysterin, parthenin, an α-methylene-γ-lactone sesquiter-pene, tetraneurins A,B,C,D, phenylacetonitrile, hexacosannol, myricyl alcohol, galactose, glucose, 6-hydroxy-kaemferol-3,7-dimethylether, kaempferol, quercetagetin-3,7-dimethyl ether a quercetin-3-O-glyco-side, betulin, ursolic acid a saponin composed of oleanolic acid and glu-cose, campestrol, β-sitosterol, stigmasterol **(plant),** histamine (0.35%), parthenin, caffeic, chlorogenic, p-hydroxybenzoic, p-anisic, vanilic, salicylic, gentisic, neo-chlorogenic and protocatechuic acids **(roots)** **(Anonymous, 1948-76a; Kamal & Mathur, 1991).**
Biological Activity	It causes allergic contact dermatitis.

Uses in Literature	Used in India as an analgesic, emmenagogue, febrifuge and tonic; and for dysentery and nasal blockage **(Ambasta, 1986; Bennet, 1986; Jain, 1991; Kaushik & Dhiman, 2000; Pande et al., 2006; Parrotta, 2001; Rana et al., 2003; Ranjan et al., 2000; Rastogi & Mehrotra, 1995a).**

Paspalum distichum L. (Plate No. 29E)

Syn.   *P. brachiatum* Trin. *ex* Nees.; *P. desressum* Steud.; *P. didactylum* Saltzen.; *P. digitaria* Poir.; *P. distachyon* Willd. *ex* Doll; *P. fernandesianum* Colla; *P. genicolatam* Heyne; *P. kleinianum* Presl; *P. littorale* Br.; *P. longiflorum* Retz.; *P. maculosum* Trin.; *P. michauxianum* Kunth; *P. notatum* flugge; *P. obtusatum* Nees; *P. obtusifolium* Trin.; *P. repens* Ic. Roxb

Family   Poaceae

Vern.	Chini - ghas.
Distribution	N.W. India, Sunderbunds, Malacca, Andaman Islands, warm countries.
Description	Annual or perennial grasses with creeping stem, narrow distichous, 15 cm long leaves, geminate spikes, ellipsoid or lanceolate spikelets, acute membranous rachis and terminal inflorescesnce. Flowers bisexual. Rachilla glabrous, Glumes greatly reduced, 3 nerved. Lemma glabrous, acute, awnless. Palea present. Stamens 3. Fruit a caryopsis.
Flowering & Fruiting	August-October.
Habitat Ecology	Occurs in ditches and marshes along with muddy coasts; **Ali Khad -** 580m; **Matwana Khad -** 610 m.
Material Examined	EBH-WL-1051; 27.09.2008.
Part Used	Whole Plant.
Folk Uses	Plant used extensively as source of **fodder** for live-stock and considered a good **soil binder**.
Uses in Literature	Used earlier as a soil binder, valuable pasture grass and for making mats (rhizomes, stolons) **(Bews, 1979; Bor, 1973; Das et al., 1996; Jain, 1991; Mehra, 1982; Uphof, 2001).**

Pennisetum lanatum **Klotzsch (Plate No. 29F)**

Syn. *P. sericeum* Munro; *P. nepalense* Griseb
Family　Poaceae

Vern.	Baru.
Distribution	W. Himalaya, (Kashmir–Garhwal) : 2,350-3,000 m, W. Tibet.
Description	Stout woody perennials with creeping rootstocks, tall stem upto 1m and flat leaves. Involucres pedicelled, densely imbricate. Spikelets solitary, scabrid. Anthers 3-nerved without hairs.
Flowering & Fruiting	June-October.
Habitat Ecology	Moist shady places in grassy meadows; **Lyond Khad -** 617 m.
Material Examined	EBH-WL-1037; 28.09.2008.
Part Used	Whole Plant.
Folk Use	Plant regarded as a valuable **fodder** grass.
Use in Literature	Useful earlier as a soil binder **(Bor, 1973; Hooker, 1872-97)**.

Phragmites karka **(Retz.) Trin. ex Steud. (Plate No. 30A)**

Syn.　*P. bifaria* Wight Herb.; *P. nepalensis* Nees *ex* Steud. *Arundo karka* Retz. *A. donax* Herb. *Trichoon karka* Roth.

Family　Poaceae

Vern.	Baru.
Distribution	Throughout India from Punjab to Myanmar.
Description	Tall perennial grasses with stems upto 3 m. Leaves close, bifarious, linear, coriaceous. Panicle 20 cm long, erect, oblong. Spikelets 0.4 cm, pedicels capillary smooth. Glumes glabrous, 3. Palea linear oblong. Seeds brown, light weight.
Flowering & Fruiting	May-June.
Habitat Ecology	Common throughout India near water sources; **Papral Khad -** 740m.
Material Examined	EBH-WL-1233; 06.05.2008.
Parts Used	Whole Plant. Roots. Rhizome.
Folk Uses	3-5 g powdered roots and rhizome given once daily on empty stomach with luke warm water for curing **diabetes**. Plant used as a good material for **thatching** purposes.

Uses in Literature	Used as animistic, diuretic, diaphoretic, ornamental, sieve for winnowing; and for thatching purposes **(Biswas & Calder, 1984; Jain, 1991; Kumar & Narain, 2010; Rana et al., 2003; Siwakoti & Varma, 1996).**

Phyllanthus urinaria L. (Plate No. 30B)

Syn.   *P. leprocarpus* Wight; *P. alatus* Blume; *P. cantoniensis* Hornem.; *P. mucronatus* Heynh.; *P. muricatus* Madr.; *P. polyphyllus* Madr.; *P. echinatus* Ham

Family   Euphorbiaceae

Vern.	Pooin anvalah.

Hindi, Sanskrit and Regional Names

Hindi	Hazarmani, Lalbhuinanvalah;
Sans.	Adhyanda, Ajata, Ajuta, Amala, Aphala, Aruha, Bahupatra;
B.	Hazar mani;
G.	Kharsadabonyaanmali, Khasadabonyaansari;
Kan.	Kempu nela nelli;
Mal.	Chirukizhukanelli, Chukannakizhanelli;
Mar.	Lalmundajanvali;
Tam.	Shivappunelli;
Tel.	Ettacuirika.

Distribution	Throughout India.
Description	Diffusely branched low erect herbs with stipular lanceolate leaves which are glaucous beneath. Flowers shortly pedicelled. Fruits echinate. Seeds transversely furrowed.
Flowering & Fruiting	July-September.
Habitat Ecology	Moist areas; **Barthin Khad -** 585 m; **Gamrola** - 528 m.
Material Examined	EBH-WL-1052; 08.09.2007.
Part Used	Roots. Stems. Leaves. Fruits.
Folk Uses	Decoction of stems and leaves used for **dyeing** cotton black. Fruits useful to **check thirst, anaemia** and **asthma**. 1-3g of its powdered roots given with cow's milk to children suffering from disturbed **sleep**.
Chemical Constituents	Reported to contain germacrene D, aromadendrene, cis-α –bisabolene, β-cubebene, γ-muurolene, 1,2,4a, 5, 6, 8a-hexahydro-4, 7-dimethyl-1-(1-methylethyl)-naphthalene **(leaves, stem).**

Uses in Literature	Recorded so far as astringent, diuretic, spasmolytic and tonic; and for bone fracture, dropsy, dysentery, flatulence, fractures, gastroenteritis, jaundice, joint pain, lactation, leucoderma, muscular pain, paralysis, piles, rheumatism, snake bite and wounds **(Ambasta, 1986; Awasthi & Goel, 2000; Barua et al., 2000; Bhattacharyya, 2000; Bora, 2000; Jain, 1991; Khan & Khanum, 2005; Kirtikar & Basu, 1984; Lindley, 1981; Panda & Das, 2000; Pande et al., 2006; Parrotta, 2001; Prajapati et al., 2006; Rao et al., 2000; Rastogi & Mehrotra, 1995b; Retnam & Martin, 2006; Samwatsar & Diwanji, 2000; Sharma, 2000).**

Physalis longifolia Nutt. (Plate No. 30C)
Family   Solanaceae

Vern.	Popti.
Distribution	Throughout India.
Description	Annual herbs. Leaves alternate. Pedicels axillary. Calyx campanulate. Corolla purple, lurid yellow. Ovary 2-celled. Berries globose. Seeds smooth.
Flowering & Fruiting	July-October.
Habitat Ecology	In paddy fields near water sources; **Karyal Khad** - 615 m.
Material Examined	EBH-WL-1053; 15.008.2008.
Part Used	Whole Plant.
Folk Use	Plant considered good for **fodder** purposes.
Uses in Literature	Reported to be described for fever, eczema and as purgative **(Duthie, 1973; Prajapati et al., 2006).**

Physalis minima L. (Plate No. 30D)
Syn.   *P. minima* L. var. *indica.*; *P. pubescens sensu* Wight
Family   Solanaceae

Vern.	Heui.

English, Hindi, Sanskrit and Regional Names

Eng.	Country gooseberry;
Hindi	Bandhapariya, Cirapoti, Tulati pati;
Sans.	Cirapotha, Mrdu kuncika;
Kan.	Gudde hannu;

Mal.	Sodakku thakkali;
Tam.	Thol thakkali;
Tel.	Kupani.
Distribution	Throughout India; as a weed upto 2,300 m on cultivated areas.
Description	Small pubescent herbs growing upto to 160 cm. Leaves simple, alternate, ovate with serrated margins. Flowers yellow, solitary, axillary. Berries globular, many seeded.
Flowering & Fruiting	June-August.
Habitat Ecology	Occasional in waste places; **Raul Khad -** 646 m.
Material Examined	EBH-WL-1086; 02.08.2008.
Parts Used	Leaves. Fruits.
Folk Uses	Leaf paste good for **fever**. Fruits **eaten** by children; also fruits roasted in coconut oil given for **treating spleen disorders**.
Chemical Constituents	A new ergostane-type steroid- with aminimum, 5-methoxy-6,7-methylenedioxyflavone and 5, 6,7-trimethoxyflavone, quercetin-3-O-galactoside isolated from **leaves**.
Biological Activity	Aerial parts diuretic.
Uses in Literature	Used as colic, diuretic, purgative and tonic; and for antifertility, ascites, bronchitis, burning sensation, cough, diabetes, dysuria, gastropathy, swelling of joints, pruritus, erysipelas, stranguary, splenomegaly and ulcers **(Ansarali & Sivadasan, 2009; Bajpayee & Dixit, 1996; Dager & Dager, 1996; Hosagoudar & Henry, 1996a; Jadhav, 2009; Jain, 1991; Kaushik & Dhiman, 2000; Parkash & Aggarwal, 2010; Prajapati et al., 2006; Rana et al., 2003; Rao & Henry, 1995; Rastogi & Mehrotra, 1995b; Retnam & Martin, 2006; Saini, 1996a; Singh et al., 1996b; Solanki et al., 2007).**

Plectranthus coetsa Buch.-Ham. ex D. Don (Plate No. 30E)

Syn.   *Ocimum coetsa* Spreng

Family   Lamiaceae

Vern.	Saru.
Distribution	Temperate and Subtropical Himalaya.
Description	Tall, strong-smelling undershrubs with ovate leaves and cymes in laxfid. Corolla blue. Nutlets oblong.
Flowering & Fruiting	October-November.

Habitat Ecology	Common in exposed places; **Sauli Khad -** 765 m.
Material Examined	EBH-WL-1095; 22.11.2008.
Part Used	Leaves.
Folk Use	Decoction of leaves good against **headache** (10-15 ml, twice daily for 3 days).
Chemical Constituents	Diterpenoids - coetsin A and coetsin B characterized from **leaves**.
Use in Literature	Earlier used for fever (**Rana et al., 2003**).

Plumbago zeylanica L. (Plate No. 30F)

Syn.   *P. auriculata* Blume; *Thela alba* Lour

Family   Plumbaginaceae

Vern.	Cheeta.
English, Hindi, Sanskrit and Regional Names	
Eng.	Ceylon leaderwort, White flowered leaderwort;
Hindi	Chita, Chitarak, Chitawar, Chiti, Chitra;
Sans.	Agni, Agnimata, Anala;
B.	Chita, Chitruk, Sufaid;
G.	Chitara, Chitrak, Chitro;
Kan.	Chitramula, vahini;
Mal.	Tumpukotuveli;
Mar.	Chitraka, Chitramala;
Oriya	Chitramulo, Krishanu;
P.	Chitrak;
Tam.	Adigarradi, Akkini;
Tel.	Agnimata, Chitramulamu.
Distribution	Throughout India. Wild in W. Peninsula & Bengal.
Description	Rambling perennial herbs having clasping stem and white flowers. Leaves alternate, entire. Flowers in long spikes, rachis glandular. Capsule included, membranous.
Flowering & Fruiting	October-March.
Habitat Ecology	Near water channels; **Auhr Khad -** 610 m.
Material Examined	EBH-WL-1097; 18.02.2009.

Part Used	Whole Plant. Roots. Leaves.
Folk Uses	Decoction of whole plant given in the morning on empty stomach to cure **fever** due to urine infection. Stem used as **brush** to clean the teeth. Poultice of leaves considered good for **scabies**; its infusion causes **abortion** (5-8 g, twice daily, 8 days). Leaves grounded with that of *Vitex altissima*, asafoetida, musk, pepper and garlic given orally in ephemeral **fever**. Root of the plant promotes the **appetite**. Paste of the root applied as poultice for **abscess**; its infusion good for **influenza, black water fever** and **rheumatism** (3-5 g, thrice daily, 3-5 days). Also, powdered roots boiled in milk given to relieve **muscular pain**. Milky juice applied on **scabies** and **unhealthy ulcers**.
Chemical Constituents	**Roots** possess plumbagin, 3-chloroplumbagin, -3,3- biplumbagin a new binaphthaquinone-chtranone-zeylinone, isozaflinone, ellipti-none and droserene.
Uses in Literature	Known so far as an appetizer, uterine stimulant; and for abortion, abscess, asthma, bite, boils, dental carries, diarrhoea, dysentery, fever, headache, jaundice, leprosy, leucorrhoea, maggot infected sores, piles, ring worm, rheumatism, scabies, skin diseases, stomach trouble, snake bites, white spots and wounds **(Alagesboopathi et al., 2000; Ambasta, 1986; Balasubramanian & Prasad, 1996; Bhogaonkar & Kanerkar, 2007; Biswas et al., 2010; Dash & Misra, 2000; Jain, 1991; Jain et al., 2010; Kaushik & Dhiman, 2000; Khan & Khanum, 2005; Khare & Khare, 2000; Kirtikar & Basu, 1984; Kumar, 2002; Lindley, 1981; Maheshwari et al., 1996; Panda & Das, 2000; Pande et al., 2006; Prajapati et al., 2006; Rana et al., 2003; Rao & Henry, 1995; Rastogi & Mehrotra, 1991; Reddy & Raju, 2000; Rout & Panda, 2010; Samwatsar & Diwanji, 2000; Saxena et al., 1997; Singh et al., 1996b; Siwakoti & Siwakoti, 2000; Sood et al., 2009b; Subramaniam, 2000; Swami & Gupta, 1996; Yasodamma et al., 2009; Watt, 1972).**

Poa supina Schrad. (Plate No. 31A)

Family   Poaceae

Vern.	Chirua.
Distribution	W. Europe to The Himalaya and Russia.
Description	Perennials having 25 cm long culms. Leaves basal. Ligule eciliate, membranous. Leaf blades flat. Inflorescence a panicle with smooth branches. Spikelets solitary, rachilla internodes smooth. Glumes persistent. Lower glumes oblong, upper ones elliptic. Fertile lemma ovate, membranous. Lodicules 2, membranous. Anthers 3. Fruit a caryopsis.

Flowering & Fruiting	August-September.
Habitat Ecology	In Grassy meadows; **Landy Khad -** 615 m; **Ali Khad -** 580 m.
Material Examined	EBH-WL-1098; 12.08.2007.
Part Used	Whole Plant.
Folk Use	Useful in the area as a good **fodder** grass.
Use in Literature	Locals regard it as a fodder grass (**Bor, 1973; Kala, 2004**).

Pogostemon plectranthoides Desf. (Plate No. 31B)

Syn.   *Origanum benghalense* Burm.f.; *P. benghalense* (Burm.f.) Kuntze; *O. indicum* Roth.; *Mentha secunda* Roxb.; *M. fruticulosa* Roxb

Family   Lamiaceae

Vern.	Bhaerda.
Regional Names	
B.	Bakoha, Jui, Jui-lata;
Bo.	Pangla;
Oriya	Badagandhuri, Dumobadotoko, Poksunga;
Tel.	Gondripulu, Kusurijang.
Distribution	W. Himalaya upto 900 m.
Description	Pubescent shrubs having membranous, irregularly toothed leaves. Flowers whitish purple, crowded in spikes. Bracts foliaceous. Nutlets smooth.
Flowering & Fruiting	February-April.
Habitat Ecology	Wastelands, roadsides; **Ali Khad -** 580 m.
Material Examined	EBH-WL-1091; 06.04.2008.
Parts Used	Leaves. Flowers.
Folk Uses	Half tsp of powdered leaves and flowers (1:1 ratio) good for various **skin troubles**; twice daily with cold water for 12-15 days.
Chemical Constituents	Essential oil from **fresh leaves** yields triacontane, phytol, vanillin and 4-hydroxy-4-mehtyl-2-pentanone.
Biological Activity	Leaves show antifungal activity against *Helminthosporium sativum*.
Uses in Literature	Used in India as an antidote (scorpion-sting, snake-bite), drink and styptic; and for cuts, haemorrhages and wounds in animals (**Aminuddin & Girach, 1991; Brahmam & Saxena, 1996; Chopra et al., 1956; Jain, 1991; Pande et al., 2006; Parrotta, 2001; Rana et al., 2003; Rao &**

Henry, 1995; Rastogi & Mehrotra; 1993; Sood & Thakur, 2004; Yoga-narasimhan, 1996).

Polygala arvensis **Willd. (Plate No. 31C)**

Syn. *P. prostrata* Willd.; *P. rothiana* W. & A. Prodr.; *P. tranquebarica* Mart.; *P. glaucoides* Wight

Family Polygalaceae

Vern.	Bachhu.
Distribution	Throughout India, from Punjab to Pegu, W. Peninnsula.
Description	Erect annual herbs. Leaves orbicular-oblong. Racemes truncate. Flowers pendulous. Seeds silky.
Flowering & Fruiting	July-December.
Habitat Ecology	Abundant in grasslands; **Raul Khad** – 765m; **Karyal Khad -** 615 m.
Material Examined	EBH-WL-1080; 18.11.2008.
Parts Used	Whole Plant. Roots. Leaves.
Folk Uses	Root poultice applied to cure **bone fracture**. 3-5 g paste of the whole plant given as an **antidote** for snake bite; also its poultice applied in the region of bite for **better** results. 15-25 ml decoction of leaves given twice daily for 10-15 days for curing **asthma** and other **bronchial disorders;** its paste applied for suppuration of boils and **skin inflammations.** Also, its extract with that of *Clerodendrum viscosum* L. (1:1 ratio) given on alternate day for curing **malaria.**
Chemical Constituents	**Roots** contain arctigenin 4'- gentiobioside and quercetin-3-rutinoside, roots and aerial parts contain arctiin (arctigenin-4'-glucoside), afzelin (kaempferol-3-rhamnoside) and myricetin-3-rhamnoside (**Jain** *et al.,* **1991).**
Uses in Literature	Used earlier for asthma, chronic bronchitis, catarrhal affections, cough, dizziness, fever, nervous disorders, paralysis, rheumatism and swollen joints (**Banerjee, 2000; Hooker, 1872-97; Jain, 1991; Kirtikar & Basu, 1984; Pande et al., 2006; Rana et al., 2003; Retnam & Martin, 2006; Sood & Thakur, 2004; Tarafder et al., 1997).**

Polygonum barbatum L. (Plate No. 31D)

Syn.   *P. horemanni* Meissn.; *P. marmoramae* Herb.; *P. fluviatile* Herb.; *P. rivulare* Koen.; *P. stagninum* Buch.-Ham. *ex* Meissn

Family   Polygonaceae

Vern.	Katur.
Regional Names	
B.	Bekhunjubaz;
Mal.	Vellutamodelamukku;
Mar.	Dhaktasheral;
P.	Narri;
Tam.	Atalari;
Tel.	Kondamalle, Niruganneru.
Distribution	Throughout the warmer parts of India.
Description	Erect herbs upto 90 cm tall. Branches stout. Racemes spiciform, thick. Bracts ciliate. Nutlets trigonous.
Flowering & Fruiting	April-July.
Habitat Ecology	Common along water sources, wet places; **Panyala Khad -** 810 m.
Material Examined	EBH-WL-1084; 02.07.2007.
Parts Used	Whole Plant. Seeds.
Folk Uses	Powdered plant given for **throat infection**; 1 tsp twice a day for 3-5 days; Plant also used as **fish poison**. Aerial plant parts used as **fodder**. Sap from pounded leaves applied for **healing wounds**. Seeds good for **relieving colic pain** (20-30 ml of its decoction, thrice daily, 3 days).
Uses in Literature	Used in India as an astringent, cicatrizant, colic, emetic, fish poison, fodder, pot herb, purgative and tonic; and for checking concepton, cooling, pain, snake-bite, stomach ache and wounds **(Ambasta, 1986; Chopra et al., 1956; Das et al., 1996; Jain, 1991; Kaushik & Dhiman, 2000; Khanna et al., 1996; Kumar & Narain, 2010; Lindley, 1981; Pande et al., 2006; Rana et al., 2003; Singh & Rao, 2003; Uphof, 2001).**

Polygonum barbatum L. sub sp. *gracile* Dansar. (Plate No. 31E)

Syn. *P. posumbu* Ham.; *P. donii* Wall.; *P. caespitosum* Blume

Family Polygonaceae

Vern.	Rohini.
Distribution	Temperate and Subtropical Himalaya: Sikkim - Nepal.
Description	Herbs with creeping stem, membranous ciliated leaves. Racemes erect, variable in length. Nuts 3-gonous, very small.
Flowering & Fruiting	April-July.
Habitat Ecology	Common along water sources; **Raul Khad -** 765 m.
Material Examined	EBH-WL-1255; 21.5.2007.
Parts Used	Aerial Plant parts.
Folk Use	Aerial plant parts given as **fodder**.
Use in Literature	Used in India as astringent (**Lindley, 1981**).

Polygonum donii Meisn. (Plate No. 31F)

Syn. *P. serrulatum* var. *donii* (Meisn.) Hook.; *P. pubescens* Bl.; *Persicaria pubescens* (Bl.) Hara

Family Polygonaceae

Vern.	Safed Rohini.
Distribution	Low hill plains of N. India, from Assam and Bengal to the Indus.
Description	Glabrous procumbent stems upto 90 cm tall. Stipules tubular. Flowers white, in slender drooping racemes. Nuts smooth.
Flowering & Fruiting	April-July.
Habitat Ecology	Common along water sources; **Gobind Sagar Lake -** 490 m; **Raul Khad -** 765 m; **Sir Khad -** 620 m.
Material Examined	EBH-WL-1085; 28.07.2008.
Part Used	Whole Plant.
Folk Use	2-3 g powdered plant given twice a day with cold water for 5-6 days to **cure cough**.
Use in Literature	Considered useful for sores and bites of insects and snakes (**Chopra et al., 1956**).

Polygonum glabrum Willd. (Plate No. 32A)

Syn.   *P. poiretii* Meisn

Family   Polygonaceae

Vern.	Senu.
Regional Names	
Ass.	Bihagni, Bihlangani, Larborna, Patharua;
Tam.	Atalari.
Distribution	Tropical Asia, Africa, and America.
Description	Glabrous herbs having petioled lanceolate leaves and membranous stipules. Racemes slender. Stamens 5-8. Nuts orbicular.
Flowering & Fruiting	January-September.
Habitat Ecology	In ditches; **Raul Khad -** 646 m.
Material Examined	EBH-WL-5010; 24.05.2008.
Parts Used	Whole Plant. Leaves.
Folk Uses	3-5 g infusion of leaves considered good against **colic pain** and **jaundice** (once daily, 10 days).
Uses in Literature	Employed as a cure for 'stitch in the side', fever and unlocking bones **(Kirtikar & Basu, 1984; Kumar & Narain, 2010; Shiddamallayya et al., 2010)**.

Polygonum lapathifolium L. (Plate No. 32B)

Family   Polygonaceae

Vern.	Rohini.
Distribution	Indo- Malaya region.
Description	Erect annual herbs. Leaves ovate-lanceolate, acuminate at apex, entire. Flowers axillary in paniculate racemes. Perianth greenish. Nuts ovoid, compressed, smooth.
Flowering & Fruiting	November-February.
Habitat Ecology	Marshy, irrigated fields or water-logged places; **Raul Khad -** 646 m.
Material Examined	EBH-WL-1276; 22.01.2008.
Part Used	Whole Plant.

Folk Use	Fomentation of plant beneficial in **chronic ulcers**.
Uses in Literature	Earlier known to be used in cold, coughs, gravel and hemorrhoidal tumours **(Stubbendiek et al., 1994)**.

Polygonum minus Huds. (Plate No. 32C)

Syn.   *P. posumbus* Wall.; *P. tenellum* Blume; *P. hypostictum* Miq.; *P. banca* Herb.; *P. strictum* All

Family   Polygonaceae

Vern.	Bishkhulti.
Distribution	Throughout hotter parts of India, ascending Himalaya to 2,000 m.
Description	Herbs with creeping, much branched stems. Leaves sessile, linear, oblong-lanceolate. Stipules strigose truncate. Racemes erect. Flowers minute. Nuts orbicular.
Flowering & Fruiting	January-December.
Habitat Ecology	Near ponds, lakes; **Ali Khad -** 580 m; **Raul Khad -** 646 m.
Material Examined	EBH-WL-1064; 05.11.2007.
Part Used	Leaves.
Folk Uses	Leaves used as **fodder**. Decoction of leaves given for relief against **indigestion**.
Uses in Literature	Used in India for diarrhoea, infection and as fodder **(Biswas & Calder, 1984; Rana et al., 2003)**.

Polygonum plebejjum Br. Prodr. (Plate No. 32D)

Syn.   *P. dryandri* Syst.; *P. aviculare* Don Prodr.; *P. herniarioides* Del.; *P. roxburghii* Meissn

Family   Polygonaceae

Vern.	Drabu, Muthisag.

Hindi and Regional Names

Hindi	Raniphul;
Sans.	Mastyakshi;
B.	Chemti sag, Dubia sag;
G.	Zinako okhard;

Oriya	Muthi saga.
Distribution	Throughout tropical India.
Description	Diffusely branched prostrate herbs having terete grooved branches and internodes usually shorter than the leaves. Stipules hyaline, short lacerate. Nuts rhomboid, smooth and shining.
Flowering & Fruiting	October-April.
Habitat Ecology	Near water sources, ponds, lakes; **Sir Khad** - 620 m; **Sauli Khad** - 765 m.
Material Examined	EBH-WL-1065; 02.04.2008.
Parts Used	Whole Plant. Root.
Folk Uses	15-20 ml decoction of the root taken on empty stomach for 10 days against **bowel** complaints and **pneumonia**. 2-4 g powdered plant given thrice daily for 3 days, for curing **amoebic dysentery** and **diarrhoea**.
Chemical Constituents	Polysaccharide complex isolated from **plant** consisted mainly of galacturonic acid and arabinose; galactose, glucose, xylose and rhamnose were other sugars of water-soluble polysaccharides.
Biological Activity	Plant exhibits antibacterial activity.
Uses in Literature	Employed earlier for checking baldness, diarrhoea, dysentery, fever, lung diseases, pneumonia, promoting lactation and stomach trouble **(Alagesboopathi et al., 1999; Anonymous, 1948-76a; Banerjee & Ghora, 1996; Chopra et al., 1956; Jain, 1991, 1997; Kaul, 1997; Kaushik & Dhiman, 2000; Kumar & Narain, 2010; Lalramnghinglova, 1996, 2003; Maliya, 2009; Panda & Das, 1999; Pande et al., 2006; Rastogi & Mehrotra, 1990; Sen & Behera, 2009).**

Polygonum pulchrum Blume (Plate No. 32E)

Syn.   *P. orientale* Wall.; *P. ochreatum* Houtt

Family   Polygonaceae

Vern.	Kyagnu.
Distribution	Common throughout India.
Description	Erect annual herbs upto 1m high. Leaves elliptic-ovate, lanceolate. Flowers in paniculate racemes. Bracts orbicular. Perianth white. Stamens 5. Nuts orbicular.
Flowering & Fruiting	October-March.

Habitat Ecology	Along the stream banks; **Raul Khad -** 646 m.
Material Examined	EBH-WL-1278; 22.02.2008.
Part Used	Whole Plant.
Folk Use	Decoction of plant given for **curing blood disorders.**
Uses in Literature	Earlier used for arthritis, blood disorder, dropsy, gout, oedema, rheumatism and swellings **(Burkill, 1935).**

Polygonum serrulatum Lag. (Plate No. 32F)

Syn. *P. flaccidum* Roxb.; *P. mile* Wall.; *P. rapte* Herb

Family Polygonaceae

Vern.	Kathua.
Distribution	Plains and low hills of N. India, from Assam and Bengal to the Indus.
Description	Herbs having prostrate stem, glabrous branches and peduncles, and subsessile linear leaves which are glabrous or sparsely hairy beneath. Racemes slender erect. Bracts glabrous, strongly ciliate, Perianth eglandular. Stamens 5-8. Nuts trigonous.
Flowering & Fruiting	June-October.
Habitat Ecology	Moist shady places; **Sir Khad -** 620 m.
Material Examined	EBH-WL-1066; 02.08.2007.
Part Used	Leaves.
Folk Use	Paste of leaves applied on **infected sores.**
Uses in Literature	Leaves used as antiseptic for cuts and wounds, sting and snakebites **(Biswas & Calder, 1984; Kapur & Nanda, 1996b; Kirtikar & Basu, 1984; Rana et al., 2003; Singh et al., 1996a).**

Potamogeton crispus L. (Plate No. 33A)

Syn. *P. crenulatus* Don; *P. tuberosus* Roxb

Family Naiadaceae

Vern.	Jalaz.
Distribution	Plains of India and Temperate Himalaya; Kashmir - Bhutan.
Description	Aquatic herbs with dichotomous compressed stem and alternate, narrowed, amlexicaulate leaves. Stipules small, caducous. Peduncles

long. Spike very short. Flowers very small. Drupelets ovoid, compressed, ribs entire or toothed.

Flowering & Fruiting   December-June.

Habitat Ecology   Fresh water tanks, pools and other water-logged areas; **Ali Khad -** 580 m.

Material Examined   EBH-WL-1091; 22.05.2008.

Part Used   Whole Plant.

Folk Use   Used as **food** for fish.

Use in Literature   Reported earlier for skin diseases **(Arber, 2003; Biswas & Calder, 1984; Hooker, 1872-97; Kapur & Srivastava, 1996).**

Potamogeton pectinatus L. (Plate No. 33B)

Syn.   *P. marinus* Ham.; *Ruppia subsessilis* Thw

Family   Naiadaceae

Vern.   Jalaz.

Distribution   Plains of India, the Himalaya.

Description   Aquatic herbs with densely distichously branched filiform stem and 5-nerved, acute, opaque leaves. Flowers whorled. Drupelets dimidate, obovoid, hardly beaked.

Flowering & Fruiting   October-January.

Habitat Ecology   Fresh water pools, low land waters; **Ali Khad -** 580 m.

Material Examined   EBH-WL-1258; 12.12.2008.

Part Used   Whole Plant.

Folk Use   Used as **food** for fish in aquaculture.

Use in Literature   Earlier used for fever **(Biswas & Calder, 1984; Kapur & Nanda, 1996a).**

Pteris cretica L. (Plate No. 33C)

Syn.   *Pteris nervosa* Thunb

Family   Pteridaceae

Vern.   Bahupatre.

English Names

    Eng.   Ribbon fern, Cretan brake.

Distribution	Plains – 3,000 m.
Description	Ferns having thick scaly rhizomes and dimorphic fronds. Fertile fronds erect but sterile ones bending backwards. Stipes and rachis thick, stramineous. Lamina imparipinnate, lanceolate. Pinnae 3-6 pairs; lanceolate veins parrallel, forked. Costae glabrous. Sori marginal. Indusia membranaceous, continuous. Spores brown with levigate undulate exine.
Flowering & Fruiting	May-October.
Habitat Ecology	Warmer parts, open and shady locations, margins in the forests; **Kothi Khad -** 810 m; **Ali Khad -** 580 m; **Sir Khad -** 620 m.
Material Examined	EBH-WL-1160; 02.09.2008.
Parts Used	Young Shoots.
Folk Uses	Young shoots **edible**. Plant cultivated as an evergreen **ornamental** fern in houses.
Chemical Constituents	Found to contain flavone-o-glycoside, luteolin 7- o –β-sophoroside, luteolin 7-o-β- gentiobioside, 2β, 6β, 16α-trihydroxy-ent-kaurane, pterosins and F from **aerial parts (Imperato, 1994; Murakami et al., 1986).**
Uses in Literature	Known earlier as antibacterial (fronds) and edible (rhizomes eaten as a kind of arrowroot) **(Ambasta, 1986; Brickell, 1996; Khare, 1996; May, 1978; Rastogi & Mehrotra, 1990; Singh, 2003c; Singh & Viswanathan, 1996).**

Pupalia lappacea **Juss. (Plate No. 33D)**

Syn. *Achyranthes lappacea* L.; *A. patula* L.

Family Amaranthaceae

Vern.	Sena.
Distribution	Throughout India; W. Tropical Himalaya from Kashmir-Kumaon.
Description	Straggling large-sized undershrubs with terete branches and lanceolate leaves. Petioles long. Sepals 3-nerved. Flowers in close clusters. Spikes long. Bracts ovate. Fruits yellow with ellipsoid seed.
Flowering & Fruiting	July-September.
Habitat Ecology	Near water sources; **Gobind Sagar Lake -** 490 m.
Material Examined	EBH-WL-1266; 12.08.2008.
Part Used	Leaves.
Folk Use	Decoction of leaves with black pepper given to check **stomach ache** (10ml, twice daily, 3 days).

Reinwardtia indica Dumort. (Plate No. 33E)

Syn.   *R. trigyna* Planch.; *Linum trigynum* Roxb.; *L. repens* Don; *Macrolinum trigynum* Reichard; *Kittelocharis trigyna* Alef.

Family   Linaceae

Vern.	Basant.
Hindi and Regional Names	
Hindi	Basanti;
Bo.	Abai;
P.	Balbasant, Basant, Gudbatal, Karkun, Kaur;
U.P.	Basant.
Distribution	Hilly parts of India; Punjab eastwards to Sikkim, ascending to 2,000 m.
Description	Tufted glabrous undershrubs upto 90 cm tall. Leaves slender. Flowers yellow. Sepals distinct. Petals yellow. Style 3. Capsules small, globose. Seeds reniform.
Flowering & Fruiting	February-April.
Habitat Ecology	Open slopes, shady localities; **Ali Khad -** 580 m; **Sauli Khad -** 620 m; **Tatoh Khad -** 710 m.
Material Examined	EBH-WL-1106; 02.04.2008.
Parts Used	Flowers. Root.
Folk Uses	Flowers used for offering to **appease** deities. Decoction of root along-with a pinch of black pepper given for treating **jaundice** and **cold** (15-20 ml, twice a day, 7 days).
Uses in Literature	Used earlier for founder diseases among cattle paralysis and wounds infested with maggots **(Ambasta, 1986; Chopra et al., 1956; Jain, 1991; Kirtikar & Basu, 1984; Pande et al., 2006; Rana et al., 2003; Sood & Thakur, 2004; Sood et al., 2009b; Watt, 1972).**

Rhynchoglossum obliquum Blume (Plate No. 33F)

Syn.   *R. blumei* DC.; *Wulfenia obliqua* Wall.; *Lexotis intermedia* Benth.; *L. obliqua* Br.

Family   Gesneriaceae

Vern.	Dushkanda.
Distribution	Throughout India.
Description	A succulent herb upto 50 cm. Flowers bluish-white, in long racemes. Bracteoles filiform. Capsules included, 2-valved.
Flowering & Fruiting	July-August.
Habitat Ecology	Damp places; **Tatoh Khad -** 710 m.
Material Examined	EBH-WL-1107; 12.07.2007.
Part Used	Whole Plant.
Folk Use	Paste of the plant applied as poultice against various **skin problems**.
Uses in Literature	Reported to cure skin diseases such as sores or boils between the toes **(Jain, 1991; Jain et al., 1991; Kumar, 2002; Sood & Thakur, 2004).**

Ricinus communis L. (Plate No. 34A)

Syn.   *R. inermis* Jacq.; *R. lividus* Jacq.; *R. speciosus* Burm.; *R. spectabilis* Blume; *R. viridis* Willd.; *Croton spinosus* L.

Family   Euphorbiaceae

Vern.	Arand.

English, Hindi, Sanskrit and Regional Names

Eng.	Castor, Castor-oil plant, Castor seed;
Hindi	Arand, Arend, Erandi, Rand;
Sans.	Eranda, Ruvuka;
Ass.	Eri;
B.	Bheranda;
Bo.	Erendi;
G.	Diveligo;
Kan.	Haralu;
Mal.	Avanallku;

Mar.	Erandi;
Oriya	Bheronta, Chitroka;
P.	Aneru, Arand;
Tam.	Amankku, Kottai muthu;
Tel.	Amadam, Amudamuchettu, Eramudapu.
U.P.	Rend.
Distribution	Common in India; naturalised near habitations.
Description	Evergreen small trees having glaucous shoots, green or reddish, linear leaves and stout erect racemes. Capsules long, globosely oblong. Seeds oblong, smooth, mottled, black.
Flowering & Fruiting	March-May.
Habitat Ecology	Frequent on wastelands, roadsides, human dwellings, drier areas; **Ali Khad -** 580 m; **Tatoh Khad -** 710 m.
Material Examined	EBH-WL-1079; 16.04.2008.
Parts Used	Fresh Plant. Roots. Stems. Leaf. Seeds.
Folk Uses	Dried stems and seeds constitute a highly **enriched feed**. A piece of stem bark tied as an **amulet** on neck of children to arrest vomiting. Leaves occasionally **fed** to cattle; its poultice applied to **boils** and **sores.** Non-mulberry silkworms, *Philosoma cynthia* are **fed** on its leaves. Leaves warmed with mustard oil used for massaging the body for **toning** and relief against **body pains** and **rheumatic joints** of woman after delivery. Leaf paste also applied externally to cure **headache, itches, boils** and **swellings**. Also, about 100-150 ml extract of leaf taken orally as a remedy for **opium** poisoning. Decoction of roots prescribed in **rheumatism** (5-10 ml, once daily, 8-10 days). Paste of seed kernels with neem leaves and turmeric powder applied externally after bath to cure **smallpox** and **chickenpox** and **pustule** formations. Seed oil used as lamp **illuminant, hair oil** and as **lubricants**.
Chemical Constituents	Found to contain ricinine and 1-methyl-3-cyano-4-methoxy-2-pyridone **(seeds, leaves)**, arachdic, chlorogenic, oleic, palmitic, ricinoleic, stearic and dihydrogenic acids; hexadecanoic, hydrocyanic and uric acid **(oil)**.
Biological Activity	Anticancerous and hypoglycaemic activities confirmed.
Uses in Literature	Reportedlly used earlier as abortifacient, anticancer, antiprotozoal, diuretic, laxative, purgative, tonic; and for allergic problems, bodyache, boils, diarrhoea, dysentery, eruptions, gastroenteritis, gonorrhoea, gums, hypoglycaemic, gout, irregular menstruation, jaundice, lactation and peritonitis, paralysis, rheumatism and sores **(Abraham,**

1997; Ambasta, 1986; Ansarali & Sivadasan, 2009; Bhattacharyya, 2000; Bhogaonkar & Kanerkar, 2007; Barua et al., 2000; Bora, 2000; Choudhury et al., 2008; Henry, 1999; Huidrom, 1996; Jadhav, 2009; Jain, 1991; Jain & Singh, 1997; Jain et al., 2010; Jamir et al., 2008; Khan & Khanum, 2005; Khanna et al., 1996; Khare & Khare, 2000; Kharkongor & Joseph, 1997; Kumar, 2002; Lalramnghinglova, 1996; Lindley, 1981; Mahato & Mahato, 1996; Maheshwari et al., 1996; Manandhar, 1996a; Meena & Yadav, 2010; Panda, 1996; Pande, et al., 2006; Parrotta, 2001; Prajapati et al., 2006; Rama Rao et al., 2008; Rana et al., 2003; Ranjan et al., 2000; Rastogi & Mehrotra, 1990; Rao & Henry, 1995; Retnam & Martin, 2006; Rout & Panda, 2010; Saini, 1996b; Satapathy, 2008; Singh, 2000; Singh & Srivastava, 2000; Singh et al., 1996b; Siwakoti & Siwakoti, 2000; Sood et al., 2009b; Subramaniam, 2000; Thakor, 2009; Tiwari & Tiwari, 1996; Watt, 1972).

Roylea cinerea (D. Don) Baill. (Plate No. 34B)

Syn. *Ballata cinerea* Don; *R. elegans* Wall.; *R. calycina* (Roxb.) Briq.; *Phlomis calycina* Roxb

Family   Lamiaceae

Vern.	Kadvya.
Hindi and Regional Names	
Hindi	Patkarru;
P.	Kaur, Kauri.
Distribution	Subtropical W. Himalaya; Kashmir - Kumaon.
Description	A tall hoary undershrub. Leaves petioled, ovate, deeply toothed. Inflorescence compound. Flowers zygomorphic. Fruiting calyx long. Corolla white or pinkish. Nutlets 4.
Flowering & Fruiting	May-October.
Habitat Ecology	Frequent on rocky slopes; **Raul Khad -** 646 m.
Material Examined	EBH-WL-1087; 22.09.2008.
Part Used	Leaves.
Folk Uses	Decoction of leaves used as a bitter **tonic** and **febrifuge**. Leaves toasted with mustard oil applied externally for **itching, softening** swollen parts and **rheumatism**.
Chemical Constituents	**Leaves** contain betulin, β-sitosterol, stigmasterol, cetyl alcohol, glucose, fructose, arabinose and palmitic, stearic, oleic, gallic, oxallic and tartaric acids calyone, calyenone and precalyone.

Biological Activity	Plant extract depresses CNS and relaxed skeletal muscle activities.
Uses in Literature	Known to be used as a blood purifier, snuff in tonsils, tonic; and for fever, headache, malarial fever, jaundice and pimples **(Ambasta, 1986; Gupta, 1981; Jain, 1964, 1991; Pande et al., 2006; Parkash & Aggarwal, 2010; Rana et al., 2003; Rastogi & Mehrotra, 1993; Sood & Thakur, 2004; Sood et al., 2009b).**

Rubus ellipticus Sm. (Plate No. 34C)

Syn.   *R. flavus* Ham.; *R. gowry-phul* Roxb.; *R. gowrae-phul* Roxb.; *R. paniculatus* Moore; *R. sessilifolius* Miq

Family   Rosaceae

Vern.	Akhae
English, Hindi and Regional Names	
Eng.	Himalayan yellow raspberry;
Hindi	Anchhu, Hinsalu;
Ass.	Jotelupoka;
Kash.	Gouriphal;
P.	Akhi.
Distribution	Temperate and Subtropical Himalaya from Sirmaur - Sikkim.
Description	Shrubs having stout branches, stout prickles and coriaceous dark green, petioled leaflets. Stipules subulate. Panicles small, many flowered, pedicels short. Bracts setaceous. Fruits globose, golden yellow, succulent, stone rugose.
Flowering & Fruiting	January-May.
Habitat Ecology	Moist shady localities, slopes, nallahs; **Sir Khad -** 620 m; **Karyal Khad** - 615 m.
Material Examined	EBH-WL-1233; 10.04.2008.
Parts Used	Fruits. Root.
Folk Uses	Fruits **edible.** Decoction of root good for **cough;** 2 tsp twice a day till relief. Pounded roots also given to cattle for promoting **lactation**.
Chemical Constituents	Octacosanol, β-sitosterol, its glucoside, octacosanoic, ursolic and rubitic acids isolated from **roots.**
Biological Activity	Plant CNS depressant and spasmolytic.
Uses in Literature	Used so far for constipation, diarrhoea, dysentery, fever, gastralgia, malaria, stomach ache, throatache, vomiting, whooping cough, worms;

and as an edible (fruits) **(Abraham, 1997; Ambasta, 1986; Jain 1991; Jain et al., 1991; Jamir et al., 2008; Kumar, 2002; Panda, 1996; Pande et al., 2006; Rana et al., 2003; Rao & Henry, 1995; Sood & Thakur, 2004; Sood et al., 2009b; Watt, 1972).**

Ruellia patula Jacq. (Plate No. 34D)

Syn. *Dipteracanthus patulus* (Jacq.) Nees

Family   Acanthaceae

Vern.	Baan.
Distribution	Throughout India.
Description	Erect, densely pubescent herbs upto 45 cm with sub-sessile flowers in clusters. Corolla purple, tubular. Tube ventricose. Seeds hairy.
Flowering & Fruiting	January-April.
Habitat Ecology	Frequent along water sources, rock crevices; **Sauli Khad -** 765 m; **Marsand Khad -** 610 m.
Material Examined	EBH-WL-1263; 31.01.2008.
Part Used	Roots.
Folk Use	3 g per dose of powdered roots given with cow's milk for 8-10 days as a **tonic** for children in the age group of 5-10.
Use in Literature	Used for cooling, high cholesterol and urinary problems **(Sood & Thakur, 2004).**

Rumex hastatus D. Don (Plate No. 34E)

Family   Polygonaceae

Vern.	Ambi.
Regional Names	
P.	Amla, Amlora, Ami, Khattimal, Malori-gha.
Distribution	Common in Himalaya from Kashmir - Kumaon.
Description	Glaucous polygamous herbs with stout erect branches arising from a shrubby base. Leaves hastately 3-lobed with narrow lobes. Racemes slender, panicled. Whorls a few-fid, distant. Valves orbicular notched at both ends.
Flowering & Fruiting	November-April.

Habitat Ecology	Rock crevices, moist slopes; **Barthin Khad -** 585 m.
Material Examined	EBH-WL-1089; 16.03.2008.
Parts Used	Whole Plant. Shoot. Leaves.
Folk Uses	Tender shoots or leaves **pickled.** Leaves have a refreshing acidic taste and raw eaten or used for making **chutney**; its poultice applied on forehead to **check headache** and **skin infections**. Whole plant mixed with rock salt given as an **antidote** for intoxication.
Chemical Constituent	**Root bark** yields tannin (21-23 %).
Uses in Literature	Reported so far for boils, burns, checking bleeding, cough, cuts, cooling, gastralgia, insect bite, sunstroke, wounds; and as a pot-herb **(Ambasta, 1986; Jain, 1991; Pande et al., 2006; Rana et al., 2003; Ranjan, 2000; Semwal et al., 2010; Sood & Thakur, 2004).**

Rumex nepalensis **Spreng. (Plate No. 34F)**

Syn.   *R. hamatus* Trevir.; *R. peregrius* Boiss.; *R. ramulosus* Meissn. *in* DC.; *R. roxburghianus* Schult. f.; *R. tuberosus* Roxb.; *R. uncinatus* Hort.

Family   Polygonaceae

Vern.	Jalbhangru.
Regional Names	
B.	Pahari palang;
Kash.	Palak.
Distribution	Temperate Himalaya: 1,350-4,000 m; W. Ghats.
Description	Robust herbs with stiff spreading branches and entire pointed leaves. Lower leaves long-stalked, oblong-ovate, cordate; upper sessile, smaller, lanceolate. Flowers 2-5 sexual, in whorls. Racemes leafless. Sepals fringed with comb-like, hooked teeth. Fruits trigonous nuts.
Flowering & Fruiting	May-October.
Habitat Ecology	Along nallahs, water sources, damp places; **Saryali Khad -** 670 m; **Tatoh Khad -** 710 m.
Material Examined	EBH-WL-1235; 22.10.2008.
Parts Used	Roots. Leaves.
Folk Uses	Leaves used as **vegetable.** Crushed leaf extract applied externally to **cure cuts, wounds, boils, blisters** and **syphilitic ulcers**; its infusion also given in **colic**. Root decoction given orally for **purgative** activity.

Chemical Constituents	**Roots** contain tannin (12.8%), nepodin and chrysophanic acid. 1,8-dihydroxy-3-methylanthraquinone, 1,6,8-trihydroxy-3-methylanthrquinone, 1,8-dihydroxy-6-methoxy-3-methylanthraquinone, lupeol and β-sitosterol.
Biological Activity	Plant CNS depressant.
Uses in Literature	Used as a colic, coolant, diuretic, pot herb, purgative; and for boils, cold and cough, chutneys, dyeing woollen clothes, fever, hepatitis, inscet bite, pickles, urinary, renal, rheumatism, scurvy, sore gums, stomach ache, swelling of muscles and syphilitic ulcers **(Bennet, 1986; Bhattarai et al., 2008; Chhetri, 2005; Devi, 2003; Gangwar & Ramakrishnan, 1990; Jain, 1991; Jain et al., 1991; Joshi, 2009; Kirtikar & Basu, 1984; Krishna & Singh, 1987; Lal et al., 1996; Manandhar, 1996a; Megoneitso & Rao, 1983; Nautiyal et al., 2003; Negi et al., 1999; Pande et al., 2006; Rana et al., 2003; Roy et al., 1998; Saklani & Jain, 1994; Samant et al., 1996, 2001a; Sood & Thakur, 2004).**

Rungia pectinata (L.) Nees (Plate No. 35A)

Syn. *Justicia pectinata* L.; *R. parviflora var. pectinata* Clarke

Family Acanthaceae

Vern.	Phuldali.

English, Hindi, Sanskrit and Regional Names

Eng.	Mustard tree, Satbrush, Tooth-brush tree.
Hindi	Jhak, Kharjal;
Sans.	Pindi;
B.	Jhal;
G.	Mothakhadsalo;
Kan.	Goni-mara;
Mal.	Punakapundu;
Mar.	Khakhim, Kickui, Miraj, Mirajoli, Pelu, Pilva, Rhakhan, Thorapilu;
Oriya	Kotungo, Toboto, Pilu;
Tam.	Punakapunda, Tavasumrungi;
Tel.	Pindikunda, Varagogu.

Distribution	Predominantly in N.W. Temperate Himalaya and Tropical regions of India.
Description	Profusely branched straggling annuals with blue flowers in 1-sided subsessile spikes. Leaves acute at base. Bracts dimorphic. Seeds yellow.

Flowering & Fruiting	January-February.
Habitat Ecology	Rock crevices, rare in dry areas; **Sir Khad -** 620 m.
Material Examined	EBH-WL-1094; 12.01.2009.
Part Used	Whole Plant.
Folk Uses	Plant decoction given along with honey to **treat urination problems** and for (15-20 ml, twice a day till cure) **relieving headache.**
Uses in Literature	Used as an aperient, febrifuge, laxative and refrigerant; and for bronchial disorders, contusions, dropsy, smallpox, suppuration of boils, swelling and urinary complaints **(Ambasta, 1986; Bhandary et al., 1996; Chandra, 1997; Jain, 1991; Parrotta, 2001; Rana et al., 2003; Retnam & Martin, 2006; Sood & Thakur, 2004).**

S

Saccharum munja Roxb. (Plate No. 35B)

Syn. *S. sinense* Roxb.; *S. arundinaceum* Retz.; *S. benghalense* Retz.; *S. ciliare* Anderss.; *S. exaltatum* Roxb.; *S. moonja* Royle; *S. procerum* Roxb.; *S. sara* Roxb.; *S. surpata*, Hb. Ham. *ex* Wall.

Family Poaceae

Vern.	Naal.

English, Hindi, Sanskrit and Regional Names

Eng.	Devil sugar cane, Reedy sugar cane, Wild sugar cane;
Hindi	Munja, Ramsar, Sara, Sarkanda, Sarpat, Sarpalta;
Sans.	Bahupraja, Bana, Bhadramunja, Brahmanya, Chakshuveshana, Darbhavhaya, Dridhatirna, Durmula, Ikshukanda, Gundra, Munja, Munjuli, Munnjanaka, Munjata, Ranjana, Sara, Shakrabhanga, Shara, Shiri, Sthuladarbha, Sumekhala, Tejanaka;
B.	Mucha, Ramshara, Sar, Teng;
Mal.	Tebaru, Mekhalapullu, Munja, Sarappulu;
Oriya	Katosore, Soro;
P.	Sarkanda, Kanda, Kharkanda, Sarjbar, Sarkanda;
Tam.	Ethudugirananal, Munji;
Tel.	Adavicheruku, Bramhamekhalamu, Gundra, Kondakanames, Munjagaddi, Mungamu, Nadam, Polagaddi, Ponika, Ponugu, Saramu.
Distribution	Throughout the plains and low hills of India, Sri Lanka, China.
Description	Tall perennial grasses possessing solid glabrous stem below the panicle. Pedicels glabrous. Spikelets slightly heteromorphous; glumes equal, membranous, lower glumes of sessile spikelet hairy on the back; the upper ones glabrous, both glumes of pedicelled spikelet hairy, the hairs at least 4mm long. Lower lemma oblong, elliptic, hairy on the back, upper ones ovate-lanceolate, ciliate on the margins, acute or very shortly awned.

Flowering & Fruiting	January-April.
Habitat Ecology	Moist areas; **Ali Khad -** 580 m; **Gobind Sagar Lake -** 490m.
Material Examined	EBH-WL-1239; 05.04.2009.
Part Used	Root. Stems.
Folk Uses	Plants due to its extensive root system regarded as an effective **soil-binder.** Smoke of burnt root considered **beneficial** for women after **delivery.** Its stem used for making **'kalams'** (writing pens).
Uses in Literature :	So far described as acrid, aphrodisiac, coolant; and for blood troubles, burning sensations, erysipelas, eye diseases, source of a fibres 'munj', ropes for manufacturing baskets, mats, thirst, 'tridosha' and urinary complaints **(Bhat Gopalakrishna & Nagendran, 2001; Bor, 1973; Cauis, 2003; Kirtikar & Basu, 1984; Pande et al., 2006; Rastogi & Malhotra, 1991; Uphof, 2001).**

Saccharum spontanium L. (Plate No. 35C)

Syn. *S. aegypticum* Willd.; *S. bengalense* Boga.; *S. biflorum* Forsk.; *S. caducum* Palisotii & *speciosissimum* Tausch; *S. canaliculatum* Roxb.; *S. chinense* Nees; *S. glaza* Reinw. *ex* Blume; *S. insulare* Brongn.; *S. klaga* Jungh. *ex* Tijdschr.; *S. punctatum* Schum. Besch.; *S. propinquum* Steud.; *S. semidecumbens* Roxb.; *Imperata spontanea* Beauv.

Family Poaceae

Vern.	Kosha.

English, Hindi, Sanskrit and Regional Names

Eng.	Thatch grass, Wild sugar cane.
Hindi	Kagara, Kans, Kansi, Kas, Kosa, Kus;
Sans.	Ikshuganndha, Kasa, Kasah, Kasha, Kahggara;
B.	Kagara, Kas, Kans, Sarkanda;
G.	Kans, kasado, Kansa doghas;
Mal.	Kusa, Nannana;
Oriya	Chhatiaagaro, Inkoro, Kaso, Khhodi, Pothhorokhhodi;
P.	Kahi, Kanch, Kans, Sarkara;
Tam.	Achabaram, Anjani, Eruvari, Kosangam, Kuncham, Kumil, Kurbagam, Nanal, Nanarbul, Nanmugappul, Peykkarumbu, Sangabidam, Saravanum, Sarupparasi, Sarabaram, Sugattan, Suredasaram, Tittiru, Tittiruchi, Tuttam, Vedasam.

Tel.	Billugaddi, Kakicheraku, Kakivedunu, Koregadi, Rasalamu, Rellugaddi.
Distribution	Throughout the warmer parts of India and Sri Lanka, ascending to 2,000 m in the Himalaya.
Description	Plant possessing tall erect stems upto 7 m, stout rootstock and long narrow leaves having convoluted margins. Mouth of sheath wooly. Ligule membranous. Panicle 60 cm, branches whorled, spreading; branchlets fragile, joints filiform, dorsally long-ciliate. Spikelets lanceolate. Fruits caryopsis.
Flowering & Fruiting	November-December.
Habitat Ecology	Occurs along river banks. Flowers after rains; **Raul Khad -** 646 m, **Karyal Khad -** 615 m, **Sir Khad -** 620 m.
Material Examined	EBH-WL-1201; 10.11.2007.
Parts Used	Whole Plant. Leaves.
Folk Uses	Plant used for **fixing** loose soil. Leaves employed for **thatching, brooms, screes**.
Uses in Literature	Useful earlier for agalactia, asthma, burning sensation, cholera, checking pus formation, cordage, dysentery, dyspepsia, general debility, haemorrhoids, menorrhagia, paper pulp, phthisis, platting, renal, vesical calculi and strangury **(Bhat Gopalakrishna & Nagendran, 2001; Bor, 1973; Hooker, 1872-97; Jain, 1991; Kirtikar & Basu, 1984; Mudaliyar, 1921; Pande et al., 2006; Prajapati et al., 2006; Rana et al., 2003; Roy, 1984; Satya & Solanki, 2008; Siwakoti & Siwakoti, 2000; Uphof, 2001).**

Salix oxycarpa Anderss. (Plate No. 35D)

Syn.   *S. zygostemon* Boiss

Family   Salicaceae

Vern.	Beeiuns.
Distribution	W. Himalaya.
Description	Small trees. Flowers appear little before the leaves. Leaves serrated. Filaments of male flowers partly connate; female ones glabrous. Ovary pubescent. Capsule long, refescent glabrous. Seeds numerous.
Flowering & Fruiting	April-June.
Habitat Ecology	Moist shady slopes, frequent; **Sir Khad -** 620 m; **Ali Khad -** 580 m.

Material Examined	EBH-WL-1102; 08.05.2008.
Parts Used	Leaves. Root-bark;
Folk Uses	Leaf poultices good against **headache.** Fresh root-bark chewed for instant relief from **toothache.**
Uses in Literature	Used for dental caries, pyorrhoea and toothache **(Sood & Thakur, 2004).**

Saussurea heteromella **(Don)** *Hand.-Mazz.* **(Plate No. 35E)**

Syn. *S. brahuica* Boiss.; *Carduus heteromallus* Don; *Aplotaxis candicans* DC.; *A. scaposa* Edgew.; *Cirsium heteromallum* Spreng.; *Saussurea candicans* (DC.) Sch.-Bip.

Family Asteraceae

Vern.	Mathaedi.
Regional Names	
P.	Batula, Kaliziri.
Distribution	Himalaya.
Description	Robust plant upto 10 cm tall. Leaves in dense head of small capitula, surrounded completely by purple wooly hair. Inflorescence cottony. Heads long-peduncled. Receptacle bristles long. Achenes 5-angled, muricate.
Flowering & Fruiting	April-June.
Habitat Ecology	Screes, open slopes; **Barthin Khad -** 585 m.
Material Examined	EBH-WL-1103; 10.04.2009.
Parts Used	Whole Plant. Roots.
Folk Uses	Paste of the whole plant applied on the forehead for relief from **headache;** also its powdered roots (2-4 g) given every morning on empty stomach with milk to restore **vigour** and **vitality.**
Chemical Constituents	Two terpene glycosides-dihydroprotopanaxadiol-di-O-arabinoside and protopanaxanone-di-O-arabinoside-isolated from **(plant)**.
Uses in Literature	Known in India as colic, carminative; and for abdominal pain, bowel complaints, cough, curing fever, dysentery, insect bites, leucoderma, night blindness, rheumatism, skin diseases and wounds **(Ambasta, 1986; Jain, 1991; Kapur & Srivastava, 1996; Pande et al., 2006; Rana et al., 2003; Rastogi & Mehrotra, 1995a; Sood & Thakur, 2004).**

Selaginella chrysocaulos (Hook. & Grev.) Spring (Plate No. 35F)

Syn.   *Lygodium chrysocaulos* Hook. et Grev

Family   Selaginellaceae

Vern.	Sindoor.
Distribution	Throughout India.
Description	Densely tufted slender stems which are stoloniferous at base. Rhizophores brownish. Leaves heteromorphic, bright-green, acute, oblique, minutely denticulate. Strobili short. Sporophylls dimorphic, dentate, larger, oblong, acuminate. Megaspores 200-350 µm, dark brown, verrucoid. Microspores 50-55 µm, deep-orange, warty.
Flowering & Fruiting	August-October.
Habitat Ecology	Moist shady, humus rich soils along roadsides and in forest floors; **Kothi Khad -** 810 m; **Matwana Khad -** 610 m.
Material Examined	EBH-WL-1203; 15.09.2007.
Part Used	Spores.
Folk Use	Plant spores used for making **'sindhoor'** (red dye) which the Indian women put in their hair parting to indicate their marital status.
Use in Literature	Plant exhibits antibacterial activity **(Singh, 2003b; Singh & Viswanathan, 1996).**

Setaria tomentosa Kunth (Plate No. 36A)

Syn.   *S. floribunda* Spreng.; *S. nubica* Link; *S. respiciens* Hochst. *ex* Miq.; *S. verticillata* Beauv.; *Panicum aparine* Steud.; *P. adhaerens* Forsk.; *P. asperum* Lamk.; *P. floribundum* Willd. *ex* Spreng.; *P. humile* Trin.; *P. italicum* Ucria.

Family   Poaceae

Vern.	Khagori.
Distribution	Shady places throughout India; ascending the Himalaya to 2,000m.
Description	Annual herbs possessing broad scrabulous leaves and long, green panicles. Spikelets smooth and shining. Glumes 3, membranous; lowest small; second and third nearly equal; uppermost with rudimentary flower. Stamens 3. Style 2, long. Grain free.
Flowering & Fruiting	January-May.

Habitat Ecology	Damp shady places; **Sir Khad -** 620 m.
Material Examined	EBH-WL-1179; 18.04.2009.
Part Used	Whole Plant.
Folk Use	Plant used in the preparation of local **tea**.
Use in Literature	Used for fermentation **(Bor, 1973; Hajra & Baishya, 1997; Hooker, 1872-97; Jain, 1991).**

Sida acuta **Burm. f. (Plate No. 36B)**

Syn.   *S. carpinifolia* L. f. SupPlate No.; *S. carpini folia* L.; *S. lanceolata* Retz

Family   Malvaceae

Vern.	Bariara.
Hindi, Sanskrit and Regional Names	
Hindi	Bariara, Kareta, Khaenti, Paharibariara;
Sans.	Bala, Brihannagabala, Pila, Pitbereela, Rajbala;
B.	Bonmethi, Pilabarelashikar, Shvetberdakorota;
G.	Bala, Janglimethi;
Kan.	Cheruparuva;
Mal.	Malatanni, Shiruparuva;
Oriya	Sibola, Sunakhodika;
Tam.	Arivalmanaippundu, Arivalmukkan, Kayappundu, Malaidangi, Malaikkurundali, Mayirmanikkam, Ponmusutlai, Vattatiruppi;
Tel.	Chittimu, Gayapaku, Muttavapulagamu, Nelabenda, Sahadevi, Visaboddi.
Distribution	Hotter parts of India. Tropics.
Description	Shrubby with slender branches and long, lanceolate, petiolate leaves. Pedicels jointed. Calyx long, triangular, acute. Corolla yellow. Carpels 5-9 puberulous. Seeds smooth, black.
Flowering & Fruiting	September-December.
Habitat Ecology	Common weed along roadsides in wet places; **Auhr Khad -** 610 m.
Material Examined	EBH-WL-1180; 22.12.2008.
Parts Used	Root. Leaves. Young Shoots.
Folk Uses	Root paste applied as poultice on **wounds**. Paste of leaves and young shoots applied over **boils**. Juice of leaves given on empty stomach in

the morning for expelling **intestinal worms**. 5-7g powdered leaves taken with cow's milk before bed for inducing **abortion**.

Chemical Constituents   Active properties due to α-amyrin, ecdysterone, ephedrine, cryptolepine **(roots)**, pristane, phytane, hentriacontane, nonacosane, cholesterol, campesterol, stigmasterol, β-sitosterol and stigmast-7-enol **(aerial parts)**.

Uses in Literature   :   Regarded as astringent, coolant, tonic; and for abortion, blood, bile and bowel complaints, intermittent fever, intestinal worms, nervous problems, snake bite, scorpion bite and urinary diseases **(Bhogaonkar & Kanerkar, 2007; Biswas et al., 2010; Cauis, 2003; Chopra et al., 1956; Dager & Dager, 1996; Gupta, 1997; Hosagoudar & Henry, 1996a; Idu et al., 2008; Jain, 1991; Jain & Singh, 1997; Kirtikar & Basu, 1984; Lindley, 1981; Maheshwari et al., 1996; Noumi, 2010; Pande et al., 2006; Parrotta, 2001; Rahman, 2000; Rana et al., 2003; Ranjan, 1999; Rao & Henry, 1995; Rao & Negi, 1980; Rastogi & Mehrotra, 1995a; Retnam & Martin, 2006; Satapathy, 2008; Subramaniam, 2000; Shukla & Verma, 1996; Siwakoti & Varma, 1996; Thomas & Britto, 1999; Tarafder et al., 1997; Upadhye et al., 1994).**

Silene conoidea L. (Plate No. 36C)

Family   Caryophyllaceae

Vern.   Dhudu chana, Jangli chana.

Distribution   W. Himalaya.

Description   Dichotomously branched, glandular pubescent herbs with pink flowers borne in terminal panicles. Fruits enclosed in inflated calyx.

Flowering & Fruiting   March-May.

Habitat Ecology   Cultivated fields; **Lyond Khad** - 617 m; **Sir Khad -**620 m.

Material Examined   EBH-WL-1181; 24.04.2008.

Part Used   Whole Plant.

Folk Uses   Powdered plant alongwith *Vicia hirsuta* useful in the treatment of **diabetes**; 1 tsp, twice a day for 15 days. Plant juice used as drops for **ophthalmic** problems.

Uses in Literature   Known to be used as an edible, emollient, fumigant and for opthalmia **(Ambasta, 1986; Jain, 1991; Kapur & Nanda, 1996a; Sood & Thakur, 2004; Viswanathan, 1997).**

Solanum nigrum L. (Plate No. 36D)

Syn.   *S. rubrum* Miller; *S. triangulare* Lamk.; *S. villosum* Lamk.; *S. incertum* Dunal; *S. nodiflorum* Jacq.; *S. uliginosum* Blume; *S. roxburghii* Dunal; *S. miniatum* Bernh.; *S. paludosum* Dunal; *S. pterocaulon* Dunal; *S. rumphii* Dunal

Family   Solanaceae

Vern.	Kale kyanun, Patkayai.

English, Hindi, Sanskrit and Regional Names

Eng.	Black nightshade, Common nightshade, Poisonberry;
Hindi	Gurkamai, Kabaiya, Makoi;
Sans.	Bahaphala, Bahutikta, Ghanaghana, Kaka, Kakamachi, Vayasi;
Ass.	Pichkati;
B.	Gurkamai, Kakmachi, Tulidun;
G.	Piludi;
Kan.	Kakarundi, Karikachi gida;
Mal.	Manttakkali; Karintakali, Karimtakkaki;
Mar.	Ghati, Kakmachi, Mako;
P.	Mako, Kambei, Kachmach, Riaungi;
Tam.	Munatakali;
Tel.	Kachchipundu, Kachi, Kamanchi, Gajju, Chettu.
Distribution	Common throughout India.
Description	Herbaceous or suffrutescents with ovate, sinuate toothed or lobed leaves, extra-axillary peduncles and subumbelled pedicels. Corolla white, glabrous. Seeds very smooth and many. Berries globose, red or black.
Flowering & Fruiting	January-July.
Habitat Ecology	Edges of cultivated areas, wastelands; **Landy Khad -** 615 m; **Lyond Khad -** 617 m.
Material Examined	EBH-WL-1069;15.04.2009.
Parts Used	Whole Plant. Fruits.
Folk Uses	Ripe fruits used for making **jam**; also its powder (2-3 g) taken with hot milk before bed to check **piles**. Decoction of the whole plant useful for **liver disorders**; 3 tsp twice a day for 20 days.

Chemical Constituents	These are solasodine, solasoliene, tigogenin, diosgenin, solasonine, solamargine, β-solamargine **(berries)**, α-solasonine and α-solamargine **(stem, leaves)** β-solamargine contained highest tigogenin content **(leaves)**, chlorogenic **(dry fruits)**, four steroidal glycoalkaloids **(immature fruits)**.
Biological Activity	Plant CNS depressant, hypotensive and spasmolytic.
Uses in Literature	Already known in India as an antidote to opium, antiseptic, cardiotonic, diaphoretic, diuretic, emollient, expectorant, hydragogue, laxative, sedative, stomachic, pot herb, vulnerary; and for bladder, boils, cough, diarrhoea, dysentery, ear complaints, easy expulsion of foetus, eye complaints, febrifuge, goitre, heart ailments, inflammation of scrotum, jaundice, kidney, liver complaints, nostrils complaints, piles, psoriasis, rejuvenating, rheumatism, ring-worm, skin diseases, sores, sprain, swelling, testicles, throat trouble, ulcers in mouth, urinary complaints and vomiting **(Ambasta, 1986, Bajpayee & Dixit, 1996; Bhattacharyya, 1996; Biswas et al., 2010; Chatterjee & Pakrashi, 1997; Chauhan, 1999; Chopra et al., 1956; Hosagoudar & Henry, 1996a; Jain, 1991; Jha et al., 1996; Joshi, 2009; Kapur & Singh, 1996; Kaul, 1997; Khanna et al., 1996; Lalramnghinglova, 2003; Lindley, 1981; Maheshwari et al., 1996; Pande et al., 2006; Parrotta, 2001; Prajapati et al., 2006; Prakasha et al., 2010; Rana et al., 2003; Rao & Henry, 1995; Rastogi & Mehrotra, 1991; Retnam & Martin, 2006; Rout & Panda, 2010; Saini, 1996b; Shiddamallayya et al., 2010; Sood & Thakur, 2004; Swami & Gupta, 1996; Viswanathan, 1997).**

Solanum xanthocarpum Schrad. & Wendl. (Plate No. 36E)

Syn.   *S. armatum* Br.; *S. jacquinii* Willd.; *S. diffusum* Roxb.; *S. virginianum* Jacq.; *S. surrattense* Burm. f

Family   Solanaceae

Vern.	Jangli bhindi.

English, Hindi, Sanskrit and Regional Names

Eng.	Yellow-berried nightshade;
Hindi	Katai, Kateli, Ringni;
Sans.	Kantakaree, Nidigadhika;
B.	Kantikari;
G.	Bhoyaringani;
Mar.	Bhuiringhani;
Oriya	Ankranti, Bheji-behun;

P.	Kandyali, Mahori, Warumba;
Tam.	Kandankattiri, Cundung katric;
Tel.	Nela mulaka, Pinna mulaka, Vankuda.
Distribution	Common throughout India.
Description	Procumbent perennials with leaves possessing many straight spines. Flowers blue, in a few-flowered extra-axillary cymes. Berries globose, yellow when ripe.
Flowering & Fruiting	March-April.
Habitat Ecology	Waste places near water; **Landy Khad -** 615 m.
Material Examined	EBH-WL-1005; 22.03.2008.
Parts Used	Stem. Fruit. Seeds.
Folk Uses	Roasted seeds **eaten**. Powdered stem and fruits used for checking **fever** and **chest affections** (1 tsp, twice a day till relief).
Chemical Constituents	Active constituents are solasodine, solmargine, β-solamargine, solasonine, diosgenin, solasurine, β-sitosterol, stigmasteryl glucoside **(fruits)**.
Biological Activity	Aq. extract of the plant hypotensive.
Uses in Literature	So far used as an appetiser, carminative, diuretic, edible (seed, fruit), and laxative; and for asthma, blisters, boils, bronchitis, chest pain, cold, cough, dropsy, ear and eye complaints, fever, gonorrhoea, gum trouble, heart diseases, lumbago, migraine, muscular pain, pain in jaw, paralysis, piles, pyorrhoea, rheumatism, scorpion-bite, snake-bite, sore throat, sores, sterility in woman, swelling and watery eyes **(Ambasta, 1986; Aminuddin & Girach, 1991; Bhatt et al., 1999; Biswas et al., 2010; Chandra, 1997; Chakraborty et al., 2003; Chatterjee & Pakrashi, 1997; Chopra et al., 1956; Girach et al., 1997; Jha et al., 1996; Jain, 1968; 1991; Jain et al., 1994, 2010; Joshi, 2008; Kirtikar & Basu, 1984; Kurian, 1999; Maliya, 2009; Meena & Yadav, 2010; Pandey et al., 2000; Parrotta, 2001; Prajapati, 2006; Rout & Panda, 2010; Sood & Thakur, 2004; Subramaniam, 2000; Thakor, 2009; Uphof, 2001).**

Sonchus arvensis L. (Plate No. 36F)

Syn. *S. longifolius* Wall.; *S. orixensis* Roxb.; *S. wallichiana* DC.; *S. wightianus* DC.

Family Asteraceae

Vern.	Sadhi.

English, Hindi and Regional Names

Eng.	Corn sow thistle, Dindle gutweed, Hogweed, Rosemary, Swine thiste, True sow thistle;
Hindi	Sahadevi bari;
B.	Banpalong;
P.	Bhangra, Kalabhangra;
Tel.	Jangli tamaku.

Distribution   Wild and in cultivated places throughout India. Common in Himalaya.

Description   Annual milky herbs with creeping rootstock, glabrous stem and pinnatifid spinous-toothed, cauline leaves. Heads and peduncles glandular-hispid. Achenes narrow, subcompressed with thick regular ribs.

Flowering & Fruiting   May-June.

Habitat Ecology   Common in moist places; **Gobind Sagar Lake** - 490 m; **Sir Khad** - 620 m.

Material Examined   EBH-WL-1245; 05.05.2008.

Part Used   Whole Plant. Roots.

Folk Uses   Herb **relished** by horses and cattle. Poultice of the plant good against **burn, scaldings** painful and irritable **ulcers**. 1-3 powdered roots mixed with honey given twice a day for **cough**.

Uses in Literature   Useful earlier as analgesic, aphrodisiac, bitter, diuretic, galactagogue, haematinic, hypnotic, saporific, stomachic; and for asthma, bronchitis, burning, chronic fevers, cough, cure biliousness, diseases of liver, leucoderma, jaundice, headache, heart diseases, improving appetite, ophthalmia, pertusis, purifying blood, scabies, sensation and troubles of nose **(Ambasta, 1986; Asolker et al., 1992; Bhalla et al., 1996; Hooker, 1872-97; Jain, 1991; Jain et al., 2010; Khanna, et al., 1996; Kirtikar & Basu, 1984; Kumar, 2002; Lalramnghinglova, 2003; Panda, 1996; Pande et al., 2006; Rana et al., 2003; Rastogi & Mehrotra, 1993).**

Sonchus asper Hill (Plate No. 37A)

Syn.   *S. ferox* Wall.; *S. oleraceus* Wall

Family   Asteraceae

Vern.   Bhursalae.

English, Hindi and Regional Name

Eng.	Du tistle, Hare's lettuce, Hare's palace, Hare's thistle, Milk weed, Milk thistle, Sow thistle, Spiny leaves, Sow thistle, Sprout thistle, Turn sole;

Hindi	Didhi;
Mar.	Mhatara.

Distribution   Throughout India, in Himalaya upto 4,000 m.

Description   Milky herbs with stem-clasping leaves and terminal, yellow homogamous heads. Achenes compressed. Pappus copious, white.

Flowering & Fruiting   March-April.

Habitat Ecology   Along roadsides, wastelands; **Karyal Khad -** 615 m; **Kothi Khad -** 810 m.

Material Examined   EBH-WL-1070; 22.03.2009.

Parts Used   Roots. Shoot. Leaves. Seeds.

Folk Uses   Young shoots eaten as **salad**. Roasted seeds also **eaten**. Infusion of the root and leaves considered good **tonic** for goats and rabbits.

Chemical Constituents   Found to contain 15-o-β-glucopyranosyl-11β, 13-dihydrourospermal A2, ursolic acid 3, lupeol 4, β-sitosterol-3-o-glucopyranoside 5 **(roots)**.

Uses in Literature   Recorded earlier as an emollient, febrifuge, sedative tonic; and for abdomen, ailments of liver, cleansing and healing ulcers, wounds and boils **(Ambasta, 1986; Asolkar et al., 1992; Bhalla et al., 1996; Chaudhury & Neogi, 2000; Jadhav, 2009; Jain, 1991; Kirtikar & Basu, 1984; Pande et al., 2006; Rastogi & Mehrotra, 1993; Uphof, 2001).**

Sorghum halepense L. Pers. (Plate No. 37B)

Syn.   *S. capense* **Herb.**; *S. dubium* C. **Koch**; *S. giganteum* **Edgew.**; *S. saccharum* Hohen.; *S. schreberi* Ten.; *Blumenbachia halepensis* Kael.; *Trachypogon avenaceous* Nees; *Holcus halepensis* Brot.; *H. decolran* Willd.; *Andropogon halpensis* Brot.; *A. arundinaceus* Scop.; *A. controversus & dubitatus* Steud

Family   Poaceae

Vern.   Barchita.

English, Hindi and Regional Names

Eng.	Aleppo grass, Evergreen millet, Johnson grass;
Hindi	Baru;
B.	Kalamucha;
Kash.	Braham;
P.	Baru, Barwa, Braham;
Tam.	Kadu cholam;
Tel.	Gaddijanu.

Distribution	Open places throughout India, Myanmar and Sri Lanka. Most warm countries.
Description	Tall stout perennials having stoloniferous roots, erect leafy stem and pubescent nodes. Leaves broad; tip filiferous, glabrous. Ligules rounded. Panicles variable in form. Spikes having 5-7 pairs of spike-lets. Lemma empty. Caryopsis long.
Flowering & Fruiting	July-October.
Habitat Ecology	Near moist areas, wet lands; **Nalti Khad -** 650 m.
Material Examined	EBH-WL-1165; 24.10.2008.
Part Used	Whole Plant. Seeds.
Folk Uses	Powdered seeds **eaten**. Plant used as **hay** and **fodder** for cattle.
Chemical Constituents	Found to contain chlorogenic, p-coumeric, fernlic p-hydroxy benzo-ic, p-hydroxy, p-hexylactic acids, dhurrin, vanillin, taxiphyllin, p-hydroxybenzaldehyde (leaves, rhizome) and p-coumaric acid (plant).
Uses in Literature	Useful as edible (seeds) and fodder **(Bor, 1973; Cauis, 2003; Chandra, 1997; Duthie, 1888; Jain, 1991; Mehra, 1982; Rana et al., 2003; Roy, 1984; Satya & Solanki, 2008; Uphof, 2001).**

Spilanthes acmella L. var. *oleracea* Jacq. (Plate No. 37C)

Syn.   *Spilanthes ciliata* H.B.K.; *Verbesina acmella* L.; *Pseudo acmella* L.

Family   Asteraceae

Vern.	Bada-karkara.
English, Hindi, Sanskrit and Regional Names	
Eng.	Para cress; Brazil cress;
B.	Roshaniya.
Distribution	Throughout India, Nepal, Sri Lanka and Brazil.
Description	Diffuse herbs rooting at nodes with terete stem and ovate-acute leaves having rounded base which are serrated at the margins. Heads yellow, subglobose, conical, axillary, solitary and rayed. Achenes black, oblong, laterally compressed.
Flowering & Fruiting	October-March.
Habitat Ecology	Wet or marshy places; **Gobind Sagar Lake -** 490 m.
Material Examined	EBH-WL-1246; 28.02.2009.
Parts Used	Roots. Leaves. Flowers.

Folk Uses	Decoction of the root considered a good **purgative**. A bath with decoction of the leaves good for **rheumatism** whereas its lotion applied for curing **scabies** and **soriasis**; A small piece of lint dipped in decoction of its leaves also applied on the gums, 3-4 times a day, for reducing **pain** and **swelling**. Flowers chewed for **freshness** of mouth.
Chemical Constituents	**Roots** possess a new triterpenoid saponin characterized as olean-12-en-3-O-β-D-galactopyranosyl (1→4)-α-L-rhamnopyranoside.
Uses in Literature	Described so far as an acrid, sialagogue, stimulant; and for affections of throat, gums, headaches, paralysis of the tongue and toothache **(Asolkar et al., 1992; Bhalla et al., 1996; Chaudhury & Neogi, 2000; Choudhury et al., 2008; Henry, 1999; Jain, 1991; Kirtikar & Basu, 1984; Lindley, 1981; Pande et al., 2006; Parrotta, 2001; Rastogi & Mehrotra, 1995a; Singh et al., 1996a; Tiwari & Tiwari, 1996; Uphof, 2001).**

Spilanthes paniculata Wall. ex DC. (Plate No. 37D)

Syn.   *S. acmella* var. *paniculta* C.B. Clarke

Family   Asteraceae

Vern.	Chota karkara.
English, Hindi, Sanskrit and Regional Names	
Eng.	Toothache plant;
Ass.	Pirazha;
Kan.	Hemmugulu, Vanamugali;
Mal.	Kuppa manjel;
Mar.	Akkalkara, Pipulka;
P.	Akarkarha, Pokarmul;
Tel.	Maratimagga, Maratiteuga.
Distribution	Throughout India, ascending the Himalaya 1,700 m.
Description	Branched tall herbs having numerous panicled heads. Achenes strongly margined, sparsely scabrid. Pappus bristles 1-2.
Flowering & Fruiting	October-March.
Habitat Ecology	Wet or marshy places; **Ali Khad -** 580 m.
Material Examined	EBH-WL-1183; 03.04.2009.
Parts Used	Whole Plant. Leaves. Flowers.
Folk Uses	Pungent flowers chewed for relief in **throat affections, paralysis of tongue** and **toothache**. Leaves given as **demulcent**. 15-20ml juice of the plant given regularly to mother as a **galactagogue**. Infusion of plant considered good against **dysentery**.

Chemical Constituents	Sitosterol, glucoside, stigmasterol, stearic acid and tetratriacontanoic acid characterized from **aerial parts**.
Uses in Literature	Employed earlier as antiscorbutic, digestive, diuretic, lithotriptic, odontalgic, purgative, sialagogue, stimulant, tonic; and for caries and inflammation of jaw-bones, psoriasis, rheumatism and scabies **(Ambasta, 1986; Bora, 2000; Barua et al., 1999; Hajra & Baishya, 1997; Jain, 1991; Kirtikar & Basu, 1984; Kumar, 2002; Pande et al., 2006; Uphof, 2001).**

Stellaria media (L.) Vill. (Plate No. 37E)

Syn.   *Alsine media* L.; *Alsinella wallichiana* Benth.; *S. monogyna* Don

Family   Caryophyllaceae

Vern.		Padyala.
English and Regional Names		
	Eng.	Chickweed;
	Ass.	Morolia.
Distribution		Temperate Himalaya. N. India.
Description		Much-branched, flaccid herbs upto 60 cm with lower leaves stalked and upper ones sessile. Flowers in terminal cymes. Capsule ovoid.
Flowering & Fruiting		March-April.
Habitat Ecology		Weed of cultivated areas, damp places, shady localities; **Sir Khad -** 620 m.
Material Examined		EBH-WL-1067; 15.03.2007.
Part Used		Whole Plant.
Folk Uses		Plant used as a **pot herb** and **purgative** both for humans (10 g) and livestock (250 g). Paste applied on forehead to **cure headache** and **giddiness**.
Chemical Constituents		**Herb** contains triterpenoid saponinis, coumarins, flavonoids, caroxylic acids and vitamin C.
Biological Activity		Plant antiinflammatory.
Uses in Literature		Useful earlier for boils, bone fracture, burns, digestive disorders, eczema, erysipelas, expelling intestinal threadworms, eye diseases, itching, haemorrhoids, inflammation in rheumatic joints, renal, respiratory diseases, snake bite, swelling, ulcer, wounds and as a vegetable **(Arora, 1997; Chandra, 1997; Chopra et al., 1956; Jain, 1991; Khanna et al., 1996; Pande et al., 2006; Prajapati et al., 2006; Rastogi & Mehrotra, 1995a).**

Strobilanthes atropurpurens Nees (Plate No. 37F)
Family   Acanthaceae

Vern.	Kaseru.
Distribution	Throughout India.
Description	Erect succulent perennials upto 90 cm. Leaves glabrous, 8 cm long; petiole winged. Lamina of lower leaves elliptic-oblong whereas that of flowering shoot ovate-lanceolate. Flowers blue, solitary. Calyx linear-oblong. Corolla 4 cm long, glabrous. Anthers muticous. Capsule oblong. Seed 4, glabrous, hairy.
Flowering & Fruiting	October-February.
Habitat Ecology	Near water sources; **Sauli Khad -** 765 m.
Material Examined	EBH-WL-1068; 22.02.2009.
Parts Used	Whole Plant. Flowers.
Folk Uses	Flowers given with sugar in **cough**. Juice of plant considered a good **antiseptic** for cuts and wounds.
Uses in Litrature	Used to cure cuts and wounds, digestive problems, fever, headache and skin infections **(Deokota & Chhetri, 2009).**

Strobilanthes dalhousianus Clarke (Plate No. 38A)
Syn.   *S. pentstemonoides* T. Anders.; *Goldfussia dalhousiana* Nees
Family   Acanthaceae.

Vern.	Gathban.
Distribution	N. W. Himalaya from Kashmir - Kumaon.
Description	Suberect shrubs having hairy stems and elliptic acuminate leaves. Bracts scarcely caducous. Corolla glabrous, purple. Calyx hairy.
Flowering & Fruiting	July-October.
Habitat Ecology	Frequent on slopes along Nallahs; **Tatoh Khad -** 710 m.
Material Examined	EBH-WL-1204; 12.10.2008.
Part Used	Whole Plant.
Folk Use	Paste of the plant in mustard oil applied as a poultice for treating **rheumatic joints**.
Use in Literature	Reported earlier for rheumatic pains **(Hooker, 1872-97; Sood & Thakur, 2004).**

Taraxacum officinale **Webb. (Plate No. 38B)**

Syn.   *T. dens-leonic* Desf.; *Leontodon taraxacum* L

Family   Asteraceae

Vern.   Kadavi, Sunachari.

English, Hindi, Sanskrit and Regional Names

Eng.	Dandelion, Fortune-teller, Miek gowan;
Hindi	Barau, Dudhal, Kanphul;
Sans.	Dugdhapheni;
B.	Pitachumki;
G.	Pathardi;
Kan.	Kaddusevent Hi;
Kash.	Hand;
Mar.	Undakanti;
Tel.	Patri.

Distribution   Throughout the Himalaya; Temperate and Cold regions.

Description   Scapigerous milky herbs possessing sessile oblanceolate, entire toothed leaves. Heads solitary. Achenes oblong-obovoid or narrow. Pappus copious.

Flowering & Fruiting   March-November.

Habitat Ecology   Grassy meadows, roadsides, damp places; **Papral Khad -** 740 m.

Material Examined   EBH-WL-1310; 18.11.2007.

Part Used   Roots. Leaves.

Folk Uses   1-2 g powdered leaves given twice a day for three days for stomach **ailments**. Juice of roots applied to **remove warts**.

Chemical Constituents   These are scopoletin, esculetin **(aerial parts)**, taraxacin, acryrtalline, taraxacerin, an acrid resin and starch **(roots)**.

Uses in Literature	So far reported as an antirheumatic, antiscorbutic, aperient, blood purifier, cholagogue, diuretic, hepatic stimulant, mild laxative, salads, stomachic, tonic, vegetable; and for atonic dyspepsia and oligurea, bowel complaints, cholecystis, dysentery, gallstones, jaundice, preparation of beer, soups, stout and wines **(Ambasta, 1986; Hooker, 1872-97; Jain, 1991; Kaul, 1997; Kumar, 2002; Pande et al., 2006; Prajapati et al., 2006; Rana et al., 2003; Rastogi & Mehrotra, 1991; Sood & Thakur, 2004; Wangchuk et al., 2008).**

Thalictrum reniforme Wall. (Plate No. 38C)

Syn.   *T. neurocarpum* Royle; *T. chelidonii* var. *reniforme* Hook

Family   Ranunculaceae

Vern.	Chaksu.
Hindi and Regional Names	
Hindi	Mamira, Pilijari, Pinjari, Shuptak;
B.	Gurbiani;
Kash.	Cheitra;
P.	Chireta, Chitramul, Gurbiani, Keraita, Mamira, Pashmaran, Phalijari.
Distribution	Found throughout the Himalaya: 1,700-2,700 m. Khasia Hills, Myanmar and Siam.
Description	Much-branched perennial herbs upto 2.5 m with leaf sheaths expanding into adnate stipules and small greenish-white flowers in panicles. Sepals deciduous. Achenes pubescent, ribs prominent.
Flowering & Fruiting	June-August.
Habitat Ecology	Grassy situations, damp places; **Karyal Khad** -615 m; **Marsand Khad** - 610 m.
Material Examined	EBH-WL-1247; 19.07.2008.
Parts Used	Root. Whole Plant.
Folk Uses	Roots chewed to get instant relief from **toothache**. Powdered plant in combination with black pepper (4:1) used for curing **leucorrhoea**; 1 tsp twice a day for 10 days.
Chemical Constituents	**Roots** yield both alkaloids and flavonoids, whereas the **aerial parts** contain only alkaloids.
Uses in Literature	Used earlier for cataract and relieving pain; and as general antidote and antimalarial **(Jain, 1991; Kala, 2004; Kirtikar & Basu, 1984; Pande et al., 2006; Sood & Thakur, 2004; Wangchuk et al., 2008).**

Thevetia neriifolia Juss. ex Steud. (Plate No. 38D)

Syn.   *T. peruviana* (Pers.) K. Schum.; *Cerbera peruviana* Pers.; *Cascabela thevetia* (L.) Lip.

Family   Apocynaceae

Vern.	Kaner.

English, Hindi, Sanskrit and Regional Names

Eng.	Bastard oleander, Exile oleander, Lucky nut tree, Yellow oleander;
Hindi	Pila, Pila kanair;
Sans.	Ashvamaraka, Nakharatila, Sidhapushpa;
B.	China-karab, Kokil-phul;
Bo.	Pilakaner, Zardakunel;
G.	Pilakanir;
Kan.	Kadukasi, Kanogalu;
Mal.	Pachchaarali;
Mar.	Pivalakanhera;
Oriya	Konyar-phul;
Tam.	Pachaiyalari, Tiruvachippu;
Tel.	Pachchaganeru;
U.P.	Pila-kaner.

Distribution	Throughout India.
Description	Shrubs. Leaves linear, lanceolate. Flowers funnel-shaped, yellow, few in a terminal clusters. Fruits black. Drupes angular, poisonous.
Flowering & Fruiting	July-September.
Habitat Ecology	Forests. Shrubberies; **Auhr Khad -** 610 m.
Material Examined	EBH-WL-1256; 22.08.2008.
Part Used	Fruits.
Folk Use	Fruits **edible.**
Chemical Constituents	Main ingredients are arabinose galactose, glucose, galacturonic acid and xylose **(fruits).**
Biological Activity	Cardiotonic and isotropic effects and musculotropic activities confirmed.

Uses in Literature	Recorded earlier as an abortifacient, acronarcotic poison, alexetric, antiperiodic, bitter cathartic, bitter tonic, cardio tonic against pentobarbital induced heart failure, emetic, febrifuge, laxative, ornamental (beads), poison insecticide, (vet); and for children diseases, criminal poisoning of cattle, dropsy, gonorrhoea, heart diseases, impotence, intermittent fevers, leucoderma, menstrual disorder, rheumatism, skin diseases, snake bite, swelling (vet), swelling of foot, tumours and worms **(Ambasta, 1986; Bhandary et al., 1996; Chauhan, 1999; Chopra et al., 1956; Das & Agarwal, 1991; Huidrom, 1996; Jain, 1991; Jha et al., 1996; Kirtikar & Basu, 1984; Kurian, 1999; Mishra et al., 1996; Said, 1997; Saini, 1996a; Sanyal, 1994; Sharma, 2003; Sharma & Sood, 1997; Singh & Pandey, 1996; Siwakoti & Verma, 1996; Watt, 1972).**

Trichodesma indicum (L.) Lehm. (Plate No. 38E)

Syn. *T. perfoliatum* Wall.; *T. hirsutum* Edgew.; *Borago spinulosa* Roxb.; *B. indica* L

Family   Boraginaceae

Vern.	Rukhadi.

Hindi, Sanskrit and Regional Names

Hindi	Chhota-kulpha, Ratmundiya;
Sans.	Adhapushpi, Adhomukha, Avakpushpi, Darvika, Dhenujivha, Gandhapushpika, Golomi, Romalu, Sarasa;
B.	Choota-kulpha, Ratmundiya;
Kash.	Nilakrai, Ratisurkh;
Mar.	Lahanakalpa;
P.	Kallri-buti, Ratmandu;
Tam.	Kazuthai-tumal;
Tel.	Guvva-gutti.
Distribution	Throughout India.
Description	A coarse branched annual, 15-45 cm tall, possessing bulbous-based hair and stem-clasping leaves. Flowers pale-blue, in a terminal few-flowered cymes. Fruits pyramidal with the persistent style, 4-ribbed.
Flowering & Fruiting	September-December.
Habitat Ecology	Open moist slopes; **Lyond Khad -** 617 m.
Material Examined	EBH-WL-1248; 02.10.2008.
Part Used	Whole Plant.

Folk Uses	One tsp of the whole plant decoction given at a time to **cure stomach ache**. Paste of the plant **heals wounds**.
Chemical Constituents	Constituents isolated are: hexacosane, ethyhexacosanoate, 21,24-hexacosadienoic acid ethylester **(leaves)**, linoleic, linolenic, oleic, palmitic and staeric acids **(seeds)**.
Uses in Literature	So far described as a blood purifier, depurative, diuretic, edible (leaves), emollient and tonic for brain; and for arthralgia, diarrhoea, dysentery, dysmenorrhoea, dyspepsia, eczema, expelling worms in animals, expulsion of the foetus, fevers, inflammations, leprosy, ophthalmopathy, skin diseases, sores, stomach ache, strangury and swelling **(Ambasta, 1986; Chatterjee & Pakrashi, 1997; Chopra et al., 1956; Jain, 1991; Parrotta, 2001; Prajapati et al., 2006; Sood & Thakur, 2004).**

Tridax procumbens L. (Plate No. 38F)

Family   Asteraceae

Vern.	Pugru.
English, Hindi and Regional Names	
Eng.	Coatbuttons, Mexican daisy;
Hindi	Pardesi langri;
Kan.	Gabbu sanna savanthi;
Tam.	Vettukkaaya-thalai;
Tel.	Raavanaasuruditalkali.
Distribution	Native of S. America. More or less throughout India.
Description	Hispid procumbent herbs with simple, lanceolate - ovate leaves having acute base and coarsely serrated margin. Flowers yellow, in terminal heads. Achenes hairy; pappus white.
Flowering & Fruiting	July-August.
Habitat Ecology	Moist rocky slopes; **Gobind Sagar Lake -** 490 m; **Sir Khad -** 620 m.
Material Examined	EBH-WL-1074; 12.07.2007.
Parts Used	Whole Plant. Leaves.
Folk Uses	Aerial plant parts used as **fodder**. Infusion of the whole plant prescribed for **stomach ache**; ½ tsp diluted in 1 cup of water given twice daily for 3-4 days. Fresh extract of leaves applied on cuts and wounds to **check bleeding**.

Chemical Constituents	**Leaves** contain fumaric acid whereas **flowers** yield glucoluteolin, luteolin, isoquercetin and quercetin.
Biological Activity	Plant hypotensive.
Uses in Literature	Used against blisters, boils, cuts and wounds, diarrhoea, dysentery, eczema, eye diseases, fever, leprosy, scorpion bite, sores, stomach ache, stone in urine bladder and toothache **(Ambasta, 1986; Ansarali & Sivadasan, 2009; Bhogaonkar & Kanerkar, 2007; Dwarakan & Ansari, 1996; Jadhav, 2009; Jain, 1991; Maheshwari et al., 1996; Mishra, 2008; Parrotta, 2001; Patil, 2009; Rana et al., 2003; Rao & Henry, 1995; Rout & Panda, 2010; Samwatsar & Diwanji, 1996a; Sood & Thakur 2004).**

Trifolium resupinatum L. (Plate No. 39A)

Family   Fabaceae

Vern.	Perseen.
English and Hindi Names	
Eng.	Persian clover;
Hindi	Shaftal.
Distribution	Common in Subtropical regions.
Description	Herbaceous annuals. Stipules adnate to the petioles. Flowers purplish, in dense axillary heads. Pods included.
Flowering & Fruiting	May-June.
Habitat Ecology	Damp places, marshy localities, wastelands; **Gobind Sagar Lake -** 490 m.
Material Examined	EBH-WL-1264; 20.05.2009.
Part Used	Whole Plant.
Folk Use	Plant used as green **fodder** for livestock.
Chemical Constituents	**Leaves** reported to contain arginine, histidine, methionine, tryptophan and total lysine.
Use in Literature	Known to be used earlier as a fodder **(Anonymous, 1976b; Uphof, 1968).**

Trigonella pubescence **Baker (Plate No. 39B)**

Family   Fabaceae

Vern.	Jangli methi.
Distribution	Common in various parts of India.
Description	Strongly-scented annuals having three leaflets and entire stipules. Flowers yellow, sessile, axillary. Pods many-seeded, beaked.
Flowering & Fruiting	March-May.
Habitat Ecology	Cultivated fields; Near water channels; **Sir Khad -** 620 m.
Material Examined	EBH-WL-1205; 05.03.2008.
Part Used	Whole Plant.
Folk Use	Considered a good **fodder** for livestock.
Chemical Constituents	**Plant** contains l-tryptophan, saponins, caumarin, fenugrukins, nicotinic acid, sapogenins and nicotinic acid.

Uraria picta (Jacq.)DC. (Plate No. 39C)

Family   Fabaceae

Vern.	Dabra.

Hindi, Sanskrit and Regional Names

Hindi	Dabra, Prisniparni, Prishtaparni;
Sans.	Prasniparni;
B.	Sankarjata;
G.	Pilvan;
Tam.	Sittirapaladai.
Distribution	Sub-Himalayan tract from Kashmir–W. Bengal and Assam.
Description	Erect branched perennial herbs having blotched white leaves upto 30 cm long. Leaflets linear, oblong, obtuse. Flowers purple in terminal racemes. Pods 3-6 jointed.
Flowering & Fruiting	July-November.
Habitat Ecology	Wet grasslands, waste places; **Barthin Khad -** 585 m.
Material Examined	EBH-WL-1280; 12.09.2008.
Part Used	Root.
Folk Use	Root extract given thrice a day for 3-4 days to cure **snake bite**.
Chemical Constituents	Alkaloids abramine, abromasterol, digitonide, isoflavanones, 5, 7-di-hydroxy-2-methoxy -3,4 – methylene dioxyisolavanones, 5,7-dihydroxy-2'3- dimethoxy -7-(5-hydroxychromen-7yl) isoflavanone isolated from **roots**.
Uses in Literature	Useful earlier as an antiseptic; and for snake bite, mouth sores, cold, fever and gonorrhoea **(Biswas et al., 2010; Jain, 1991; Jain & Singh, 1997; Pande et al., 2006; Parrotta, 2001; Prajapati et al., 2006).**

Urena lobata L. (Plate No. 39D)

Syn. *U. cana* Wall.; *U. palmata* Roxb

Family Malvaceae

Vern.	Badi-dredae.

Hindi, Sanskrit and Regional Names

Hindi	Bachita, Bahata;
Sans.	Vanabhenda;
B.	Benochra;
Mal.	Udiram, Uran, Vatto;
Mar.	Rantupkada, Vanabendha;
Oriya	Bilokapasia, Jotyahola;
Tam.	Ottatti;
Tel.	Peddabenda.
Distribution	Throughout the hotter parts of India.
Description	Erect herbs upto 10 cm with angled leaves and pink flowers in clusters. Capsules covered with blunt spines.
Flowering & Fruiting	July-September.
Habitat Ecology	Frequent on moist shady places; **Auhr Khad -** 610 m.
Material Examined	EBH-WL-1077; 19.08.2008.
Parts Used	Seeds. Whole Plant.
Folk Uses	1-2 g powdered seeds given twice a day with milk for 3-5 days for checking **flatulence**. Whole plant boiled in 'sesame' oil and applied externally for **joint pains** till **cure**.
Chemical Constituents	Chief constituents are mangiferin and quercetin **(aerial parts)**, stigmasterol and β-sitosterol **(whole plant)**.
Biological Activity	Plant hypotensive and CNS depressant.
Uses in Literature	Recorded in India for cordage, cough, fever, diarrhoea, dysentery, hyperacidity, hydrophobia, lumbago and windy colic; and as an expectorant, diuretic, gargle for aphthae and sore throat **(Ambasta, 1986; Biswas et al., 2010; Chatterjee & Pakrashi 1997; Chopra et al., 1956; Dager & Dager, 1996; Hosagoudar & Henry, 1996a; Jain, 1991; Joshi, 2009; Kaushik & Dhiman, 2000; Kirtikar & Basu, 1984; Kumar, 2002; Lalramnghinglova, 1996; Lindley, 1981; Maheshwari et al., 1996; Pan-**

de et al., 2006; Pandey et al., 1996; Parrotta, 2001; Rama Rao et al., 2008; Rana et al., 2003; Rao & Henry, 1995; Rastogi & Mehrotra, 1993; Shukla & Verma, 1996; Sinha et al., 1996; Siwakoti & Varma, 1996; Uniyal, 1989; Upadhye et al., 1994; Watt, 1972).

Urtica dioica L. (Plate No. 39E)

Family   Urticaceae

Vern.	Kogsi.
English and Hindi Names	
Eng.	Nettle, Stinging nettle;
Hindi	Bichhu booti.
Distribution	Kashmir-shimla.
Description	Erect stout grooved herbs with stinging hairs. Leaves cordate, long-stalked, toothed. Flowers green, dioecious, clustered on drooping axillary panicles. Achenes flat, included in the persistent perianth.
Flowering & Fruiting	June-October.
Habitat Ecology	Moist slopes, wastelands; **Tatoh Khad -** 710 m.
Material Examined	EBH-WL-1248; 19.08.2008.
Parts Used	Whole Plant. Shoots. Leaves.
Folk Uses	Tender shoots and leaves lopped for **chutany, vegetable** and **cattle fodder**. Poultice of the plant efficacious in **chronic painful joints**. Necklace of pieces of roots worn on Saturday to ward off any ill effect of **black magic**.
Chemical Constituents	Found to contain histamine, 5-hydroxytryptamine, acetylcholine, glycoprotein, serine-o-galactoside glycopeptide, arabinose, vitamin A and C, iron, calcium, magnesium, potassium **(leaves)**.
Biological Activity	Antiviral, aphrodisiac, CNS-depressant, cytotoxic, depurative, expectorant, haemostatic, hypotensive, mitogenic, uterotonic and vermifuge activities axhibited.
Uses in Literature	Known to be used as an antihelminhic, antiseptic, astringent, diuretic, emenagogue, haemostatic, lithotriptic, pot-herb; and for asthma, bladder stone, bleeding, boils, bronchosis, congestion, constipation, consumption, cramp, cuts and wounds, dandruff, diabetes, diarrhoea, dislocated bones, dog bites, edema, enhanced milk flow, fever, fibers, goitre, gonorrhoea, gout, hay fever, haematuria, headache, herpes, jaundice, kidney troubles, leucorrhoea, menorrhagia, nephri-

tis, paralysis, rheumatism, sciatica, sprain, swelling, throat diseases, toothache, urticaria and wounds **(Agarwal, 2003, Ambasta, 1986; Arora & Pandey, 1996; Aswal & Mehrotra, 1994; Bhogaonkar & Kanerkar, 2007; Chopra et al., 1956; Devi, 2003; Duke et al., 2002; Ganai & Nawachoo, 2003; Jain, 1991; Kala & Rawat, 2001; Kaul, 1997; Khare, 2004; Kirtikar & Basu, 1984; Lal et al., 1996; Lindley, 1981; Malla & Chhetri, 2009; Manandhar, 1996a; Negi et al., 1999; Pande et al., 2006; Panthi & Chaudhary, 2003; Prajapati et al., 2006; Rana et al., 2003; Said, 1997; Saklani & Jain, 1994; et al., 1996, 2001a; Sarin 1990; Semwal et al., 2010; Sharma, 2003; Sood & Thakur, 2004; Uphof, 1968; Yonzone et al., 1996; Viswanathan, 1997).**

Vernonia cinerea (L.) Less. (Plate No. 39F)

Syn. *V. albicans* DC.; *V. abbreviata* DC.; *V. laxiflore* Less.; *V. physalifolia* DC.; *V. parviflora* Reinw.; *Conyza cinerea* L

Family Asteraceae

Vern.	Sahdaiya.

English, Hindi, Sanskrit and Regional Names

Eng.	Ash-coloured flea bane;
Hindi	Dandotpala, Sahadevi, Sadodi, Sadori;
Sans.	Dandotpala, Devasasha, Devika, Gandhavalli, Govandani, Saha, Sahadeva, Vishamajvaranashini, Vishadeva;
B.	Kalajira, Kukshim, Kuksim;
G.	Sadodi, Sedardi, Shedardi;
Kan.	Sahadevi;
Mal.	Puvankuruntal;
Mar.	Osari, Sadadi;
P.	Sahadevi;
Tam.	Puvamkurundal, Sahadevi, Sirashengalanis;
Tel.	Garitikamma, Gharitikamini.
Distribution	Throughout India, ascending to 2,700 m in the Himalaya, Khasia and Peninsular regions.
Description	Erect perennials with slender, grooved and ribbed stem and membranous, coriaceous leaves possessing variable petiole. Heads open, flat-topped and in corymbs. Peduncles slender, corolla pubescent. Achenes with dirty white pappus.
Flowering & Fruiting	October-March.

Habitat Ecology	Common throughout the wastelands, moist places; **Ali Khad -** 580 m; **Sir Khad -** 620 m; **Sauli Khad -** 765 m.
Material Examined	EBH-WL-1071; 23.03.2009.
Parts Used	Whole Plant. Leaves. Roots. Flowers.
Folk Uses	One tsp of the plant paste mixed with honey taken twice a day for 3 days against **gastric complaints**. Plant juice taken orally on empty stomach to cure **rheumatism**. 15-25ml decoction of leaves, flowers or roots good against **fever** (thrice a day till cure). 3-5g powdered leaves taken with cow's milk before bed to **check constipation**.
Chemical Constituents	β-amyrin acetate, β-amyrin benzoate, lupeol and its acetate, β-sitosterol, stigmasterol and α-spinasterol, 24-hydroxytaraxer-14-ene isolated from **plant**.
Uses in Literature	Known so far as alexiphoretic, antihelminthic, astringent, diaphoretic, stomachic, strangury, tonic; and for asthma, bronchitis, cunjunctivitis, dropsy, removing kidney stones, piles and spasm of the bladder **(Ansarali & Sivadasan, 2009; Barua et al., 2000; Bhalla et al., 1996; Chandra, 1997; Chaudhury & Neogi, 2000; Dash & Misra, 1999b; Goud et al., 2000; Hooker, 1872-97; Jain, 1991; Jain et al., 2010; Khanna et al., 1996; Kirtikar & Basu, 1984; Mahato & Mahato, 1996; Pande et al., 2006; Parrotta, 2001; Prajapati et al., 2006; Rana et al., 2003; Rastogi & Mehrotra, 1995b; Retnam & Martin, 2006; Samwatsar & Diwanji, 1999; Singh, 1999; Singh & Srivastava, 2000; Siwakoti & Siwakoti, 1999; Thomas & Britto, 1999; Thakor, 2009; Viswanathal et al., 2006).**

Veronica anagalis-aquatica L. (Plate No. 40A)

Syn.   *V. angallis* L

Family   Scrophulariaceae

Vern.	Sadevi.
Hindi Name	
Hindi	Titlokia.
Distribution	Throughout the greater parts of India.
Description	Erect or decumbent-ascending annual herbs having glabrous stems and sessile, lanceolate-oblong, entire leaves with cordate base. Flowers white, in lax and axillary racemes. Pedicels filiformous. Capsules compressed, gland-ciliate.
Flowering & Fruiting	November-January.

Habitat Ecology	Near water streams; **Sir Khad -** 620 m.
Material Examined	EBH-WL-1078; 24.12.2008.
Part Used	Whole Plant.
Folk Uses	Bruised herb applied for healing **burns**. Gargles of its root decoction good for **throat infection**.
Chemical Constituents	Benzoic, p-hydroxybenzioc, protocatechuic, caffeic, vanillic, ferulic, isoferulic and p-coumaric acids, rhinanthin, aucuboside, catalpol and its acyl derivatives, glucose, fructose and sucrose have been identified **(herb)**.
Uses in Literature	Described earlier for boils, burns, epilepsy, fever, scurby, impurity of blood, scrofulous affections, skin diseases, headache, healing burns, ulcers, whitlows, mitigation of swollen piles; and as antiscorbutic **(Ambasta, 1986; Chopra et al., 1956; Jain, 1991; Kaushik & Dhiman, 2000; Kumar & Narain, 2010; Pande et al., 2006; Rana et al., 2003; Rastogi & Mehrotra, 1993).**

Vicia sativa L. (Plate No. 40C)

Family   Fabaceae

Vern.	Bada Kaer, Roothi.
English, Hindi and Regional Names	
Eng.	Common vetch;
Hindi	Akra, Ankra;
B.	Ankari;
Oriya	Choni, Rothi.
Distribution	Throughout India.
Description	Annual herbs having glabrous or obscurely downy suberect stems. Stipules small, deeply toothed. Inflorescence in pairs on short peduncles. Flowers solitary, in clusters. Corolla red-blue. Pods glabrescent, long, 8-10 seeded.
Flowering & Fruiting	January-February.
Habitat Ecology	Grassy hill slopes; **Gobind Sagar Lake -** 620m.
Material Examined	EBH-WL-1072; 04.02.2009.
Part Used	Whole Plant. Seeds.
Folk Use	Plant used as a green **feed** for livestock. Grounded seeds used for **culinary** preparations like 'pakoras'.

Chemical Constituent	**Leaves** contain total sterols, free sterols (1.22), esterified sterols rich in 7-dehydrostigmasterol; (0.18), sterol glucosides (0.063), vicianine and HCN **(seeds)**.
Biological Activity	Seeds show specific haemagglutinating activity.
Uses in Literature	So far reported as astringent, detergent, edible (seeds), fodder of high protein value, pot herb; and for diarrhoea **(Ambasta, 1986; Kaul, 1997; Rana et al., 2003; Rastogi & Mehrotra, 1995b; Roy et al., 1998; Sharma & Rana, 2005; Singh & Kumar, 2000a).**

Viola pilosa **Blume (Plate No. 40D)**

Syn.   *V. serpens* Wall. *ex* Roxb

Family   Violaceae

Vern.	Banapsha.
Hindi and Regional Names	
Hindi	Banafsha, Thungtu;
P.	Banafsha.
Distribution	India, hilly districts.
Description	Small, stoloniferous glabrous herbs with toothed stipules, lilac flowers and racemose inflorescence. Bracteoles 2. Fruits a capsule. Seed carunculate.
Flowering & Fruiting	February-April.
Habitat Ecology	Moist shady slopes; **Karyal Khad -** 615 m.
Material Examined	EBH-WL-1163; 06.04.2009.
Parts Used	Aerial Plant Parts. Flowers. Root.
Folk Uses	Decoction of leaves, stem and flowers effectively cures **cough, cold** and **chest affections**; 2 tsp thrice a day for 7-10 days. Flowers **eaten** as such and also used for **flavouring** tea. Decoction of root good for **vaginal discharges** (10-15ml, twice a day, 15 days).
Chemical Constituents	Therapeutic properties due to rutin, violin, saliculic acid **(herb)**, rutin, quercetin, violanthin, violaxanthin, p-hydroxycinnimic acid, delphinidin, C-glycoside-saponin and 15-cis-violaxanthin **(flowers)**.
Biological Activity	Aq. extract of flower exhibits antibiotic properties.
Uses in Literature	Recorded in India as an antipyretic, demulcent, diaphoretic, emetic, emollient, febrifuge and purgative; and for biliousness, cold, cough and lung diseases **(Ambasta, 1986; Aswal, 1996; Chauhan, 1999; Kala & Rawat, 2001; Krishna & Singh, 1987; Lal et al., 1996; Nautiyal, 1981; Pande et al., 2006; Shah, 1997; Sood & Thakur, 2004).**

Withania somnifera Dunal (Plate No. 40E)

Family   Solanaceae

Vern.	Ashvagandha.

English, Hindi, Sanskrit and Regional Names

Eng.	Winter cherry;
Hindi	Asgand, Punir;
Sans.	Ashvagandha, Asvakandika, Ashvarodha, Pillivendramu, Varahakarni;
Kan.	Viremadddinagaddi, Kiremallinagida;
Mal.	Amukkuram;
Mar.	Askandha, Kanchuki;
Oriya	Asugandha;
P.	Ak, Aksan, Asgand, Asgandnagori, Isgand;
Tam.	Amukkira, Asubam, Asuvagandhi;
Tel.	Asvagandhi, Dommadolu, Pennerce, Pillivendramu.
Distribution	Throughout drier subtropical India; frequent in west.
Description	Thinly woolly unarmed erect shrubs having round branches, entire leaves and hermaphrodite, greenish flowers. Corolla 3-6. Stamens 5. Ovary 2-celled. Berries globose, many seeded.
Flowering & Fruiting	September-October.
Habitat Ecology	Wastelands, common near wet places; **Gobind Sagar Lake -** 490 m.
Material Examined	EBH-WL-1164; 28.09.2007.
Parts Used	Roots. Leaves. Berries. Tubers.
Folk Uses	Bruised green berries rubbed to cure **ringworm** in both humans and animals. Paste of its root applied for **snake bite** and **scorpion-sting**. 15-20 ml decoction of the tuber taken thrice daily till **cure for asthma**

and **bronchitis**. Poultice of leaves good against **tumours, tuberculous glands, painful swellings** and **sore eyes**.

Chemical Constituents

Properties associated due to 5, 20α-dihydroxy-6α, 7α-epoxy-1-oxowitha-2, 24-dienolide **(roots)**, nine new steroidal lactones-withanolides E,F,G,H,I,J,K,L and M- **(leaves)**; seven of these characterized as 20-hydroxy-1-oxo-20 R, 22 R- witha- 2,5,8 (14), 24- tetraenolide, 20, 27 –dihyroxy-1-oxo- 20 R, 22 R- witha- 3,5,8(14), 24-tetraenolide (withanolid I), 17, 20- dihydroxy-1-oxo-20s, 22 R- witha-2,5,8 (14), 24-tetraenolide (withanolide J), 17, 20-dihyroxy-1-oxo-20s, 22R-with a-2,5,14,24-tetraenolide (withanolide L) and 17,20-dihydroxy-1-oxo-14, 15α-epoxy-20s, 22R- with a- 2,5,24-trienolide (withanolide M) and another withanolide-WS-1 **(seeds)**.

Biological Activity

CNS active, antihepatotoxic, antibacterial, antiviral, bradycardic, diuretic, hepatoprotective, hyponotic, phagocytotic, proteolytic, sedative and vermifugal activities exhibited.

Uses in Literature

Reportedly described as adaptogenic, antiarthritic, antiinflammatory, antispasmodic, antitumourous, hypotensive, hypnotic, rejuvenatory, respiratory stimulant; and for arthritis, asthma, brabycardic, cataract, eczema, itches, rectal diseases, tuberculosis, urinary troubles and white patches **(Ambasta, 1986; Bhatt et al., 1999; Bhogaonkar & Kanerkar, 2007; Biswas et al., 2010; Jain, 1991; Patil et al., 2007; Parabia & Pathak, 2007; Kamble et al., 2010; Kaushik & Dhiman, 2000; Kirtikar & Basu, 1984; Khan & Khanum, 2005; Pande et al., 2006; Parrotta, 2001; Prajapati et al., 2006; Prakasha et al., 2010; Rana et al., 2003; Rastogi & Mehrotra, 1991; Retnam & Martin, 2006; Rout & Panda, 2010; Saini, 1996a; Shukla & Verma, 1996; Yoganarasimhan, 1996, Yasodamma et al., 2009).**

Xanthium strumarium L. (Plate No. 40F)

Syn. *X. indicum* DC.; *X. roxburghii* Wallr.; *X. discolor* Wallr.; *X. brevirostre* Wallr.; *X. orientale* Blume

Family Asteraceae

Vern.	Chinjadoo.

English, Hindi, Sanskrit and Regional Names

Eng.	Bur-weed, Cockle bur;
Hindi	Banakara, Chhotagokhru;
Sans.	Arishta, Chanda, Itara, Pitapushpi;
Ass.	Agara;
B.	Banokra;
Bo.	Shankeshvara;
G.	Gadriyun;
H.	Bankra, Chhotgokhru;
Kash.	Lannetsuru;
Mal.	Buah anjang;
Mar.	Dumundi, Sankeshwara;
P.	Chirru, Gudal;
Tam.	Marlumutta;
Tel.	Marulamatangi, Parsvapu.

Distribution	Throughout hotter parts of India.
Description	Coarse unarmed annual herbs with 3-lobed leaves and greenish-white heads in terminal axillary racemes. Achenes clothed with strong hooked spines.
Flowering & Fruiting	September-November.

Habitat Ecology	Open slopes, wastelands, roadsides; **Matwana Khad -** 610 m; **Marol Khad -** 610 m.
Material Examined	EBH-WL-1073; 16.11.2008.
Parts Used	Aerial Parts. Seeds.
Folk Uses	An infusion of aerial plant parts applied for treating **rheumatism**. Seeds eaten raw to **cure** headache.
Chemical Constituents	Found to contain sesquiterpene lactones, xanthin, 4-oxo-bedfordia acid, hydroquinone, caffeoylquinic acids **(aerial parts)** and hydroquinone, choline, iodine **(seed)**.
Biological Activity	Roots hypoglycaemic, antiinflammatory and anti-tumourous.
Uses in Literature	Herb reported earlier as diaphoretic, diuretic, emollient, pot herb, sedative and sudorific; and for antimicrobial activity, boils, cancer wounds, cooling, cuts and wounds, eye diseases, headache, herpes, inflammatory swellings, leucorrhoea, malaria, menorrhagia, night blindness, piles, rheumatism, ringworm, scrofula, snake-bite, toothache, ulcers and urinary complaints **(Ambasta, 1986; Bhalla et al., 1996; Bhogaonkar & Ahmed, 2007; Chatterjee & Pakrashi, 1997; Chauhan, 1999; Hajra & Baishya, 1997; Henry, 1999; Hosagoudar & Henry, 1996b; Hooker, 1872-97; Jain, 1991; Jain et al., 2010; Khan & Khanum, 2005; Khanna et al., 1996; Kirtikar & Basu, 1984; Meena & Yadav, 2010; Pande et al., 2006; Parrotta, 2001; Patil et al., 2007; Prajapati et al., 2006; Rana et al., 2003; Rastogi & Mehrotra, 1991; Singh et al., 1996a).**

CHAPTER 3

Conclusion and Prospects

The overall diversity comprises the use of 203 plant species belonging to 156 genera under 66 families with predominance of polypetalous group (Table 37) by the wetland populace of district Bilaspur, Himachal Pradesh (Figure. 5; Table 2). Of the taxa delineated the floristic composition is dominated by dicots with bulk share of having 73.89% representation (150 species, 116 genera; Tables 3, 5), followed by monocots (42 species, 32 genera; Table 6), pteridophytes (10 species, 7 genera: *Adiantum, Ampelopteris, Cheilanthes, Christella, Equisetum, Pteris, Marsilea*; Table 4) and bryophytes (1 species, 1 genus: *Marchantia*; Figure. 8). Among these, 174 species (85.71%) of herbs, 15 shrubs (7.38%), 10 climbers (4.92%), 9 undershrubs (4.43%), 6 trees (2.95%) and 2 lianas (0.98%) constitute the entire gamut of ethnobotanical wetland diversity (Figures. 6, 7; Tables 10-15), of which 15 species occur as wild and under cultivation (Table 8). Of these, 196 species have been recorded as wetland hydrophytes (WL), followed by 4 submerged anchored hydrophytes (SA), 2 suspended hydrophytes (SH) and 1 emergent hydrophytes (EA) (Table 19). Based on analysis of their life period, 120 species are classified into annuals (62.50%), 63 as perennials (32.81%), 3 as annuals/perennials (1.56%), 4 as biennials (2.08%) and 2 as perennials/annuals (1.04%) (Tables 7, 20). Phenologically, a large majority of collected taxa produce flowers and fruits predominantly during summers (51.23%), followed by spring (21.67%) and rainy (19.70%) season (Figures. 10,11; Table 18). Relatively, the monocots and dicots ratio is 1 : 3.57 (Figure. 8; Table 3). The dicotyledonous families are over 4 times larger than the monocotyledonous ones, the genera and species are more than 3.5 times the monocots (Figures. 8, 9). Correspondingly, the proportion of species belonging to monocots to dicots is 1 : 3.57, of genera 1 : 3.62 and of families is 1 : 4.8. However, the ratio of the total number of genera to species is low, i.e., 1 : 0.98 in comparison to information for the whole India, i.e., 1 : 7 (Hooker, 1872-1897), lending further support to the view point of Good (1964) that within the same floral region, flora of smaller areas and remote islands have lower genus-species ratio as ecologically diversity tends to be smaller. In overall, the maximum diversity of ethnobotanical usages of wetland families is represented by Asteraceae (21 species), Poaceae (19 species), Fabaceae (15 species), Polygonaceae (11 species), Lamiaceae (10 species), Acanthaceae (9 species), Amaranthaceae (8 species), Convolvulaceae (6 species), Cyperaceae (6 species), Euphorbiaceae (6 species), Solanaceae (6 species), Malvaceae (4 species), Naiadaceae (4 species), Urticaceae (4 species), etc., (Tables 31, 32). Relatively, the top fourteen families account for 63.54 % of the recorded species diversity from the study area (Figures. 12-14). Likewise, the relative percentage in taxa for predominant dicotyledonous families varies from 10.34 (Asteraceae) to 0.49 (28 families; Table 33) whereas for monocots it is 9.35 (Poaceae) to 0.49 (Cannaceae,

Liliaceae) (Table 34). Moreover, out of total 66 wetland families used presently, 5 families, viz., Araceae (2 species), Ceratophyllaceae (1 species), Hydrocharitaceae (1 species), Naiadaceae (4 species) and Marsileaceae (1 species) with predominace of monocots (Tables 9, 16, 17) are purely aquatic, i.e., all the species belonging to these families are aquatic. With regard to utilitarian genera, *Polygonum* (9 species), *Ipomoea* (5 species), *Amaranthus* (4 species), *Euphorbia* (4 species) and *Chenopodium* (3 species) for dicots (Figure. 15; Table 35) and *Cyperus* (5 species, Cyperaceae) for monocots (Figure. 16; Table 36) are predominantly employed by the wetland populace of the region, and overall accounting for 14.77% of the reported diversity. From the view point of fruit types, capsule is the most common type, followed by achenes (19.79%), nutlets (14.58%), caryopsis (9.89%), pods (9.89%), berries (6.25%), drupe (3.12%), etc. (Figure. 17; Tables 21-30).

Table 2   Total Number of Ethnobotanically Collected Wetland Species, Genera and Families of District Bilaspur (H.P.) Under Various Divisions of Plant Kingdom

S. No.	Divisions	Genera	Species	Families
1.	Dicots	116	150	48
2.	Monocots	32	42	**10**
3.	Bryophytes	1	1	1
4.	Pteridophytes	7	10	7

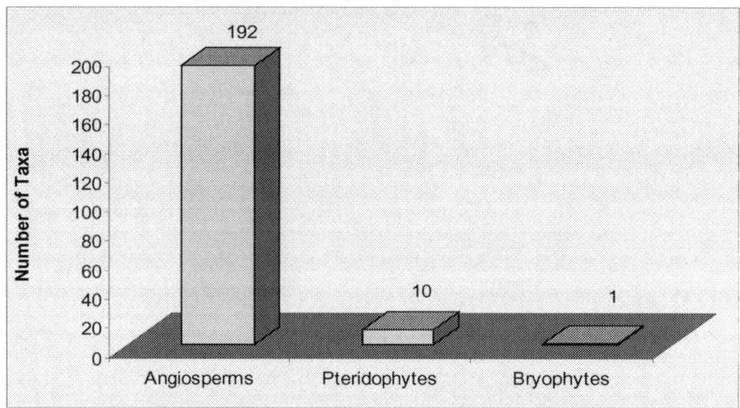

Figure 5 Histogram showing various divisions of ethnobotanically used wetland plants of District Bilaspur.

Table 3   Comparative Number of Wetland Genera, Species and Families of Monocots and Dicots Used by the Local Populace Inhabiting Nearby Areas of District Bilaspur

S.No.	Divisions	Genera	Species	Families
1.	Dicots	116	150	48
2.	Monocots	32	42	10

Table 4 Total Number of Plants Used by the Locals Inhabiting Fringe Areas of Wetlands in District Bilaspur

S.No.	Plant Kingdom	Number
1	Angiosperms	192
2	Bryophytes	1
3	Pteridophytes	10

Table 5 Ethnobotanically Important Wetland Dicots of District Bilaspur

Plants	Family
Abelmoschus crinitus Wall.	Malvaceae
Abrus precatorius L.	Fabaceae
Abutilon indicum L. Sweet	Malvaceae
Achyranthes aspera L.	Amaranthaceae
Aerva sanguinolenta (L.) Blume	Amaranthaceae
Aeschynomene aspera L.	Fabaceae
Ageratum conyzoides L.	Asteraceae
Ajuga bracteosa Wall. *ex* Benth.	Lamiaceae
Amaranthus gangeticus L. Syst.	Amaranthaceae
Amaranthus paniculatus L.	Amaranthaceae
Amaranthus tricolor (L.) var. *gangeticus* (L.) Fiori	Amaranthaceae
Amaranthus viridis L.	Amaranthaceae
Andrographis paniculata (Burm. f.) Wall. *ex* Nees	Acanthaceae
Anisomeles indica (L.) O. Kuntze	Lamiaceae
Argemone mexicana L.	Papaveraceae
Artemisia indica Waldst. & Kit.	Asteraceae
Artemisia scoparia Waldst. & Kit.	Asteraceae
Bacopa monnieri (L.) Pennell	Scrophulariaceae
Barleria cristata L.	Acanthaceae
Begonia picta Sm.	Begoniaceae
Bidens pilosa L.	Asteraceae
Boehmeria platyphylla Don	Urticaceae
Bryophyllum calycinum Salisb.	Crassulaceae
Calamintha umbrosum (M.B.) C. Koch	Lamiaceae
Cannabis sativa L.	Cannabinaceae
Capsella bursa-pastoris (L.) Medik.	Brassicaceae
Cardiospermum halicacobum L.	Sapindaceae
Cassia absus L.	Fabaceae
Cassia occidentalis L.	Fabaceae
Centella asiatica L.	Apiaceae
Ceratophyllum demersum L.	Ceratophyllaceae
Chenopodium album L.	Chenopodiaceae
Chenopodium ambrosiodes L.	Chenopodiaceae
Chenopodium murale L.	Chenopodiaceae
Cissampelos pareira L.	Menispermaceae
Clematis gouriana Roxb.	Ranunculaceae
Convolvulus arvensis L.	Convolvulaceae

Plants	Family
Conyza bonariensis L.	Asteraceae
Conyza stricta Willd.	Asteraceae
Crotolaria alata Buch.-Ham.	Fabaceae
Crotolaria mysorensis Roth	Fabaceae
Cryptolepis buchanani Roem. & Schult.	Asclepiadaceae
Cucumis pubescens Willd.	Cucurbitaceae
Cyathocline purpurea (Don) Kuntze	Asteraceae
Datura stramonium L.	Solanaceae
Debregeasia hypoleuca Wedd.	Urticaceae
Dicliptera roxburghiana Nees	Acanthaceae
Emilia sonchifolia (L.) DC.	Asteraceae
Eupatorium adenophorum Spreng.	Asteraceae
Euphorbia geniculata Ort. *ex* Boiss.	Euphorbiaceae
Euphorbia helioscopia L.	Euphorbiaceae
Euphorbia hirta L.	Euphorbiaceae
Euphorbia parviflora L.	Euphorbiaceae
Ficus hispida L.f.	Moraceae
Ficus roxburghii Wall.	Moraceae
Fragaria indica Andr.	Rosaceae
Fragaria nubicola Lindl.	Rosaceae
Fumaria indica (Hausskn.) Pugsley.	Fumariaceae
Galium aparine L.	Rubiaceae
Geranium nepalense Sweet	Geraniaceae
Girardinia heterophylla Decne.	Urticaceae
Gnaphalium pensylvanicum Willd.	Asteraceae
Gomphrena celosioides Mart.	Amaranthaceae
Hedera helix Clarke	Araliaceae
Impatiens balsamina L.	Balsaminaceae
Ipomoea cairica L.	Convolvulaceae
Ipomoea carnea Facq.	Convolvulaceae
Ipomoea muricata Jacq.	Convolvulaceae
Ipomoea nil (L.) Roth	Convolvulaceae
Ipomoea pestigridis L.	Convolvulaceae
Justicia simplex Don	Acanthaceae
Lactuca dissecta Don	Asteraceae
Lannea coromandelica (Houtt.) Merr.	Anacardiaceae
Lantana camara L.	Verbenaceae
Lathyrus aphaca L.	Fabaceae
Leea crispa Willd.	Leeaceae
Lepidagathis cuspidata Nees	Acanthaceae
Lespedeza sericea Miq.	Fabaceae
Lindernia ciliata Colsm.	Scrophulariaceae
Macrotyloma unifloruma (Lam.) Verdc.	Fabaceae
Martynia annua L.	Martyniaceae
Medicago denticulata Willd.	Fabaceae
Melothria heterophylla Cogn.	Cucurbitaceae
Mentha longifolia L.	Lamiaceae
Mentha piperita L.	Lamiaceae
Micromeria biflora (Buch.-Ham.) Benth.	Lamiaceae
Mirabilis jalapa L.	Nyctaginaceae
Momordica dioica Roxb. *ex* Willd.	Cucurbitaceae

Plants	Family
Mucuna pruriens DC.	Fabaceae
Nasturtium officinale R. Br.	Brassicaceae
Nerium indicum Mill.	Apocynaceae
Ocimum basilicum L.	Lamiaceae
Oroxylum indicum Vent.	Bignoniaceae
Oxalis corniculata L.	Oxalidaceae
Parthenium hysterophorus L.	Asteraceae
Phyllanthus urinaria L.	Euphorbiaceae
Physalis longifolia Nutt.	Solanaceae
Physalis minima L.	Solanaceae
Plectranthus coetsa Buch.-Ham. *ex* D. Don	Lamiaceae
Plumbago zeylanica L.	Plumbaginaceae
Pogostemon plectranthoides Desf.	Lamiaceae
Polygala arvensis Willd.	Polygalaceae
Polygonum barbatum L.	Polygonaceae
Polygonum barbatum L. sub sp. *gracile*Dansar.	Polygonaceae
Polygonum donii Meisn.	Polygonaceae
Polygonum glabrum Willd.	Polygonaceae
Polygonum lapathifolium L.	Polygonaceae
Polygonum minus Huds.	Polygonaceae
Polygonum plebejjum Br. Prodr.	Polygonaceae
Polygonum pulchrum Blume	Polygonaceae
Polygonum serrulatum Lag.	Polygonaceae
Pupalia lappacea Juss.	Amaranthaceae
Reinwardtia indica Dumort.	Linaceae
Rhynchoglossum obliquum Blume	Gesneriaceae
Ricinus communis L.	Euphorbiaceae
Roylea cinerea Baill.	Lamiaceae
Rubus ellipticus Sm.	Rosaceae
Ruellia patula Jacq.	Acanthaceae
Rumex hastatus Don	Polygonaceae
Rumex nepalensis Spreng.	Polygonaceae
Rungia pectinata (L.) Nees	Acanthaceae
Salix oxycarpa Anderss.	Salicaceae
Saussurea heteromala (Don) Hand.-Mazz.	Asteraceae
Sida acuta Burm. f.	Malvaceae
Silene conoidea L.	Caryophyllaceae
Solanum nigrum L.	Solanaceae
Solanum xanthocarpum Schrad. & Wendl.	Solanaceae
Sonchus arvensis L.	Asteraceae
Sonchus asper Hill	Asteraceae
Spilanthes acmella L. var. *oleracea* Jacq.	Asteraceae
Spilanthes paniculata Wall. *ex* DC.	Asteraceae
Stellaria media L.	Caryophyllaceae
Strobilanthes atropurpurens Nees	Acanthaceae
Strobilanthes dalhousianus Clarke	Acanthaceae
Taraxacum officinale Wigg	Asteraceae
Thalictrum reniforme Wall.	Ranunculaceae
Thevetia neriifolia Juss. *ex* Steud.	Apocynaceae
Trichodesma indicum Br.	Boraginaceae
Tridax procumbens L.	Asteraceae

Plants	Family
Trifolium resupinatum L.	Fabaceae
Trigonella pubescence Edgew.	Fabaceae
Uraria picta Desf.	Fabaceae
Urena lobata L.	Malvaceae
Urtica dioica L.	Urticaceae
Vernonia cinerea (L.) Less.	Asteraceae
Veronica anagalis-aquatica L.	Scrophulariaceae
Vicia sativa L.	Fabaceae
Viola pilosa Blume	Violaceae
Withania somnifera Dunal	Solanaceae
Xanthium strumarium L.	Asteraceae

Table 6 Ethnobotanically Important Wetland Monocots of District Bilaspur

Plants	Family
Acorus calamus L.	Araceae
Apluda mutica Auct.	Poaceae
Blyxa auberti Rich.	Hydrocharitaceae
Bromus catharticus Vahl	Poaceae
Canna indica L.	Cannaceae
Coix lachryma - jobi L.	Poaceae
Colocasia antiquorum Schott	Araceae
Commelina diffusa Burm. f.	Commelinaceae
Commelina paludosa Burm.	Commelinaceae
Costus speciosus Sm.	Zingiberaceae
Curcuma longa Wall.	Zingiberaceae
Cymbopogon citratus Stapf	Poaceae
Cymbopogon martinii Stapf	Poaceae
Cynodon dactylon (L.) Pers.	Poaceae
Cyperus compressus L.	Cyperaceae
Cyperus distans L.f.	Cyperaceae
Cyperus flabelliformis Rottb.	Cyperaceae
Cyperus iria L.	Cyperaceae
Cyperus rotundus L.	Cyperaceae
Dichanthium annulatum Hack.	Poaceae
Digitaria griffithii (Hook.f.) Henn.	Poaceae
Dioscorea belophylla Voigt. *ex* Haines.	Dioscoreaceae
Dioscorea bulbifera L.	Dioscoreaceae
Echinochloa frumentacea Link	Poaceae
Eleusine indica Gaertn.	Poaceae
Eriophorum comosum Wall.	Cyperaceae
Gloriosa superba L.	Liliaceae
Hedychium spicatum Buch.-Ham. *ex* Sm.	Zingiberaceae
Heteropogon contortus (L.) Beauv. *ex* Roem. & Schult.	Poaceae
Hydrilla verticillata (L.f.) Royle	Hydrocharitaceae
Najas graminea Dd.	Naiadaceae
Najas indica (Willd.) Cham.	Naiadaceae
Paspalum distichum L.	Poaceae
Pennisetum lanatum Klotzsch	Poaceae
Phragmites karka Roxb.	Poaceae

Plants	Family
Poa supina Schrad.	Poaceae
Potamogeton crispus L.	Naiadaceae
Potamogeton pectinatus L.	Naiadaceae
Saccharum munja L.	Poaceae
Saccharum spontanium L.	Poaceae
Setaria tomentosa Kunth	Poaceae
Sorghum halepense Wall.	Poaceae

Table 7   Growth Habits, Life Forms and Fruit Types of Ethnobotanically Important Wetland Angiospermic Plants of District Bilaspur

Name	Life Form	Growth Habit	Fruit Type
Abelmoschus crinitus Wall.	WL	A	Capsule
Abrus precatorius L.	WL	A	Pods
Abutilon indicum L. Sweet	WL	P	Capsule
Achyranthes aspera L.	WL	A	Capsules
Acorus calamus L.	WL	P	Berries
Aerva sanguinolenta (L.) Blume	WL	P	Capsules
Aeschynomene aspera L.	WL	P	Pods
Ageratum conyzoides L.	WL	A	Achenes
Ajuga bracteosa Wall. *ex* Benth.	WL	A	Nutlets
Amaranthus gangeticus L. Syst.	WL	A	Capsules
Amaranthus paniculatus L.	WL	A	Capsules
Amaranthus tricolor (L.) var. *gangeticus* (L.) Fiori	WL	A	Capsules
Amaranthus viridis L.	WL	A	Capsules
Andrographis paniculata (Burm.f.) Wall. *ex* Nees	WL	A	Capsules
Anisomeles indica (L.) O. Kuntze	WL	A	Nutlets
Apluda mutica Auct.	WL	P	Caryopsis
Argemone mexicana L.	WL	A	Capsules
Artemisia indica Waldst. & Kit.	WL	P	Achenes
Artemisia scoparia Waldst. & Kit.	WL	A	Achenes
Bacopa monnieri (L.) Pennell	WL	P/A	Capsules
Barleria cristata L.	WL	P	Capsules
Begonia picta Sm.	WL	A	Capsules
Bidens pilosa L.	WL	A	Achenes
Blyxa auberti Rich.	SH	A	Capsules
Boehmeria platyphylla Don	WL	P	Achenes
Bromus catharticus Vahl	WL	A	Caryopsis
Bryophyllum calycinum Salisb.	WL	P	Follicles
Calamintha umbrosum (M.B.) C. Koch	WL	P	Nutlets
Canna indica L.	WL	P	Capsule
Cannabis sativa L.	WL	A	Achenes
Capsella bursa-pastoris (L.) Medik.	WL	A	Pods
Cardiospermum halicacobum L.	WL	A	Capsules
Cassia absus L.	WL	A	Pods
Cassia occidentalis L.	WL	B	Pods
Centella asiatica L.	WL	A/P	Cremocarp
Ceratophyllum demersum L.	SH	P/A	Nutlets
Chenopodium album L.	WL	A	Nut
Chenopodium ambrosiodes L.	WL	A	Nut

Name	Life Form	Growth Habit	Fruit Type
Chenopodium murale L.	WL	A	Nut
Cissampelos pareira L.	WL	P	Drupe
Clematis gouriana Roxb.	WL	P	Achene
Coix lachryma - jobi L.	WL	A	Caryopsis
Colocasia antiquorum Schott	WL	B	Berry
Commelina diffusa Burm. f.	WL	A	Capsule
Commelina paludosa Burm.	WL	A	Capsules
Convolvulus arvensis L.	WL	A	Capsules
Conyza bonariensis L.	WL	P	Cypsella
Conyza stricta Willd.	WL	A	Cypsella
Costus speciosus Sm.	WL	A	Capsules
Crotolaria alata Buch.-Ham.	WL	P	Pods
Crotolaria mysorensis Roth	WL	A	Pods
Cryptolepis buchanani Roem. & Schult.	WL	P	Pods
Cucumis pubescens Willd.	WL	A	Berry
Curcuma longa Wall.	WL	B	Capsules
Cyathocline purpurea (Don) Kuntze	WL	A	Achenes
Cymbopogon citratus Stapf	WL	P	Caryopsis
Cymbopogon martinii Stapf	WL	P	Caryopsis
Cynodon dactylon (L.) Pers.	WL	P	Caryopsis
Cyperus compressus L.	WL	A	Achenes
Cyperus distans L.f.	WL	A	Achenes
Cyperus flabelliformis Rottb.	WL	P	Achenes
Cyperus iria L.	WL	A	Achenes
Cyperus rotundus L.	WL	A	Achenes
Datura stramonium L.	WL	A	Capsules
Debregeasia hypoleuca Wedd.	WL	P	Achenes
Dichanthium annulatum Hack.	WL	P	Caryopsis
Dicliptera roxburghiana Nees	WL	A	Capsules
Digitaria griffithii (Hook.f.) Henn.	WL	A	Caryopsis
Dioscorea belophylla Voigt. *ex* Haines.	WL	A	Capsules
Dioscorea bulbifera L.	WL	A	Capsules
Echinochloa frumentacea Link	WL	A	Caryopsis
Eleusine indica Gaertn.	WL	A	Caryopsis
Emilia sonchifolia (L.) DC.	WL	A	Achenes
Eriophorum comosum Wall.	WL	A	Nuts
Eupatorium adenophorum Spreng.	WL	A	Achenes
Euphorbia geniculata Ort. *ex* Boiss.	WL	A	Capsules
Euphorbia helioscopia L.	WL	A	Capsules
Euphorbia hirta L.	WL	A	Capsules
Euphorbia parviflora L.	WL	A	Capsules
Ficus hispida L.f.	WL	P	Achenes
Ficus roxburghii Wall.	WL	P	Achenes
Fragaria indica Andr.	WL	A	Achenes
Fragaria nubicola Lindl.	WL	A	Achenes
Fumaria indica (Hausskn.) Pugsley.	WL	A	Capsule
Galium aparine L.	WL	A	Capsule
Geranium nepalense Sweet	WL	A	Capsules
Girardinia heterophylla Decne.	WL	P	Achenes
Gloriosa superba L.	WL	A	Capsules
Gnaphalium pensylvanicum Willd.	WL	A	Achenes
Gomphrena celosioides Mart.	WL	A	Capsules

Name	Life Form	Growth Habit	Fruit Type
Hedera helix Clarke	WL	P	Berry
Hedychium spicatum Buch.-Ham. *ex* Sm.	WL	B	Capsules
Heteropogon contortus (L.) Beauv. *ex* Roem. & Schult.	WL	P	Caryopsis
Hydrilla verticillata (L.f.) Royle	EA	A	Capsules
Impatiens balsamina L.	WL	A	Capsules
Ipomoea cairica L.	WL	P	Capsules
Ipomoea carnea Facq.	WL	P	Capsules
Ipomoea muricata Jacq.	WL	A	Capsules
Ipomoea nil (L.) Roth	WL	A	Capsules
Ipomoea pestigridis L.	WL	A	Capsules
Justicia simplex Don	WL	A	Capsules
Lactuca dissecta Don	WL	A	Achenes
Lannea coromandelica (Houtt.) Merrill	WL	P	Drupe
Lantana camara L.	WL	P	Drupe
Lathyrus aphaca L.	WL	A	Pods
Leea crispa Willd.	WL	P	Berries
Lepidagathis cuspidata Nees	WL	P	Capsules
Lespedeza sericea Miq.	WL	P	Pods
Lindernia ciliata Colsm.	WL	A	Capsules
Macrotyloma unifloruma (Lam.) Verdc.	WL	A	Pods
Martynia annua L.	WL	A	Pods
Medicago denticulata Willd.	WL	A	Pods
Melothria heterophylla Cogn.	WL	P	Berry
Mentha longifolia L.	WL	P	Nutlets
Mentha piperita L.	WL	P	Nutlets
Micromeria biflora (Buch.-Ham.) Benth.	WL	P	Nutlets
Mirabilis jalapa L.	WL	A	Nutlets
Momordica dioica Roxb. *ex* Willd.	WL	A	Berry
Mucuna pruriens DC.	WL	A	Pods
Najas graminea Dd.	SA	A	Achenes
Najas indica (Willd.) Cham.	SA	A	Achenes
Nasturtium officinale R. Br.	WL	A	Pods
Nerium indicum Mill.	WL	P	Capsule
Ocimum basilicum L.	WL	A	Nutlets
Oroxylum indicum Vent.	WL	P	Capsules
Oxalis corniculata L.	WL	A	Capsules
Parthenium hysterophorus L.	WL	A	Achenes
Paspalum distichum L.	WL	A	Caryopsis
Pennisetum lanatum Klotzsch	WL	P	Caryopsis
Phragmites karka Roxb.	WL	P	Caryopsis
Phyllanthus urinaria L.	WL	A	Capsules
Physalis longifolia Nutt.	WL	A	Berries
Physalis minima L.	WL	A	Berries
Plectranthus coetsa Buch.-Ham. *ex* D. Don	WL	A	Nutlets
Plumbago zeylanica L.	WL	P	Capsules
Poa supina Schrad.	WL	P	Caryopsis
Pogostemon plectranthoides Desf.	WL	P	Nutlets
Polygala arvensis Willd.	WL	A	Capsules
Polygonum barbatum L.	WL	A	Nutlets
Polygonum barbatum L. sub sp. *gracile* Dansar.	WL	A	Nutlets
Polygonum donii Meisn.	WL	A	Nutlets
Polygonum glabrum Willd.	WL	A	Nutlets

Name	Life Form	Growth Habit	Fruit Type
Polygonum lapathifolium L.	WL	A	Nutlets
Polygonum minus Huds.	WL	A	Nutlets
Polygonum plebejjum Br. Prodr.	WL	A	Nutlets
Polygonum pulchrum Blume	WL	A	Nutlets
Polygonum serrulatum Lag.	WL	A	Nutlets
Potamogeton crispus L.	SA	A	Drupe
Potamogeton pectinatus L.	SA	A	Drupe
Pupalia lappacea Juss.	WL	P	Capsules
Reinwardtia indica Dumort.	WL	P	Capsules
Rhynchoglossum obliquum Blume	WL	A	Capsules
Ricinus communis L.	WL	P	Capsules
Roylea cinerea Baill.	WL	P	Nutlets
Rubus ellipticus Sm.	WL	P	Achenes
Ruellia patula Jacq.	WL	A	Capsules
Rumex hastatus Don	WL	A	Nuts
Rumex nepalensis Spreng.	WL	A	Nuts
Rungia pectinata (L.) Nees	WL	A	Capsules
Saccharum munja L.	WL	P	Caryopsis
Saccharum spontanium L.	WL	P	Caryopsis
Salix oxycarpa Anderss.	WL	P	Capsules
Saussurea heteromala (Don) Hand.-Mazz.	WL	P	Achenes
Setaria tomentosa Kunth	WL	A	Caryopsis
Sida acuta Burm.f.	WL	P	Capsules
Silene conoidea L.	WL	A	Capsules
Solanum nigrum L.	WL	A	Berries
Solanum xanthocarpum Schrad. &Wendl.	WL	P	Berries
Sonchus arvensis L.	WL	A	Achenes
Sonchus asper Hill	WL	A	Achenes
Sorghum halepense Wall.	WL	P	Caryopsis
Spilanthes acmella L.var. *oleracea* Jacq.	WL	A/P	Achenes
Spilanthes paniculata Wall. *ex* DC.	WL	A	Achenes
Stellaria media L.	WL	A	Capsules
Strobilanthes atropurpurens Nees	WL	P	Capsules
Strobilanthes dalhousianus Clarke	WL	P	Capsules
Taraxacum officinale Wigg	WL	A	Achenes
Thalictrum reniforme Wall.	WL	P	Achenes
Thevetia neriifolia Juss. *ex* Steud.	WL	P	Drupe
Trichodesma indicum Br.	WL	A	Nutlets
Tridax procumbens L.	WL	A	Achenes
Trifolium resupinatum L.	WL	A	Pods
Trigonella pubescence Edgew.	WL	A	Pods
Uraria picta Desf.	WL	P	Pods
Urena lobata L.	WL	A	Capsules
Urtica dioica L.	WL	A	Achenes
Vernonia cinerea (L.) Less.	WL	A	Achenes
Veronica anagalis-aquatica L.	WL	A/P	Capsules
Vicia sativa L.	WL	A	Pods
Viola pilosa Blume	WL	A	Capsules
Withania somnifera Dunal	WL	P	Berries
Xanthium strumarium L.	WL	A	Achenes

[Annual (A), Biennial (B), Perennial (P), Annual/ Perennial (A/P), Perennial/ Annual (P/A)]

Table 8   Wild as well as Cultivated Wetland Plants of District Bilaspur (H.P.)

Wild Plants	Plants occuring Wild and under Cultivation
Abelmoschus crinitus	*Acorus calamus*
Abrus precatorius	*Andrographis paniculata*
Abutilon indicum	*Canna indica*
Achyranthes aspera	*Cannabis sativa*
Acorus calamus	*Chenopodium album*
Adiantum capillus-veneris	*Colocasia antiquorum*
Adiantum incisum	*Cucumis pubescens*
Aerva sanguinolenta	*Cymbopogon martinii*
Aeschynomene aspera	*Echinochloa frumentacea*
Ageratum conyzoides	*Hedychium spicatum*
Ajuga bracteosa	*Mentha piperita*
Amaranthus gangeticus	*Mentha longifolia*
Amaranthus paniculatus	*Mirabilis jalapa*
Amaranthus tricolor var. *gangeticus*	*Spilanthes acmella* var. *oleracea*
Amaranthus viridis	*Withania somnifera*
Ampelopteris prolifera	
Andrographis paniculata	
Anisomeles indica	
Apluda mutica	
Argemone mexicana	
Artemisia indica	
Artemisia scoparia	
Bacopa monnieri	
Barleria cristata	
Begonia picta	
Bidens pilosa	
Blyxa auberti	
Boehmeria platyphylla	
Bromus catharticus	
Bryophyllum calycinum	
Calamintha umbrosum	
Canna indica	
Cannabis sativa	
Capsella bursa-pastoris	
Cardiospermum halicacobum	
Cassia absus	
Cassia occidentalis	
Centella asiatica	
Ceratophyllum demersum	
Cheilanthes bicolor	
Chenopodium album	
Chenopodium ambrosiodes	
Chenopodium murale	
Christella dentata	
Cissampelos pareira	
Clematis gouriana	
Coix lachryma - jobi	
Colocasia antiquorum	
Commelina diffusa	
Commelina paludosa	

Wild Plants	Plants occuring Wild and under Cultivation
Convolvulus arvensis	
Conyza bonariensis	
Conyza stricta	
Costus speciosus	
Crotolaria alata	
Crotolaria mysorensis	
Cryptolepis buchanani	
Cucumis pubescens	
Curcuma longa	
Cyathocline purpurea	
Cymbopogon citratus	
Cymbopogon martinii	
Cynodon dactylon	
Cyperus compressus	
Cyperus distans	
Cyperus flabelliformis	
Cyperus iria	
Cyperus rotundus	
Datura stramonium	
Debregeasia hypoleuca	
Dichanthium annulatum	
Dicliptera roxburghiana	
Digitaria griffithii	
Dioscorea belophylla	
Dioscorea bulbifera	
Echinochloa frumentacea	
Eleusine indica	
Emilia sonchifolia	
Equisetum arvense	
Equisetum debile	
Eriophorum comosum	
Eupatorium adenophorum	
Euphorbia geniculata	
Euphorbia helioscopia	
Euphorbia hirta	
Euphorbia parviflora	
Ficus hispida	
Ficus roxburghii	
Fragaria indica	
Fragaria nubicola	
Fumaria indica	
Galium aparine	
Geranium nepalense	
Girardinia heterophylla	
Gloriosa superba	
Gnaphalium pensylvanicum	
Gomphrena celosioides	
Hedera helix	
Hedychium spicatum	
Heteropogon contortus	
Hydrilla verticillata	
Impatiens balsamina	

Wild Plants	Plants occuring Wild and under Cultivation
Ipomoea cairica	
Ipomoea carnea	
Ipomoea muricata	
Ipomoea nil	
Ipomoea pestigridis	
Justicia simplex	
Lactuca dissecta	
Lannea coromandelica	
Lantana camara	
Lathyrus aphaca	
Leea crispa	
Lepidagathis cuspidata	
Lespedeza sericea	
Lindernia ciliata	
Macrotyloma unifloruma	
Marchantia palmata	
Marsilea minuta	
Martynia annua	
Medicago denticulata	
Melothria heterophylla	
Mentha longifolia	
Mentha piperita	
Micromeria biflora	
Mirabilis jalapa	
Momordica dioica	
Mucuna pruriens	
Najas graminea	
Najas indica	
Nasturtium officinale	
Nerium indicum	
Ocimum basilicum	
Oroxylum indicum	
Oxalis corniculata	
Parthenium hysterophorus	
Paspalum distichum	
Pennisetum lanatum	
Phragmites karka	
Phyllanthus urinaria	
Physalis longifolia	
Physalis minima	
Plectranthus coetsa	
Plumbago zeylanica	
Poa supina	
Pogostemon plectranthoides	
Polygala arvensis	
Polygonum barbatum	
Polygonum barbatum sub sp. *gracile*	
Polygonum donii	
Polygonum glabrum	
Polygonum lapathifolium	
Polygonum minus	
Polygonum plebejjum	

Wild Plants	**Plants occuring Wild and under Cultivation**
Polygonum pulchrum	
Polygonum serrulatum	
Potamogeton crispus	
Potamogeton pectinatus	
Pteris cretica	
Pupalia lappacea	
Reinwardtia indica	
Rhynchoglossum obliquum	
Ricinus communis	
Roylea cinerea	
Rubus ellipticus	
Ruellia patula	
Rumex hastatus	
Rumex nepalensis	
Rungia pectinata	
Saccharum munja	
Saccharum spontanium	
Salix oxycarpa	
Saussurea heteromala	
Selaginella chrysocaulos	
Setaria tomentosa	
Sida acuta	
Silene conoidea	
Solanum nigrum	
Solanum xanthocarpum	
Sonchus arvensis	
Sonchus asper	
Sorghum halepense	
Spilanthes acmella var. *oleracea*	
Spilanthes paniculata	
Stellaria media	
Strobilanthes atropurpurens	
Strobilanthes dalhousianus	
Taraxacum officinale	
Thalictrum reniforme	
Thevetia neriifolia	
Trichodesma indicum	
Tridax procumbens	
Trifolium resupinatum	
Trigonella pubescence	
Uraria picta	
Urena lobata	
Urtica dioica	
Vernonia cinerea	
Veronica anagalis-aquatica	
Vicia sativa	
Viola pilosa	
Withania somnifera	
Xanthium strumarium	

Table 9 Ethnobotanically Important Wetland and Aquatic Plant Resources of District Bilaspur

Wetland Plant	Aquatic Plant
Abelmoschus crinitus	*Blyxa auberti*
Abrus precatorius	*Ceratophyllum demersum*
Abutilon indicum	*Hydrilla verticillata*
Achyranthes aspera	*Marsilea minuta*
Acorus calamus	*Najas graminea*
Adiantum capillus-veneris	*Najas indica*
Adiantum incisum	*Potamogeton crispus*
Aerva sanguinolenta	*Potamogeton pectinatus*
Aeschynomene aspera	
Ageratum conyzoides	
Ajuga bracteosa	
Amaranthus gangeticus	
Amaranthus paniculatus	
Amaranthus tricolor var. *gangeticus*	
Amaranthus viridis	
Ampelopteris prolifera	
Andrographis paniculata	
Anisomeles indica	
Apluda mutica	
Argemone mexicana	
Artemisia indica	
Artemisia scoparia.	
Bacopa monnieri	
Barleria cristata	
Begonia picta	
Bidens pilosa	
Boehmeria platyphylla	
Bromus catharticus	
Bryophyllum calycinum	
Calamintha umbrosum	
Canna indica	
Cannabis sativa	
Capsella bursa-pastoris	
Cardiospermum halicacobum	
Cassia absus	
Cassia occidentalis	
Centella asiatica	
Cheilanthes bicolor	
Chenopodium album	
Chenopodium ambrosiodes	
Chenopodium murale	
Christella dentata.	
Cissampelos pareira	
Clematis gouriana	
Coix lachryma - jobi	
Colocasia antiquorum	
Commelina diffusa	
Commelina paludosa	
Convolvulus arvensis	
Conyza bonariensis	

Wetland Plant	Aquatic Plant
Conyza stricta	
Costus speciosus	
Crotolaria alata	
Crotolaria mysorensis	
Cryptolepis buchanani	
Cucumis pubescens	
Curcuma longa	
Cyathocline purpurea	
Cymbopogon citratus	
Cymbopogon martinii	
Cynodon dactylon	
Cyperus compressus	
Cyperus distans	
Cyperus flabelliformis	
Cyperus iria	
Cyperus rotundus	
Datura stramonium	
Debregeasia hypoleuca	
Dichanthium annulatum	
Dicliptera roxburghiana	
Digitaria griffithii	
Dioscorea belophylla	
Dioscorea bulbifera	
Echinochloa frumentacea	
Eleusine indica	
Emilia sonchifolia	
Equisetum arvense	
Equisetum debile	
Eriophorum comosum	
Eupatorium adenophorum	
Euphorbia geniculata	
Euphorbia helioscopia	
Euphorbia hirta	
Euphorbia parviflora	
Ficus hispida	
Ficus roxburghii	
Fragaria indica	
Fragaria nubicola	
Fumaria indica	
Galium aparine	
Geranium nepalense	
Girardinia heterophylla	
Gloriosa superba	
Gnaphalium pensylvanicum	
Gomphrena celosioides	
Hedera helix	
Hedychium spicatum	
Heteropogon contortus	
Impatiens balsamina	
Ipomoea cairica	
Ipomoea carnea	

Wetland Plant	Aquatic Plant
Ipomoea muricata	
Ipomoea nil	
Ipomoea pestigridis	
Justicia simplex	
Lactuca dissecta	
Lannea coromandelica	
Lantana camara	
Lathyrus aphaca	
Leea crispa	
Lepidagathis cuspidata	
Lespedeza sericea	
Lindernia ciliata	
Macrotyloma unifloruma	
Marchantia palmata	
Martynia annua	
Medicago denticulata	
Melothria heterophylla	
Mentha longifolia	
Mentha piperita	
Micromeria biflora	
Mirabilis jalapa	
Momordica dioica	
Mucuna pruriens	
Nasturtium officinale	
Nerium indicum	
Ocimum basilicum	
Oroxylum indicum	
Oxalis corniculata	
Parthenium hysterophorus	
Paspalum distichum	
Pennisetum lanatum	
Phragmites karka	
Phyllanthus urinaria	
Physalis longifolia	
Physalis minima	
Plectranthus coetsa	
Plumbago zeylanica	
Poa supina	
Pogostemon plectranthoides	
Polygala arvensis	
Polygonum barbatum	
Polygonum barbatum sub sp. *gracile*	
Polygonum donii	
Polygonum glabrum	
Polygonum lapathifolium	
Polygonum minus	
Polygonum plebejjum	
Polygonum pulchrum	
Polygonum serrulatum	
Pteris cretica	
Pupalia lappacea	
Reinwardtia indica	

Wetland Plant	Aquatic Plant
Rhynchoglossum obliquum	
Ricinus communis	
Roylea cinerea	
Rubus ellipticus	
Ruellia patula	
Rumex hastatus	
Rumex nepalensis	
Rungia pectinata	
Saccharum munja	
Saccharum spontanium	
Salix oxycarpa	
Saussurea heteromala	
Selaginella chrysocaulos	
Setaria tomentosa	
Sida acuta	
Silene conoidea	
Solanum nigrum	
Solanum xanthocarpum	
Sonchus arvensis	
Sonchus asper	
Sorghum halepense	
Spilanthes acmella var. *oleracea*	
Spilanthes paniculata	
Stellaria media	
Strobilanthes atropurpurens	
Strobilanthes dalhousianus	
Taraxacum officinale	
Thalictrum reniforme	
Thevetia neriifolia	
Trichodesma indicum	
Tridax procumbens	
Trifolium resupinatum	
Trigonella pubescence	
Uraria picta	
Urena lobata	
Urtica dioica	
Vernonia cinerea	
Veronica anagalis-aquatica	
Vicia sativa	
Viola pilosa	
Withania somnifera	
Xanthium strumarium	

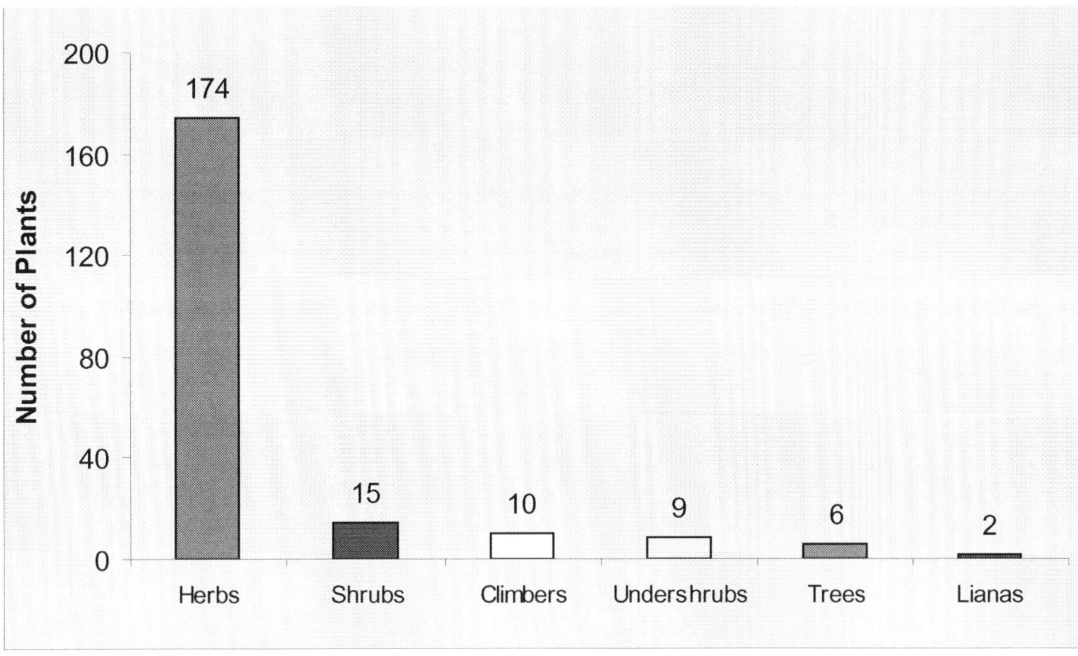

Figure 6 Histogram showing total number of wetland herbs, shrubs, climbers, undershrubs, trees and lianas used in District Bilaspur.

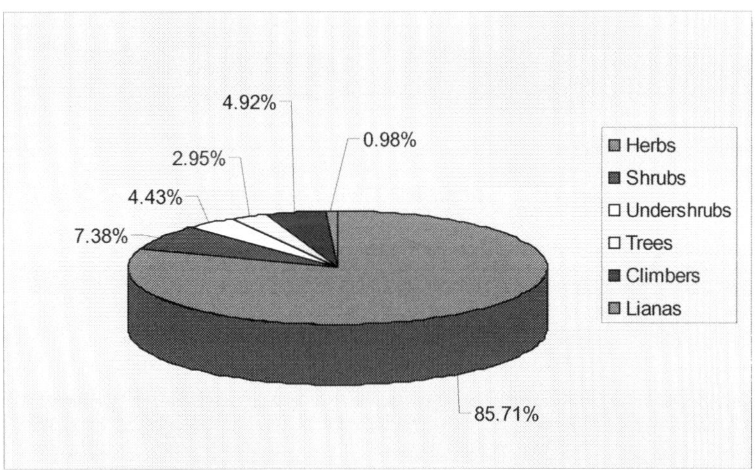

Figure 7 Relative percentage of ethnobotanically collected wetland herbs, shrubs, undershrubs, trees, climbers and lianas used in District Bilaspur.

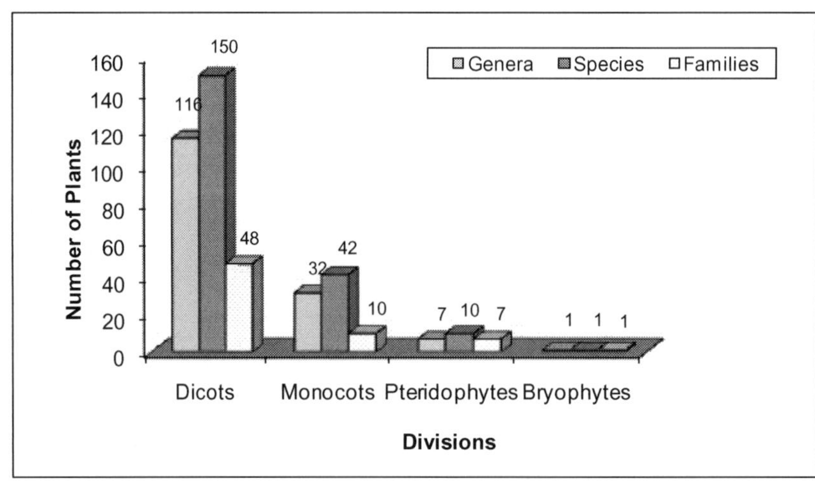

Figure 8 Histogram showing total number of ethnobotanically collected wetland species, genera and families under various divisions of plant kingdom.

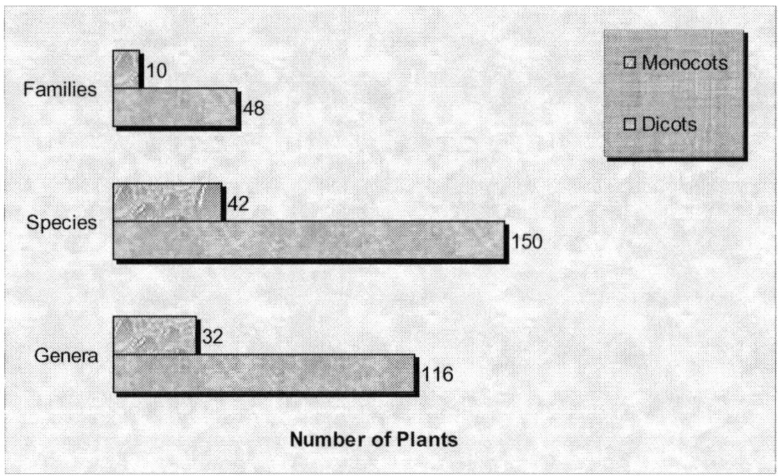

Figure 9 Histogram showing the comparative number of wetland genera, species and families of monocots and dicots used in District Bilaspur.

Table 10   Ethnobotanically Important Wetland Herbs of District Bilaspur (H.P.)

Botanical Name	Local Name/s
Abelmoschus crinitus	Beuli
Abutilon indicum	Jangli bhindi
Achyranthes aspera	Puthkanda, Lathjeera
Acorus calamus	Barae
Adiantum capillus-veneris	Vikrantaa
Adiantum incisum	Morpanki
Ageratum conyzoides	Ukal Booty, Phulnu
Ajuga bracteosa	Neelkanthi
Amaranthus gangeticus	Lal sag
Amaranthus paniculatus	Chollayi
Amaranthus tricolor var. *gangeticus*	Lal Sag
Amaranthus viridis	Cholai
Ampelopteris prolifera	Sena

Botanical Name	Local Name/s
Andrographis paniculata	Kalmegh
Anisomeles indica	Basinga
Apluda mutica	Chofki basar
Argemone mexicana	Chooly, Badi–Kandayi
Artemisia scoparia	Churisaroj
Bacopa monnieri	Jalnema
Begonia picta	Khattu
Bidens pilosa	Lamb
Blyxa auberti	Jaladhru
Bromus catharticus	Jangdu
Bryophyllum calycinum	Patharchatta
Calamintha umbrosum	Jangli-tulsi
Canna indica	Sudershan
Cannabis sativa	Bhang
Capsella bursa-pastoris	Seksi
Cardiospermum halicacobum	Fuka fucha
Cassia absus	Chakshu
Centella asiatica	Brahmi
Ceratophyllum demersum	Jalaj, Shaival
Cheilanthes bicolor	Shopa
Chenopodium album	Bathu
Chenopodium ambrosiodes	Bathu
Chenopodium murale	Baj-bhanga
Christella dentata	Kathu
Coix lachryma - jobi	Bajayanti mala
Colocasia antiquorum	Jangli Kachalu
Commelina diffusa	Kapala
Commelina paludosa	Chura
Convolvulus arvensis	Dudhua-bel
Conyza bonariensis	Shesherda
Conyza stricta	Hodari
Costus speciosus	Kedae ki challi
Crotolaria mysorensis	Sirsan
Curcuma longa	Ban-haldi
Cyathocline purpurea	Gandh-bhadra
Cymbopogon citratus	Makoda Ghas
Cymbopogon martinii	Rusha ghas, Palmarosa
Cynodon dactylon	Doob
Cyperus compressus	Tuli ghas
Cyperus distans	Ganeechi
Cyperus flabelliformis	Sola
Cyperus iria	Doob
Cyperus rotundus	Motha
Datura stramonium	Dhatura
Dichanthium annulatum	Kirdu
Dicliptera roxburghiana	Saundi
Digitaria griffithii	Ghuni
Echinochloa frumentacea	Madir
Eleusine indica	Rajputana
Emilia sonchifolia	Pooshathala
Equisetum arvense	Rugosika, Sehet bund
Equisetum debile	Fox tailed asperagus

Botanical Name	Local Name/s
Eriophorum comosum	Bagad
Eupatorium adenophorum	Kalibasuti
Euphorbia geniculata	Khabad-dudhia
Euphorbia helioscopia	Bada-dudhia
Euphorbia hirta	Bada-dudhia
Euphorbia parviflora	Lal Dudhi
Fragaria indica	Puin Aakha
Fragaria nubicola	Bhee-kaphal
Fumaria indica	Pitpapda
Galium aparine	Dhanpatri
Geranium nepalense	Bakra
Girardinia heterophylla	Badi-kogsi
Gnaphalium pensylvanicum	Sukaru
Gomphrena celosioides	Kangu
Hedychium spicatum	Ban haldi
Heteropogon contortus	Bandarpuncha
Hydrilla verticillata	Thangi, Jala
Impatiens balsamina	Teurya
Ipomoea cairica	Railway creeper
Ipomoea muricata	Ghaudan
Ipomoea nil	Ghaudan
Ipomoea pestigridis	Photial wagpadi
Justicia simplex	Juffa, Pitpapda
Lactuca dissecta	Khal
Lathyrus aphaca	Sudu
Lindernia ciliata	Gingu
Macrotyloma unifloruma	Kulthi
Marchantia palmate	Matakain
Marsilea minuta	Tripatre
Martynia annua	Kanv
Medicago denticulata	Khokani
Mentha longifolia	Jungli-pudina, Pudina
Mentha piperita	Pipermint
Micromeria biflora	Pushanbanda
Mirabilis jalapa	Shivkali
Najas graminea	Jalchida
Najas indica	Chu
Nasturtium officinale	Chuch
Ocimum basilicum	Bhabhri
Oxalis corniculata	Malora
Parthenium hysterophorus	Chikadu
Paspalum distichum	Chini ghas
Pennisetum lanatum	Baru
Phragmites karka	Baru
Phyllanthus urinaria	Pooin anvalah
Physalis longifolia	Popti
Physalis minima	Heui
Plumbago zeylanica	Cheeta
Poa supina	Chirua
Polygala arvensis	Bachhu
Polygonum barbatum	Katur
Polygonum barbatum sub sp. *gracile*	Rohini

Botanical Name	Local Name/s
Polygonum donii	Safed Rohini
Polygonum glabrum	Senu
Polygonum lapathifolium	Rohini
Polygonum minus	Bishkhulti
Polygonum plebejjum	Drabu, Muthisag
Polygonum pulchrum	Kyagnu
Polygonum serrulatum	Kathua
Potamogeton crispus	Jalaz
Potamogeton pectinatus	Jalaz
Pteris cretica	Bahupatre
Rhynchoglossum obliquum	Dushkanda
Ruellia patula	Baan
Rumex hastatus	Ambi
Rumex nepalensis	Jalbhangru
Rungia pectinata	Phuldali
Saccharum munja	Naal
Saccharum spontanium	Kosha
Saussurea heteromala	Mathaedi
Selaginella chrysocaulos	Sindoor
Setaria tomentosa	Khagori
Silene conoidea	Dhudu chana, Jangli chana
Solanum nigrum	Patkayai, Kale kyanun
Solanum xanthocarpum	Jangli bhindi
Sonchus arvensis	Sadhi
Sonchus asper	Bhursalae
Sorghum halepense	Barchita
Spilanthes acmella var. *oleracea*	Bada karkara
Spilanthes paniculata	Chota karkara
Stellaria media	Padyala
Strobilanthes atropurpurens	Kaseru
Taraxacum officinale	Kadavi, Sunachari
Thalictrum reniforme	Chaksu
Trichodesma indicum	Rukhadi
Tridax procumbens	Pugru
Trifolium resupinatum	Perseen
Trigonella pubescence	Jangli methi
Uraria picta	Dabra
Urena lobata	Badi-dredae
Urtica dioica	Kogsi
Vernonia cinerea	Sahdaiya
Veronica anagalis-aquatica	Sadevi
Vicia sativa	Bada Kaer, Roothi
Viola pilosa	Banapsha
Xanthium strumarium	Chinjadoo

Table 11   Ethnobotanically Important Wetland Shrubs of District Bilaspur (H.P.)

Botanical Name	Local Name
Aeschynomene aspera	Solu
Artemisia indica	Charmaar

Boehmeria platyphylla	Paryoon, Chamrala
Debregeasia hypoleuca	Shyaru
Ipomoea carnea	Naktibuti, Jablota
Lantana camara	Ujadu
Leea crispa	Ghangolae
Lepidagathis cuspidata	Puthkanda
Nerium indicum	Kaner, Ghanira
Pogostemon plectranthoides	Bhaerda
Rubus ellipticus	Akhae
Sida acuta	Bariara
Strobilanthes dalhousianus	Gathban
Thevetia neriifolia	Kaner
Withania somnifera	Ashvagandha

Table 12   Ethnobotanically Important Wetland Undershrubs of District Bilaspur (H P.)

Botanical Name	Local Name
Aerva sanguinolenta	Sufed-phulia
Barleria cristata	Morni
Cassia occidentalis	Badi-aeluan
Crotolaria alata	Jhunka
Lespedeza sericea	Binni
Plectranthus coetsa	Saru
Pupalia lappacea	Sena
Reinwardtia indica	Basant
Roylea cinerea	Kadvya

Table 13   Ethnobotanically Important Wetland Climbers of District Bilaspur (H.P.)

Botanical Name	Local Name/s
Abrus precatorius	Raten
Clematis gouriana	Bakerbel
Cryptolepis buchanani	Chink booti, Dudhli, Taern
Cucumis pubescens	Photnu
Dioscorea belophylla	Tardi
Dioscorea bulbifera	Dragal
Gloriosa superba	Langhi
Melothria heterophylla	Bhains
Momordica dioica	Meetha karela
Mucuna pruriens	Dryagal

Table 14   Ethnobotanically Important Wetland Lianas of District Bilaspur (H.P.)

Botanical Name	Local Name/s
Cissampelos pareira	Bhatindu
Hedera helix	Dakari

Table 15   Ethnobotanically Important Wetland Trees of District Bilaspur (H.P.)

Botanical Name	Local Name
Ficus hispida	Dabru
Ficus roxburghii	Trayambalu
Lannea coromandelica	Gujar
Oroxylum indicum	Arlu
Ricinus communis	Arand
Salix oxycarpa	Beeiuns

Table 16   Pure Aquatic Families of the Study Area

Family	No. of Species
Araceae	2
Ceratophyllaceae	1
Hydrocharitaceae	1
Naiadaceae	4
Marsileaceae	1

Table 17   Pure Aquatic Monocot Families of the Study Area

Family	No. of Species
Araceae	2
Hydrocharitaceae	1
Naiadaceae	4
Marsileaceae	1

Table 18 Flowering and Fruiting Seasons of Ethnobotanically Important Wetland Plants of District Bilaspur

Season	Plants	%age
Summer	104	51.23
Winter	15	07.38
Spring	44	21.67
Rainy	40	19.70

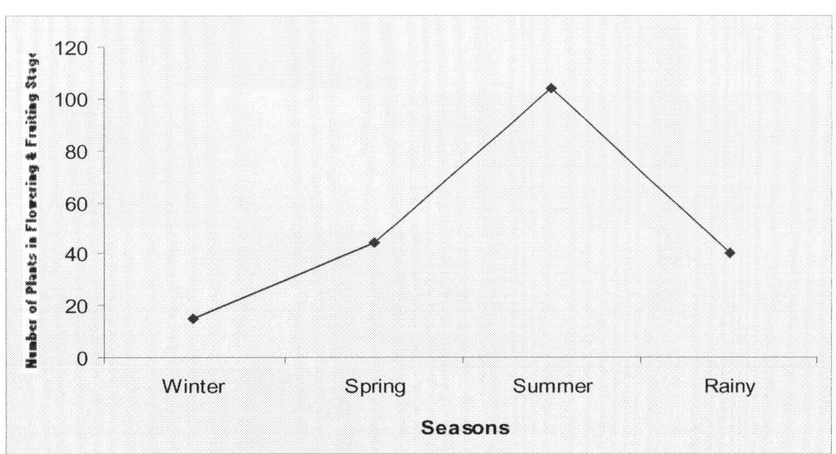

Figure 10   Flowering and fruiting seasons of wetland plants used ethnobotancally in District Bilaspur.

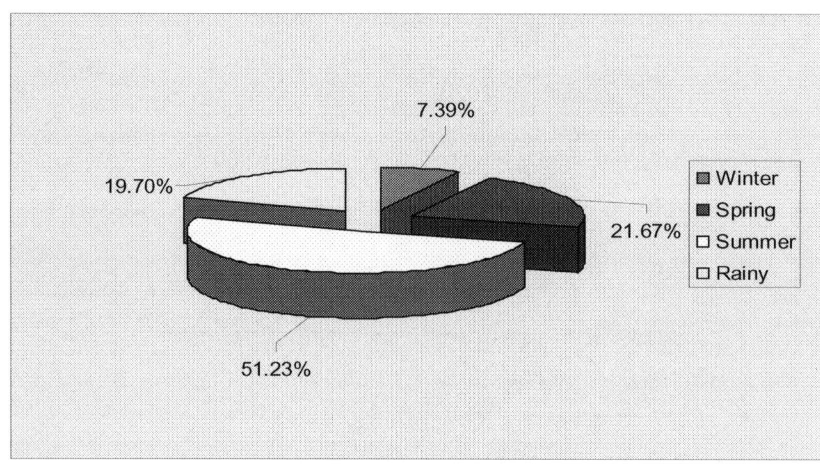

Figure 11 Pie diagram showing relative percentage of flowering and fruiting seasons of ethnobotanically collected wetland plants of District Bilaspur.

Table 19 Life Forms of Ethnobotanically Important Wetland Plants of the Study Area

Habitat	No. of Species
WL	196
SH	2
SA	4
EA	1

[Wetland Hydrophytes (WL), Suspended Hydrophytes (SH), Submerged Anchored Hydrophytes (SA), Emergent Hydrophytes (EA)]

Table 20 Growth Forms of Wetland Plant Resources of District Bilaspur

Growth Habit	No. of Species	%age of Total Species
Annual	120	62.50
Biennial	4	02.08
Perennial	63	32.81
Annual/ Perennial	3	1.56
Perennial/ Annual	2	1.04

Table 21 Fruit Types of Ethnobotanically Important Wetland Plant Resources of District Bilaspur

Fruit Types	No. of Species	%age of Fruit Type
Achenes	38	19.79
Berries	12	06.25
Capsules	66	34.37
Caryopsis	19	09.89
Cremocarp	01	00.52
Cypsella	02	01.04
Drupe	06	03.12
Follicles	01	00.52
Nutlets	28	14.58
Pods	19	09.89

Table 22   Ethnobotanically Important Wetland Angiosperms Having Achene Type of Fruits

Name	Life Form	Growth Habit
Ageratum conyzoides L.	WL	A
Artemisia indica Waldst. & Kit.	WL	P
Artemisia scoparia Waldst. & Kit.	WL	A
Bidens pilosa L.	WL	A
Boehmeria platyphylla Don	WL	P
Cannabis sativa L.	WL	A
Clematis gouriana Roxb.	WL	P
Cyathocline purpurea (Don) Kuntze	WL	A
Cyperus compressus L.	WL	A
Cyperus distans L.f.	WL	A
Cyperus flabelliformis Rottb.	WL	P
Cyperus iria L.	WL	A
Cyperus rotundus L.	WL	A
Debregeasia hypoleuca Wedd.	WL	P
Emilia sonchifolia (L.) DC.	WL	A
Eupatorium adenophorum Spreng.	WL	A
Ficus hispida L.f.	WL	P
Ficus roxburghii Wall.	WL	P
Fragaria indica Andr.	WL	A
Fragaria nubicola Lindl.	WL	A
Girardinia heterophylla Decne.	WL	P
Gnaphalium pensylvanicum Willd.	WL	A
Lactuca dissecta Don	WL	A
Najas graminea Dd.	A	A
Najas indica (Willd.) Cham.	A	A
Parthenium hysterophorus L.	WL	A
Rubus ellipticus Sm.	WL	P
Saussurea heteromala (Don) Hand.-Mazz.	WL	P
Sonchus arvensis L.	WL	A
Sonchus asper Hill	WL	A
Spilanthes acmella L.var. *oleracea* Jacq.	WL	A/P
Spilanthes paniculata Wall. *ex* DC.	WL	A
Taraxacum officinale Wigg	WL	A
Thalictrum reniforme Wall.	WL	P
Tridax procumbens L.	WL	A
Urtica dioica L.	WL	A
Vernonia cinerea (L.) Less.	WL	A
Xanthium strumarium L.	WL	A

Table 23   Ethnobotanically Important Wetland Angiosperms Having Berries

Name	Life Form	Growth Habit
Acorus calamus L.	WL	P
Colocasia antiquorum Schott	WL	B
Cucumis pubescens Willd.	WL	A
Hedera helix Clarke	WL	P
Leea crispa Willd.	WL	P
Melothria heterophylla Cogn.	WL	P
Momordica dioica Roxb. *ex* Willd.	WL	A
Physalis longifolia Nutt.	WL	A
Physalis minima L.	WL	A

Solanum nigrum L.	WL	A
Solanum xanthocarpum Schrad. & Wendl.	WL	P
Withania somnifera Dunal	WL	P

Table 24 Ethnobotanically Important Wetland Angiosperms Having Capsular Fruits

Name	Life Form	Growth Habit
Abelmoschus crinitus Wall.	WL	A
Abutilon indicum L. Sweet	WL	P
Canna indica L.	WL	P
Commelina diffusa Burm. f.	WL	A
Fumaria indica (Hausskn.) Pugsley.	WL	A
Galium aparine L.	WL	A
Nerium indicum Mill.	WL	P
Achyranthes aspera L.	WL	A
Aerva sanguinolenta (L.) Blume	WL	P
Amaranthus gangeticus L. Syst.	WL	A
Amaranthus paniculatus L.	WL	A
Amaranthus tricolor (L.) var. *gangeticus* (L.) Fiori	WL	A
Amaranthus viridis L.	WL	A
Andrographis paniculata (Burm. f.) Wall. *ex* Nees	WL	A
Argemone mexicana L.	WL	A
Bacopa monnieri (L.) Pennell	WL	A
Barleria cristata L.	WL	P
Begonia picta Sm.	WL	A
Blyxa auberti Rich.	A	A
Cardiospermum halicacobum L.	WL	A
Commelina paludosa Burm.	WL	A
Convolvulus arvensis L.	WL	A
Costus speciosus Sm.	WL	A
Curcuma longa Wall.	WL	B
Datura stramonium L.	WL	A
Dicliptera roxburghiana Nees	WL	A
Dioscorea belophylla Voigt. *ex* Haines.	WL	A
Dioscorea bulbifera L.	WL	A
Euphorbia geniculata Ort. *ex* Boiss.	WL	A
Euphorbia helioscopia L.	WL	A
Euphorbia hirta L.	WL	A
Euphorbia parviflora L.	WL	A
Geranium nepalense Sweet	WL	A
Gloriosa superba L.	WL	A
Gomphrena celosioides Mart.	WL	A
Hedychium spicatum Buch.-Ham. *ex* Sm.	WL	B
Hydrilla verticillata (L.f.) Royle	A	A
Impatiens balsamina L.	WL	A
Ipomoea cairica L.	WL	P
Ipomoea carnea Facq.	WL	P
Ipomoea muricata Jacq.	WL	A
Ipomoea nil (L.) Roth	WL	A
Ipomoea pestigridis L.	WL	A
Justicia simplex Don	WL	A
Lepidagathis cuspidata Nees	WL	P

Name	Life Form	Growth Habit
Lindernia ciliata Colsm.	WL	A
Oroxylum indicum Vent.	WL	P
Oxalis corniculata L.	WL	A
Phyllanthus urinaria L.	WL	A
Plumbago zeylanica L.	WL	P
Polygala arvensis Willd.	WL	A
Pupalia lappacea Juss.	WL	P
Reinwardtia indica Dumort.	WL	P
Rhynchoglossum obliquum Blume	WL	A
Ricinus communis L.	WL	P
Ruellia patula Jacq.	WL	A
Rungia pectinata (L.) Nees	WL	A
Salix oxycarpa Anderss.	WL	P
Sida acuta Burm.f.	WL	P
Silene conoidea L.	WL	A
Stellaria media L.	WL	A
Strobilanthes atropurpurens Nees	WL	P
Strobilanthes dalhousianus Clarke	WL	P
Urena lobata L.	WL	A
Veronica anagalis-aquatica L.	WL	A
Viola pilosa Blume	WL	A

Table 25 Ethnobotanically Important Wetland Angiosperms Having Caryopsis

Name	Life Form	Growth Habit
Apluda mutica Auct.	WL	P
Bromus catharticus Vahl	WL	A
Coix lachryma - jobi L.	WL	A
Cymbopogon citratus Stapf	WL	P
Cymbopogon martinii Stapf	WL	P
Cynodon dactylon (L.) Pers.	WL	P
Dichanthium annulatum Hack.	WL	P
Digitaria griffithii (Hook.f.) Henn.	WL	A
Echinochloa frumentacea Link	WL	A
Eleusine indica Gaertn.	WL	A
Heteropogon contortus (L.) Beauv. *ex* Roem. & Schult.	WL	P
Paspalum distichum L.	WL	A
Pennisetum lanatum Klotzsch	WL	P
Phragmites karka Roxb.	WL	P
Poa supina Schrad.	WL	P
Saccharum munja L.	WL	P
Saccharum spontanium L.	WL	P
Setaria tomentosa Kunth	WL	A
Sorghum halepense Wall.	WL	P

Table 26 Ethnobotanically Important Wetland Angiosperms Having Cremocarp and Cypsellar Fruits

Name	Life Form	Growth Habit
Centella asiatica L.	WL	A/P
Conyza bonariensis L.	WL	P
Conyza stricta Willd.	WL	A

Table 27   Ethnobotanically Important Wetland Angiosperms Having Nuts and Nutlets

Name	Life Form	Growth Habit
Ajuga bracteosa Wall. *ex* Benth.	WL	A
Anisomeles indica (L.) O. Kuntze	WL	A
Calamintha umbrosum (M.B.) C. Koch	WL	P
Ceratophyllum demersum L.	A	A
Chenopodium album L.	WL	A
Chenopodium ambrosiodes L.	WL	A
Chenopodium murale L.	WL	A
Eriophorum comosum Wall.	WL	A
Mentha longifolia L.	WL	P
Mentha piperita L.	WL	P
Micromeria biflora (Buch.-Ham.) Benth.	WL	P
Mirabilis jalapa L.	WL	A
Ocimum basilicum L.	WL	A
Plectranthus coetsa Buch.-Ham. *ex* D. Don	WL	A
Pogostemon plectranthoides Desf.	WL	P
Polygonum barbatum L.	WL	A
Polygonum barbatum L. sub sp. *gracile* Dansar.	WL	A
Polygonum donii Meisn.	WL	A
Polygonum glabrum Willd.	WL	A
Polygonum lapathifolium L.	WL	A
Polygonum minus Huds.	WL	A
Polygonum plebejjum Br. Prodr.	WL	A
Polygonum pulchrum Blume	WL	A
Polygonum serrulatum Lag.	WL	A
Roylea cinerea Baill.	WL	P
Rumex hastatus Don	WL	A
Rumex nepalensis Spreng.	WL	A
Trichodesma indicum Br.	WL	A

Table 28   Ethnobotanically Important Wetland Angiosperms Having Drupes

Name	Life Form	Growth Habit
Cissampelos pareira L.	WL	P
Lannea coromandelica (Houtt.) Merrill	WL	P
Lantana camara L.	WL	P
Potamogeton crispus L.	A	A
Potamogeton pectinatus L.	A	A
Thevetia neriifolia Juss. *ex* Steud.	WL	P

Table 29   Ethnobotanically Important Wetland Angiosperms Having Follicles

Name	Life Form	Growth Habit
Bryophyllum calycinum Salisb.	WL	P

Table 30   Ethnobotanically Important Wetland Angiosperms Having Pods

Name	Life Form	Growth Habit
Abrus precatorius L.	WL	A
Aeschynomene aspera L.	WL	P
Capsella bursa-pastoris (L.) Medik.	WL	A
Cassia absus L.	WL	A
Cassia occidentalis L.	WL	B
Crotolaria alata Buch.-Ham.	WL	P
Crotolaria mysorensis Roth	WL	A
Cryptolepis buchanani Roem. & Schult.	WL	P
Lathyrus aphaca L.	WL	A
Lespedeza sericea Miq.	WL	P
Macrotyloma unifloruma (Lam.) Verdc.	WL	A
Martynia annua L.	WL	A
Medicago denticulata Willd.	WL	A
Mucuna pruriens DC.	WL	A
Nasturtium officinale R. Br.	WL	A
Trifolium resupinatum L.	WL	A
Trigonella pubescence Edgew.	WL	A
Uraria picta Desf.	WL	P
Vicia sativa L.	WL	A

Table 31   Ethnobotanically Predominant Wetland Dicot Families of District Bilaspur

Family	No. of Species
Asteraceae	21
Fabaceae	15
Polygonaceae	11
Lamiaceae	10
Acanthaceae	9
Amaranthaceae	8
Convolvulaceae	6
Euphorbiaceae	6
Solanaceae	6
Malvaceae	4
Urticaceae	4
Chenopodiaceae	3
Cucurbitaceae	3
Rosaceae	3
Scrophulariaceae	3
Apocynaceae	2
Brassicaceae	2
Caryophyllaceae	2
Moraceae	2
Ranunculaceae	2
Anacardiaceae	1
Apiaceae	1
Araliaceae	1
Asclepiadaceae	1
Balsaminaceae	1

Family	No. of Species
Begoniaceae	1
Bignoniaceae	1
Boraginaceae	1
Cannabinaceae	1
Ceratophyllaceae	1
Crassulaceae	1
Fumariaceae	1
Geraniaceae	1
Gesneriaceae	1
Leeaceae	1
Linaceae	1
Martyniaceae	1
Menispermaceae	1
Nyctaginaceae	1
Oxalidaceae	1
Papaveraceae	1
Plumbaginaceae	1
Polygalaceae	1
Rubiaceae	1
Salicaceae	1
Sapindaceae	1
Verbenaceae	1
Violaceae	1

Table 32 Ethnobotanically Predominant Wetland Monocot Families of District Bilaspur

Family	No. of Species
Poaceae	19
Cyperaceae	6
Naiadaceae	4
Zingiberaceae	3
Araceae	2
Commelinaceae	2
Dioscoreaceae	2
Hydrocharitaceae	2
Cannaceae	1
Liliaceae	1

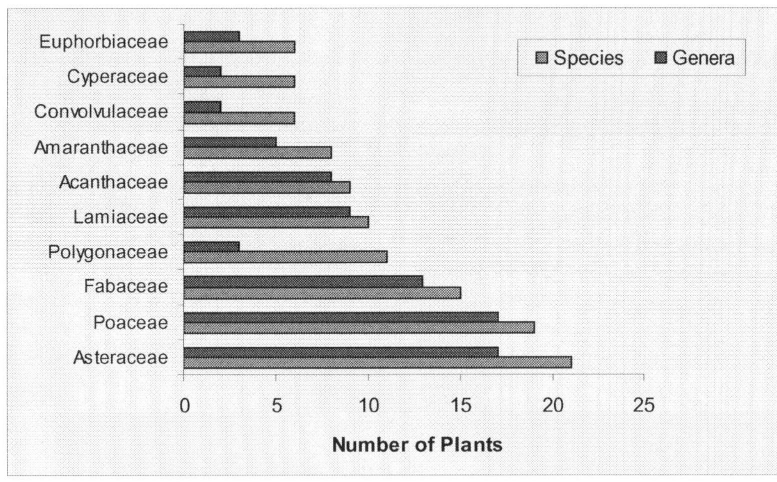

Figure 12   Relative disposition of wetland species and genera under predominant ethnobotanical families of District Bilaspur.

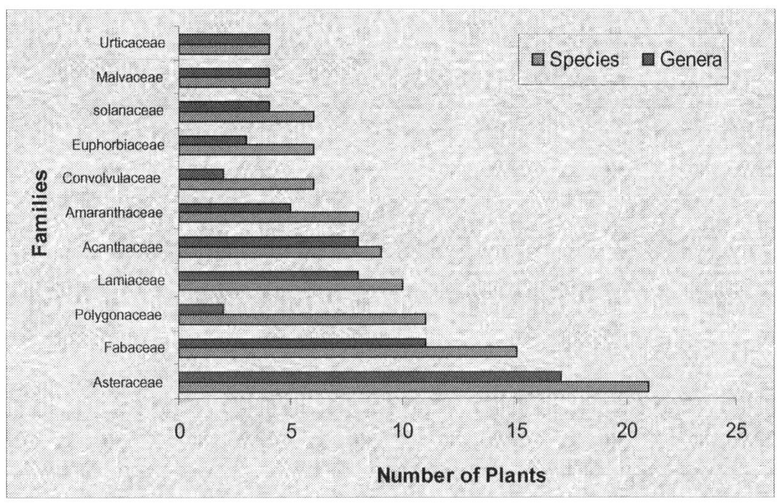

Figure 13   Predominant ethnobotanical wetland dicot families of District Bilaspur.

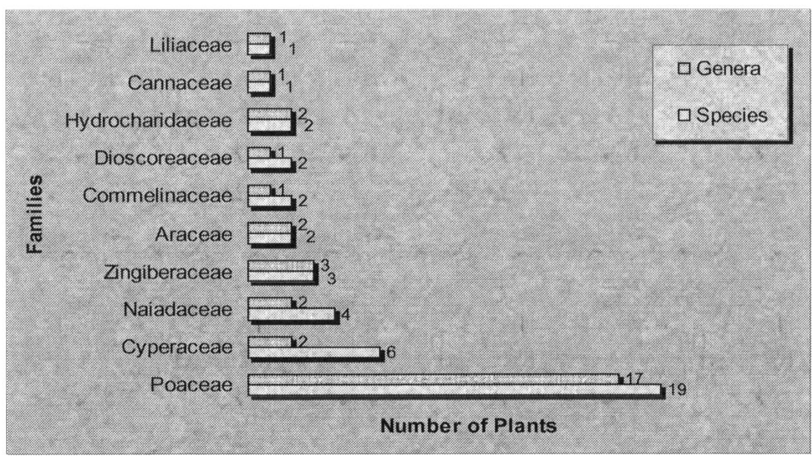

Figure 14   Predominant ethnobotanical wetland monocot families of District Bilaspur.

Table 33 Relative %age Disposition of Ethnobotanically Predominant Wetland Dicot Families of the Study Area

Family	Total No. of Species	Total % in Taxa
Asteraceae	21	10.34
Fabaceae	15	07.38
Polygonaceae	11	05.41
Lamiaceae	10	04.92
Acanthaceae	9	04.43
Amaranthaceae	8	03.94
Convolvulaceae	6	02.95
Euphorbiaceae	6	02.95
Solanaceae	6	02.95
Malvaceae	4	01.97
Urticaceae	4	01.97
Chenopodiaceae	3	01.47
Cucurbitaceae	3	01.47
Rosaceae	3	01.47
Scrophulariaceae	3	01.47
Apocynaceae	2	0.98
Brassicaceae	2	0.98
Caryophyllaceae	2	0.98
Moraceae	2	0.98
Ranunculaceae	2	0.98
Anacardiaceae	1	0.49
Apiaceae	1	0.49
Araliaceae	1	0.49
Asclepiadaceae	1	0.49
Balsaminaceae	1	0.49
Begoniaceae	1	0.49
Bignoniaceae	1	0.49
Boraginaceae	1	0.49
Cannabinaceae	1	0.49
Ceratophyllaceae	1	0.49
Crassulaceae	1	0.49
Fumariaceae	1	0.49
Geraniaceae	1	0.49
Gesneriaceae	1	0.49
Leeaceae	1	0.49
Linaceae	1	0.49
Martyniaceae	1	0.49
Menispermaceae	1	0.49
Nyctaginaceae	1	0.49
Oxalidaceae	1	0.49
Papaveraceae	1	0.49
Plumbaginaceae	1	0.49
Polygalaceae	1	0.49
Rubiaceae	1	0.49
Salicaceae	1	0.49
Sapindaceae	1	0.49
Verbenaceae	1	0.49
Violaceae	1	0.49

Table 34 Relative %age Disposition of Ethnobotanically Important Wetland Taxa Under Predominant Monocot Families of District Bilaspur

Family	Species Used	%age in Taxa
Poaceae	19	09.35
Cyperaceae	6	02.95
Naiadaceae	4	01.97
Zingiberaceae	3	01.47
Araceae	2	0.98
Commelinaceae	2	0.98
Dioscoreaceae	2	0.98
Hydrocharitaceae	2	0.98
Cannaceae	1	0.49
Liliaceae	1	0.49

Table 35 Ethnobotanically Predominant Wetland Dicot Genera of District Bilaspur

Genera	Family	No. of Species
Polygonum	Polygonaceae	9
Ipomoea	Convolvulaceae	5
Amaranthus	Amaranthaceae	4
Euphorbia	Euphorbiaceae	4
Chenopodium	Chenopodiaceae	3
Artemisia	Asteraceae	2
Cassia	Fabaceae	2
Conyza	Asteraceae	2
Crotolaria	Fabaceae	2
Ficus	Moraceae	2
Fragaria	Rosaceae	2
Mentha	Lamiaceae	2
Physalis	Solanaceae	2
Rumex	Polygonaceae	2
Solanum	Solanaceae	2
Sonchus	Asteraceae	2
Spilanthes	Asteraceae	2
Strobilanthes	Acanthaceae	2
Abelmoschus	Malvaceae	1
Abrus	Fabaceae	1
Abutilon	Malvaceae	1
Achyranthes	Amaranthaceae	1
Aerva	Amaranthaceae	1

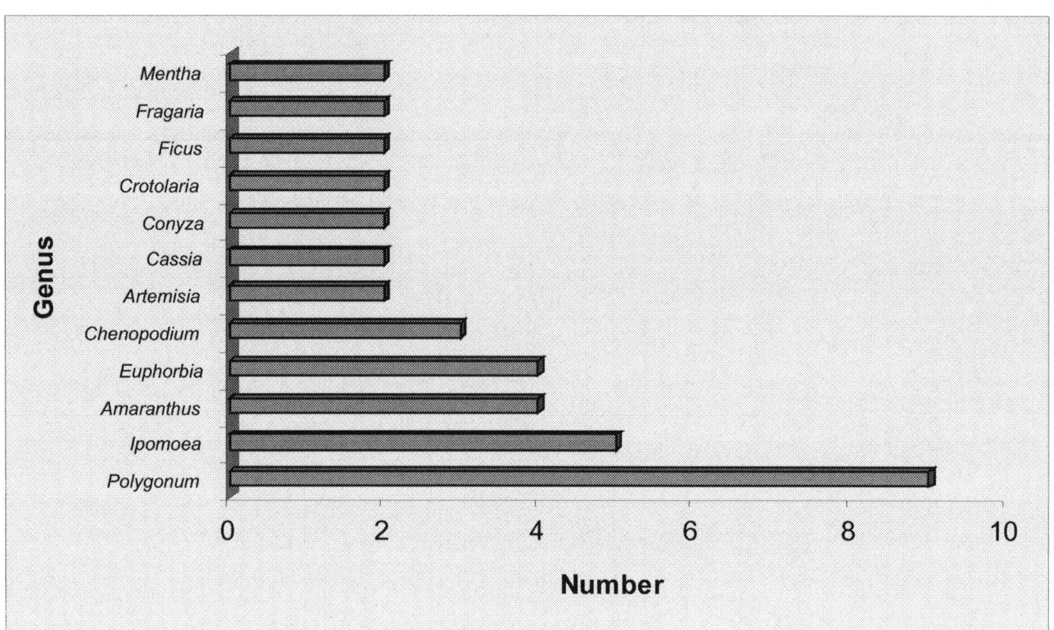

Figure 15   Predominant ethnobotanical wetland dicot genera of District Bilaspur.

Table 36   Ethnobotanically Predominant Wetland Monocot Genera of District Bilaspur

Genera	Family	No. of Species
Cyperus	Cyperaceae	5
Commelina	Commelinaceae	2
Cymbopogon	Poaceae	2
Dioscorea	Dioscoreaceae	2
Najas	Naiadaceae	2

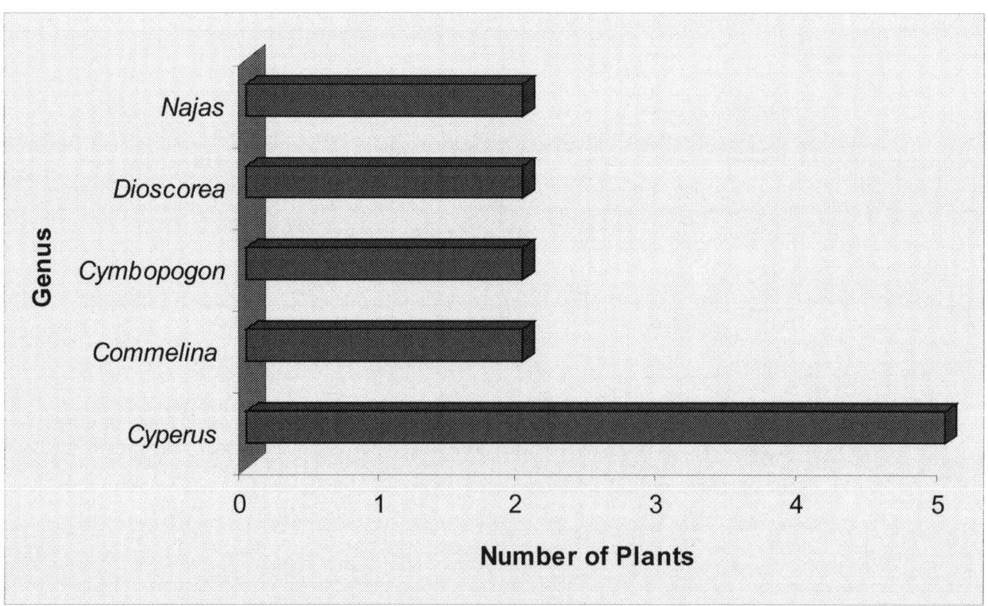

Figure 16   Predominant ethnobotanical wetland monocot genera of District Bilaspur.

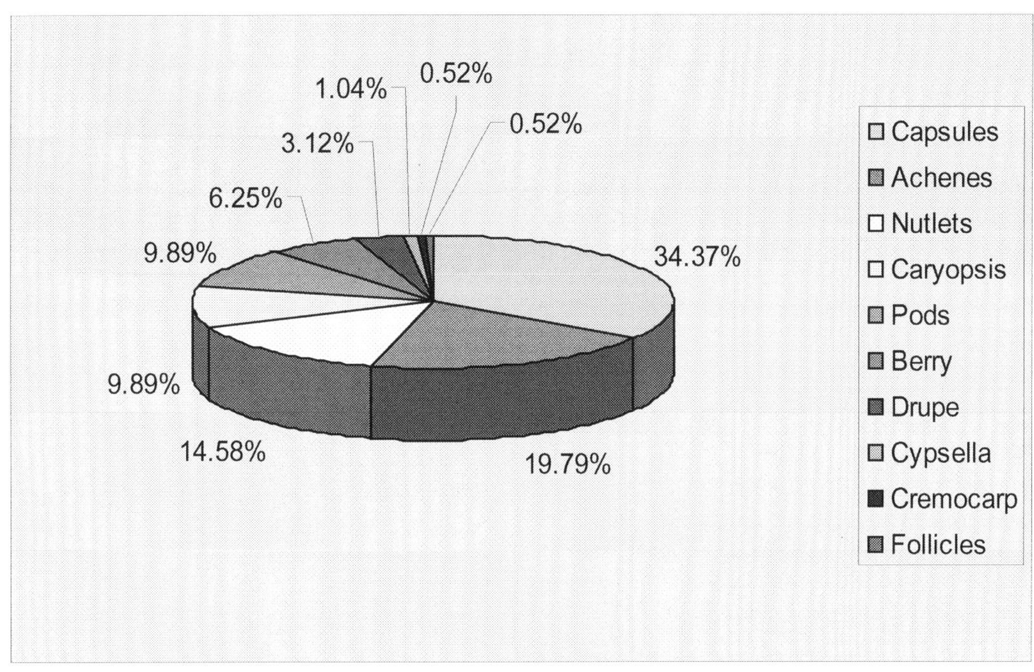

Figure 17 Relative percentage of wetland fruit types of District Bilaspur.

Table 37 Layout of Ethnobotanically Important Wetland Families of District Bilaspur as per Bentham & Hooker System of Classification (1862-1883)

Polypetalous Families	Gamopetalous Families	Monochlamdos Families	Monocot Families	Bryophytic Families	Pteridophytic Families
Anacardiaceae	Acanthaceae	Amaranthaceae	Araceae	Marchantiaceae	Adiantaceae
Apiaceae	Apocynaceae	Cannabinaceae	Cannaceae		Equisetaceae
Araliaceae	Asclepiadaceae	Ceratophyllaceae	Commelinaceae		Marsileaceae
Balsaminaceae	Asteraceae	Chenopodiaceae	Cyperaceae		Pteridaceae
Begoniaceae	Bignoniaceae	Euphorbiaceae	Dioscoreaceae		Selaginellaceae
Brassicaceae	Boraginaceae	Nyctaginaceae	Hydrocharitaceae		Sinopteridaceae
Caryophyllaceae	Convolvulaceae	Polygonaceae	Liliaceae		Thelypteridaceae
Crassulaceae	Gesneriaceae	Salicaceae	Naiadaceae		
Cucurbitaceae	Lamiaceae	Urticaceae	Poaceae		
Fabaceae	Martyniaceae	Moraceae	Zingiberaceae		
Fumariaceae	Plumbaginaceae				
Geraniaceae	Rubiaceae				
Leeaceae	Scrophulariaceae				
Linaceae	Solanaceae				
Malvaceae	Verbenaceae				
Menispermaceae					
Papaveraceae					
Oxalidaceae					
Polygalaceae					
Ranunculaceae					
Rosaceae					
Sapindaceae					
Violaceae					

During the survey of study area, it was observed that as many as 168 plant species belonging to 132 genera under 63 families were being used for curing about 68 types of ailments affecting the rural populace inhabiting fringe areas of wetlands of district Bilaspur (Appendix 3). As evidenced from the use, the maximum number of plants were used for skin ailments (36 species), followed by headaches (20), cough (18), gout/rheumatism/muscular pain (13), stomach ailments (16), toothaches (13), dysentery/diarrhoea, fever (11 each), boils/blisters/burns (10), bone fractures, bowel complaints and swellings (8 each), urinary disorders, sore throat/mouth infection (7 each), anaemia, liver disorders/jaundice, sores and ulcers (6 each), asthma/shortness of breath, body pains, bronchitis/chest affections (5 each), eye and ear ailments, epilepsy, kidney stones (4 each), cholera, cuts/wounds, flatulence, paralysis, piles, hair fall, snake bites, spermatorrhoea, (3 species each), etc., whereas incidence of other ailments/cure namely abdominal pain, antidote, antiseptic, appetiser, body massage, children ailments, diabetes, dropsy, expelling worms, eye lotion, gonorrhoea, hysteria, immunity, influenza, insect bites, insect repellent, intestinal worms, irritable ulcers, leucorrhoea, measles, memory loss, neuralgia, nose bleeding, pneumonia, convulsions, colic pain, regulating blood pressure, rickets, ringworm, slipped naval, sores, spleen disorders, tuberculosis, vaginal disorders, vomiting, weakness, etc., are less in the region as indicated relatively by the use of one or two plants for these ailments (Appendix 3). For alleviating these complaints/problems, various plant part/s used in order of preferance are: whole plant (99 species), leaves (91 species), roots (49 species), seed (31 species), stem (29 species), fruit (19 species), flower (11 species), aerial plant parts (10 species), etc. (Figure. 18; Appendix 4). commonly used families are: Asteraceae (21 species), Polygonaceae (11 species), Lamiaceae (10 species), Fabaceae (10 species), Acanthaceae (9 species), Poaceae (8 species), Amaranthaceae (6 species), Convolvulaceae (6 species), Euphorbiaceae (6 species), Solanaceae (6 species), Malvaceae (4 species), Urticaceae (4 species), Cucurbitaceae (3 species), Zingiberaceae (3 species), Rosaceae (3 species), Brassicaceae (2 species), Caryophyllaceae (2 species), Ranunculaceae (2 species), Scrophulariaceae (2 species), Thelypteridaceae (2 species), Adiantaceae (2 species), Chenopodiaceae (2 species), Cyperaceae (2 species), Equisetaceae (2 species), Moraceae (2 species), (Table 38). Likewise, commonly used genera employed for curing various ailments are : *Polygonum* (9 species), *Ipomoea* (5 species), *Amaranthus* (4 species), *Euphorbia* (4 species), *Adiantum* (2 species), *Artemisia* (2 species), *Cassia* (2 species), *Chenopodium* (2 species), *Conyza* (2 species), *Crotolaria* (2 species), *Cymbopogon* (2 species), *Cyperus* (2 species), *Equisetum* (2 species), *Ficus* (2 species), *Fragaria* (2 species), *Mentha* (2 species), *Physalis* (2 species), *Rumex* (2 species), *Solanum* (2 species), *Sonchus* (2 species), *Spilanthes* (2 species), *Strobilanthes* (2 species) (Tables 39, 40). Undeniably, the rural populace of the region has vast knowledge of the local plant wealth and preferred to modern system of medicine as these are considered cheap, safe and with no side effects. Moreover, the documentation of most of these remedies are hitherto unreported and worthy of consideration as perspective healthcare plants after phytochemical/pharmacological investigation and chemical evaluation leading to new drug compounds for the welfare of mankind.

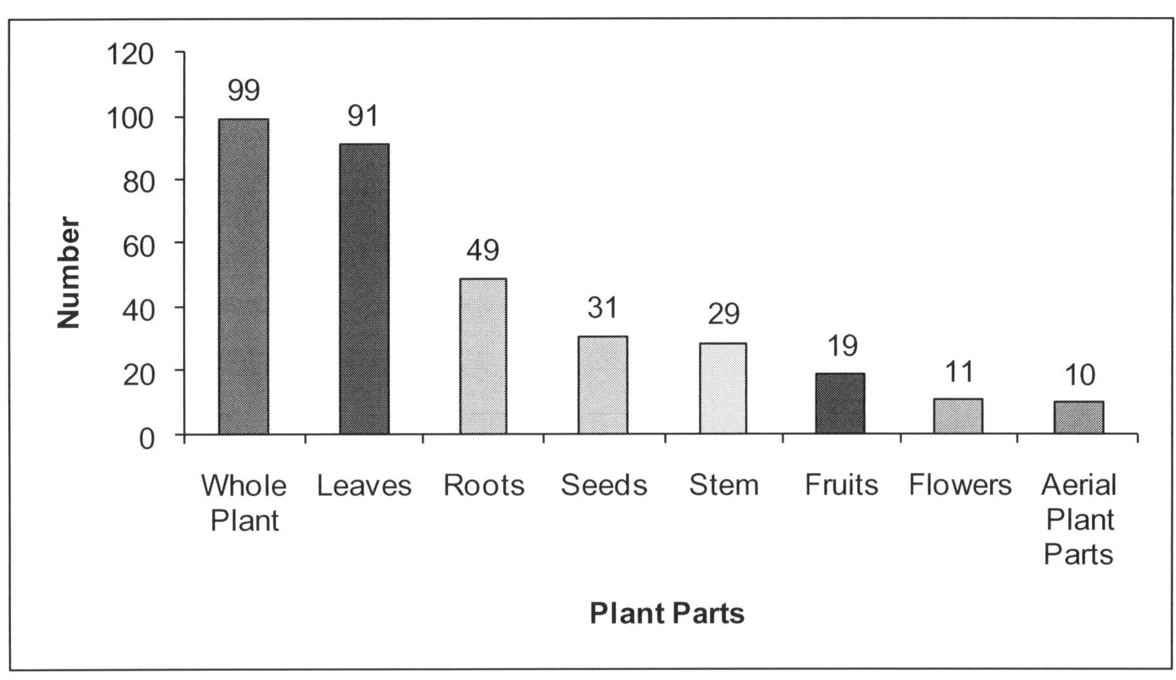

Figure 18 Histogram showing relative number of wetland plant parts used for medicinal purposes in District Bilaspur.

Table 38 Ethnobotanically Important Medicinal Plant Families of the Study Area

Family	Plant Names	Genera/ Species
Asteraceae	*Ageratum conyzoides, Artemisia indica, Artemisia scoparia, Bidens pilosa, Conyza bonariensis, Conyza stricta, Cyathocline purpurea, Emilia sonchifolia, Eupatorium adenophorum, Gnaphalium pensylvaticum, Lactuca dissecta, Parthenium hysterophorus, Saussurea heteromala, Sonchus arvensis, Sonchus asper, Spilanthes acmella* var. *oleracea, Spilanthes paniculata, Taraxacum officinale, Tridax procumbens, Vernonia cinerea, Xanthium strumarium*	17/21
Lamiaceae	*Ajuga bracteosa, Anisomeles indica, Calamintha umbrosum, Mentha longifolia, Mentha piperita, Micromeria biflora, Ocimum basilicum, Plectranthus coetsa, Pogostemon plectranthoides, Roylea cinerea*	9/10
Fabaceae	*Abrus precatorius, Cassia absus, Cassia occidentalis, Crotolaria alata, Crotolaria mysorensis, Lathyrus aphaca, Macrotyloma unifloruma, Medicago denticulata, Mucuna pruriens, Uraria picta*	8/10
Acanthaceae	*Andrographis paniculata, Barleria cristata, Dicliptera roxburghiana, Justicia simplex, Lepidagathis cuspidata, Ruellia patula, Rungia pectinata, Strobilanthes atropurpurens, Strobilanthes dalhousianus*	8/9
Poaceae	*Apluda mutica, Coix lachryma–jobi, Cymbopogon citratus, Cymbopogon martinii, Cynodon dactylon, Echinochloa frumentacea, Eleusine indica, Heteropogon contortus*	7/8
Solanaceae	*Datura stramonium, Physalis longifolia, Physalis minima, Solanum nigrum, Solanum xanthocarpum, Withania somnifera*	4/6
Malvaceae	*Abelmoschus crinitus, Abutilon indicum, Sida acuta, Urena lobata*	4/4
Urticaceae	*Boehmeria platyphylla, Debregeasia hypoleuca, Girardinia heterophylla, Urtica dioica,*	4/4

Family	Plant Names	Genera/Species
Polygonaceae	*Polygonum barbatum, Polygonum barbatum* sub sp. *gracile, Polygonum donii, Polygonum glabrum, Polygonum lapathifolium, Polygonum minus, Polygonum plebejjum, Polygonum pulchrum, Polygonum serrulatum, Rumex hastatus, Rumex nepalensis*	2/11
Amaranthaceae	*Achyranthes aspera, Amaranthus gangeticus, Amaranthus paniculatus, Amaranthus tricolor* var. *gangeticus, Amaranthus viridis, Pupalia lappacea*	3/6
Euphorbiaceae	*Euphorbia geniculata, Euphorbia helioscopia, Euphorbia hirta, Euphorbia parviflora, Phyllanthus urinaria, Ricinus communis*	3/6
Cucurbitaceae	*Cucumis pubescens, Melothria heterophylla, Momordica dioica*	3/3
Zingiberaceae	*Costus speciosus, Curcuma longa, Hedychium spicatum*	3/3
Convolvulaceae	*Convolvulus arvensis, Ipomoea cairica, Ipomoea carnea, Ipomoea muricata, Ipomoea nil, Ipomoea pestigridis,*	2/6
Rosaceae	*Fragaria indica, Fragaria nubicola, Rubus ellipticus*	2/3
Brassicaceae	*Capsella bursa-pastoris, Nasturtium officinale*	2/2
Caryophyllaceae	*Silene conoidea, Stellaria media*	2/2
Ranunculaceae	*Clematis gouriana, Thalictrum reniforme*	2/2
Scrophulariaceae	*Bacopa monnieri, Veronica anagalis-aquatica*	2/2
Thelypteridaceae	*Ampelopteris prolifera, Christella dentata*	2/2
Adiantaceae	*Adiantum capillus-veneris, Adiantum incisum*	1/2
Chenopodiaceae	*Chenopodium album, Chenopodium ambrosiodes*	1/2
Cyperaceae	*Cyperus iria, Cyperus rotundus*	1/2
Equisetaceae	*Equisetum arvense, Equisetum debile*	1/2
Moraceae	*Ficus hispida, Ficus roxburghii*	1/2
Anacardiaceae	*Lannea coromandelica*	1/1
Apiaceae	*Centella asiatica*	1/1
Apocynaceae	*Nerium indicum*	1/1
Araceae	*Acorus calamus*	1/1
Araliaceae	*Hedera helix*	1/1
Asclepiadaceae	*Cryptolepis buchanani*	1/1
Balsaminaceae	*Impatiens balsamina*	1/1
Begoniaceae	*Begonia picta*	1/1
Bignoniaceae	*Oroxylum indicum*	1/1
Boraginaceae	*Trichodesma indicum*	1/1
Cannabinaceae	*Cannabis sativa*	1/1
Cannaceae	*Canna indica*	1/1
Ceratophyllaceae	*Ceratophyllum demersum*	1/1
Crassulaceae	*Bryophyllum calycinum,*	1/1
Dioscoreaceae	*Dioscorea belophylla*	1/1
Fumariaceae	*Fumaria indica*	1/1
Geraniaceae	*Geranium nepalense*	1/1
Gesneriaceae	*Rhynchoglossum obliquum*	1/1
Hydrocharitaceae	*Blyxa auberti*	1/1
Leeaceae	*Leea crispa*	1/1
Liliaceae	*Gloriosa superba*	1/1
Linaceae	*Reinwardtia indica*	1/1
Marchantiaceae	*Marchantia palmata*	1/1
Marsileaceae	*Marsilea minuta*	1/1
Martyniaceae	*Martynia annua*	1/1
Menispermaceae	*Cissampelos pareira*	1/1
Naiadaceae	*Potamogeton crispus*	1/1
Nyctaginaceae	*Mirabilis jalapa*	1/1

Family	Plant Names	Genera/ Species
Oxalidaceae	*Oxalis corniculata*	1/1
Papaveraceae	*Argemone mexicana*	1/1
Plumbaginaceae	*Plumbago zeylanica*	1/1
Polygalaceae	*Polygala arvensis*	1/1
Rubiaceae	*Galium aparine*	1/1
Salicaceae	*Salix oxycarpa*	1/1
Sapindaceae	*Cardiospermum halicacobum*	1/1
Sinopteridaceae	*Cheilanthes bicolor*	1/1
Verbenaceae	*Lantana camara*	1/1
Violaceae	*Viola pilosa*	1/1

Table 39   Ethnobotanically Important Wetland Medicinal Plant Genera of District Bilaspur

Plant Names	Familiy	Genera
Abelmoschus	Malvaceae	1
Abrus	Fabaceae	1
Abutilon	Malvaceae	1
Achyranthes	Amaranthaceae	1
Acorus	Araceae	1
Adiantum	Adiantaceae	2
Ageratum	Asteraceae	1
Ajuga	Lamiaceae	1
Amaranthus	Amaranthaceae	4
Ampelopteris	Thelypteridaceae	1
Andrographis	Acanthaceae	1
Anisomeles	Lamiaceae	1
Apluda	Poaceae	1
Argemone	Papaveraceae	1
Artemisia	Asteraceae	2
Bacopa	Scrophulariaceae	1
Barleria	Acanthaceae	1
Begonia	Begoniaceae	1
Bidens	Asteraceae	1
Blyxa	Hydrocharitaceae	1
Boehmeria	Urticaceae	1
Bryophyllum	Crassulaceae	1
Calamintha	Lamiaceae	1
Canna	Cannaceae	1
Cannabis	Cannabinaceae	1
Capsella	Brassicaceae	1
Cardiospermum	Sapindaceae	1
Cassia	Fabaceae	2
Centella	Apiaceae	1
Ceratophyllum	Ceratophyllaceae	1
Cheilanthes	Sinopteridaceae	1
Chenopodium	Chenopodiaceae	2
Christella	Thelypteridaceae	1
Cissampelos	Menispermaceae	1
Clematis	Ranunculaceae	1

Plant Names	Familiy	Genera
Coix	Poaceae	1
Convolvulus	Convolvulaceae	1
Conyza	Asteraceae	2
Costus	Zingiberaceae	1
Crotolaria	Fabaceae	2
Cryptolepis	Asclepiadaceae	1
Cucumis	Cucurbitaceae	1
Curcuma	Zingiberaceae	1
Cyathocline	Asteraceae	1
Cymbopogon	Poaceae	2
Cynodon	Poaceae	1
Cyperus	Cyperaceae	2
Datura	Solanaceae	1
Debregeasia	Urticaceae	1
Dicliptera	Acanthaceae	1
Dioscorea	Dioscoreaceae	1
Echinochloa	Poaceae	1
Eleusine	Poaceae	1
Emilia	Asteraceae	1
Equisetum	Equisetaceae	2
Eupatorium	Asteraceae	1
Euphorbia	Euphorbiaceae	4
Ficus	Moraceae	2
Fragaria	Rosaceae	2
Fumaria	Fumariaceae	1
Galium	Rubiaceae	1
Geranium	Geraniaceae	1
Girardinia	Urticaceae	1
Gloriosa	Liliaceae	1
Gnaphalium	Asteraceae	1
Hedera	Araliaceae	1
Hedychium	Zingiberaceae	1
Heteropogon	Poaceae	1
Impatiens	Balsaminaceae	1
Ipomoea	Convolvulaceae	5
Justicia	Acanthaceae	1
Lactuca	Asteraceae	1
Lannea	Anacardiaceae	1
Lantana	Verbenaceae	1
Lathyrus	Fabaceae	1
Leea	Leeaceae	1
Lepidagathis	Acanthaceae	1
Macrotyloma	Fabaceae	1
Marchantia	Marchantiaceae	1
Marsilea	Marsileaceae	1
Martynia	Martyniaceae	1
Medicago	Fabaceae	1
Melothria	Cucurbitaceae	1
Mentha	Lamiaceae	2
Micromeria	Lamiaceae	1
Mirabilis	Nyctaginaceae	1

Plant Names	Familiy	Genera
Momordica	Cucurbitaceae	1
Mucuna	Fabaceae	1
Nasturtium	Brassicaceae	1
Nerium	Apocynaceae	1
Ocimum	Lamiaceae	1
Oroxylum	Bignoniaceae	1
Oxalis	Oxalidaceae	1
Parthenium	Asteraceae	1
Phyllanthus	Euphorbiaceae	1
Physalis	Solanaceae	2
Plectranthus	Lamiaceae	1
Plumbago	Plumbaginaceae	1
Pogostemon	Lamiaceae	1
Polygala	Polygalaceae	1
Polygonum	Polygonaceae	9
Potamogeton	Naiadaceae	1
Pupalia	Amaranthaceae	1
Reinwardtia	Linaceae	1
Rhynchoglossum	Gesneriaceae	1
Ricinus	Euphorbiaceae	1
Roylea	Lamiaceae	1
Rubus	Rosaceae	1
Ruellia	Acanthaceae	1
Rumex	Polygonaceae	2
Rungia	Acanthaceae	1
Salix	Salicaceae	1
Saussurea	Asteraceae	1
Sida	Malvaceae	1
Silene	Caryophyllaceae	1
Solanum	Solanaceae	2
Sonchus	Asteraceae	2
Spilanthes	Asteraceae	2
Stellaria	Caryophyllaceae	1
Strobilanthes	Acanthaceae	2
Taraxacum	Asteraceae	1
Thalictrum	Ranunculaceae	1
Trichodesma	Boraginaceae	1
Tridax	Asteraceae	1
Uraria	Fabaceae	1
Urena	Malvaceae	1
Urtica	Urticaceae	1
Vernonia	Asteraceae	1
Veronica	Scrophulariaceae	1
Viola	Violaceae	1
Withania	Solanaceae	1
Xanthium	Asteraceae	1

Table 40 Ethnobotanically Predominant Wetland Medicinal Plant Genera of District Bilaspur (H.P.)

Botanical Names	Familiy	Genera
Polygonum	Polygonaceae	9
Ipomoea	Convolvulaceae	5
Amaranthus	Amaranthaceae	4
Euphorbia	Euphorbiaceae	4
Adiantum	Adiantaceae	2
Artemisia	Asteraceae	2
Cassia	Fabaceae	2
Chenopodium	Chenopodiaceae	2
Conyza	Asteraceae	2
Crotolaria	Fabaceae	2
Cymbopogon	Poaceae	2
Cyperus	Cyperaceae	2
Equisetum	Equisetaceae	2
Ficus	Moraceae	2
Fragaria	Rosaceae	2
Mentha	Lamiaceae	2
Physalis	Solanaceae	2
Rumex	Polygonaceae	2
Solanum	Solanaceae	2
Sonchus	Asteraceae	2
Spilanthes	Asteraceae	2
Strobilanthes	Acanthaceae	2
Abelmoschus	Malvaceae	1
Abrus	Fabaceae	1
Abutilon	Malvaceae	1
Achyranthes	Amaranthaceae	1
Acorus	Araceae	1
Andrographis	Acanthaceae	1
Barleria	Acanthaceae	1
Dicliptera	Acanthaceae	1
Justicia	Acanthaceae	1
Lepidagathis	Acanthaceae	1
Ruellia	Acanthaceae	1
Rungia	Acanthaceae	1
Strobilanthes	Acanthaceae	1

For edibility, 45 wetland species belonging to 34 genera under 26 families have been reported to be used presently (Figure. 19; Table 41). From the analysis of the data it is evident that 9 genera and 11 species for fruits, 13 genera and 19 species for vegetables, pot herbs and chutney (Table 42), 5 genera and 5 species for pickling (Table 43), 5 genera and 5 species (*Dioscorea bulbifera, Hedera helix, Mentha longifolia, Setaria tomentosa, Viola pilosa*) as flavourants and imparting medicinal value to tea (Table 44) 2 genera and 2 species (*Adiantum capillus-veneris, Cassia occidentalis*) as substitute of tea, 2 genera and 2 species (*Oxalis corniculata, Spilanthes acmella* var. *oleracea*) as mouth freshner (Table 45), and 2 genera and 2 species (*Oxalis corniculata, Rumex hastatus*) as souring agent have been listed as alternative foods whereas 7 species belonging to 5 genera under 4 families (*Ceratophyllum demersum, Hydrilla verticillata, Ipomoea cairica, Najas graminea, Najas indica, Potamogeton crispus, Potamogeton pectinatus*) with predominance of Naiadaceae have been employed as fish food (Figure. 19; Tables 46, 78). Organ-wise analysis reveals that inhabitants of this part of Himalaya fancied leaves (*Aeschynomene aspera, Amaranthus gangeticus, Amaranthus paniculatus, Amaranthus tricolor* var. *gangeticus, Amaranthus viridis, Chenopodium album, Chenopodium ambrosiodes, Chenopodium murale, Colocasia antiquorum, Commelina diffusa, Commelina paludosa, Marsilea minuta, Mentha longifolia, Mirabilis jalapa, Oxalis corniculata, Rumex hastatus, Rumex nepalensis, Urtica dioca*) more than the other parts, mostly these are consumed raw and/or cooked. Relatively, the maximum diversity of supplementary foods for humans is represented by Amaranthaceae (4), followed by Chenopodiaceae (3), Cucurbitaceae (3), Rosaceae (3), Solanaceae (3), Polygonaceae, Fabaceae, Commelinaceae, Dioscoreaceae, Moraceae, Brassicaceae, Poaceae (2 each), Malvaceae, Thelypteridaceae, Apocynaceae, Araceae, Araliaceae, Caryophylaceae, Marsileaceae, Nyctaginaceae, Oxalidaceae, Urticaceae (1 each) (Table 41). Among the genera, *Amaranthus* (4 species) showed highest species diversity, followed by *Chenopodium* (3), *Solanum, Rumex, Fragaria, Commelina, Dioscorea, Ficus* (2 each), *Abelmoschus, Aeschynomene, Ampelopteris, Capsella bursa-pastoris, Cassia, Colocasia, Cucumis, Echinochloa, Hedera, Marsilea, Melothria, Mirabilis, Momordica, Nasturtium, Oxalis, Physalis, Rubus, Sorghum, Stellaria, Thevetia* and *Urtica* (1 each). Of the reported estimated diversity of 3900 species of food plants consumed by Indian tribals (Anonymous, 1994; Singh & Arora, 1978), it is interesting to mention that 16 new records of food plants hitherto unknown in ethnobotanical literature have been reported from the studied wetland area/s. These are: *Aeschynomene* (Fabaceae), *Amaranthus gangeticus* (Amaranthaceae), *Amaranthus tricolor* var. *gangeticus* (Amaranthaceae), *Ampelopteris prolifera* (Thelypteridaceae), *Commelina diffusa* (Commelinaceae), *Commelina paludosa* (Commelinaceae), *Cucumis pubescens* (Cucurbitaceae), *Echinochloa frumentacea* (Poaceae), *Hedera helix* (Araliaceae), *Momordica dioica* (Cucurbitaceae), *Oxalis corniculata* (Oxalidaceae), *Rubus ellipticus* (Rosaceae), *Rumex hastatus* (Polygonaceae), *Solanum nigrum* (Solanaceae), *Thevetia neriifolia* (Apocynaceae), *Urtica dioca* (Urticaceae) (Table 47). Besides, nutiritional components of as many as 10 edibles have also been enlisted (Table 48) thereby highlighting the reasons for enjoying good health, vigour and vitality by the wetland inhabitants of the region. In the scenario of expanding humanity and dwindling resources, all these plants can be utilised effectively for extending the base of our sustainable food security system as well as to enhance our chances of survival in the event of any crisis. It will be interesting to analyse these plants chemically for their nutraceutical potential. Pertinently, there is also much scope for improving the growth form of promising wetland wild edibles for domestication by using modern scientific tools, and this no doubt offers a challenging task for the agronomists.

Table 41   Wetland Edibles of District Bilaspur

Name	Parts Used	Family
Abelmoschus crinitus	Roots	Malvaceae
Aeschynomene aspera	Leaves	Fabaceae
Amaranthus gangeticus	Leaves	Amaranthaceae
Amaranthus paniculatus	Leaves	Amaranthaceae
Amaranthus tricolor var. *gangeticus*	Leaves	Amaranthaceae
Amaranthus viridis	Leaves	Amaranthaceae
Ampelopteris prolifera	Young Fronds	Thelypteridaceae
Begonia picta	Stalks, Stems	Begoniaceae
Capsella bursa-pastoris	Aerial Parts	Brassicaceae
Cassia occidentalis	Seeds	Fabaceae

Name	Parts Used	Family
Chenopodium album	Leaves	Chenopodiaceae
Chenopodium ambrosiodes	Leaves	Chenopodiaceae
Chenopodium murale	Leaves	Chenopodiaceae
Colocasia antiquorum	Leaves, Tubers	Araceae
Commelina diffusa	Leaves	Commelinaceae
Commelina paludosa	Leaves	Commelinaceae
Cucumis pubescens	Fruits	Cucurbitaceae
Dioscorea belophylla	Tubers	Dioscoreaceae
Dioscorea bulbifera	Tubers	Dioscoreaceae
Echinochloa frumentacea	Aerial Parts	Poaceae
Ficus hispida	Fruits	Moraceae
Ficus roxburghii	Fruits	Moraceae
Fragaria indica	Fruits	Rosaceae
Fragaria nubicola	Fruits	Rosaceae
Hedera helix	Berries	Araliaceae
Marsilea minuta	Leaves, Sprouts	Marsileaceae
Melothria heterophylla	Fruits	Cucurbitaceae
Mentha longifolia	Leaves	Lamiaceae
Mirabilis jalapa	Leaves	Nyctaginaceae
Momordica dioica	Young Fruits	Cucurbitaceae
Nasturtium officinale	Aerial Plant Parts	Brassicaceae
Oxalis corniculata	Leaves	Oxalidaceae
Physalis minima	Fruits	Solanaceae
Rubus ellipticus	Fruits	Rosaceae
Rumex hastatus	Leaves,Tender Shoots	Polygonaceae
Rumex nepalensis	Leaves	Polygonaceae
Setaria tomentosa	Roots	Poaceae
Solanum nigrum	Fruits	Solanaceae
Solanum xanthocarpum	Seeds	Solanaceae
Sorghum halepense	Seeds	Poaceae
Spilanthes acmella var. *oleracea*	Flowers	Asteraceae
Stellaria media	Whole Plant	Caryophyllaceae
Thevetia neriifolia	Fruits	Apocynaceae
Urtica dioca	Leaves, Tender Shoots	Urticaceae
Viola pilosa	Flowers	Violaceae

Table 42   Wetland Plants Used as Vegetables, Pot Herbs and Chutney

Plants	Family
Amaranthus tricolor var. *gangeticus*	Amaranthaceae
Amaranthus viridis	Amaranthaceae
Ampelopteris prolifera	Thelypteridaceae
Capsella bursa-pastoris	Brassicaceae
Chenopodium album	Chenopodiaceae
Chenopodium ambrosiodes	Chenopodiaceae
Chenopodium murale	Chenopodiaceae
Colocasia antiquorum	Araceae
Commelina diffusa	Commelinaceae
Commelina paludosa	Commelinaceae
Dioscorea belophylla	Dioscoreaceae

Dioscorea bulbifera	Dioscoreaceae
Marsilea minuta	Marsileaceae
Mirabilis jalapa	Nyctaginaceae
Momordica dioica	Cucurbitaceae
Nasturtium officinale	Brassicaceae
Rumex nepalensis	Polygonaceae
Stellaria media	Caryophyllaceae
Urtica dioica	Urticaceae

Table 43   Wetland Plants Used for Making Pickles

Plants	Family
Ampelopteris prolifera	Thelypteridaceae
Begonia picta	Begoniaceae
Colocasia antiquorum	Araceae
Ficus hispida	Moraceae
Rumex hastatus	Polygonaceae

Table 44   Wetland Flavourants of District Bilaspur

Name	Family
Dioscorea bulbifera	Dioscoreaceae
Hedera helix	Araliaceae
Mentha longifolia	Lamiaceae
Setaria tomentosa	Poaceae
Viola pilosa	Violaceae

Table 45   Mouth Freshners of District Bilaspur

Name	Family
Oxalis corniculata	Oxalidaceae
Spilanthes acmella var. *oleracea*	Asteraceae

Table 46   Wetland Plants Used as Fish Food in District Bilaspur

Name	Family
Ceratophyllum demersum	Ceratophyllaceae
Hydrilla verticillata	Hydrocharitaceae
Ipomoea cairica	Convolvulaceae
Najas graminea	Naiadaceae
Najas indica	Naiadaceae
Potamogeton crispus	Naiadaceae
Potamogeton pectinatus	Naiadaceae

Table 47   Wetland Plants of District Bilaspur with New Edible Uses

Name	Family
Aeschynomene aspera	Fabaceae
Amaranthus gangeticus	Amaranthaceae
Amaranthus tricolor var. *gangeticus*	Amaranthaceae
Ampelopteris prolifera	Thelypteridaceae
Commelina diffusa	Commelinaceae
Commelina paludosa	Commelinaceae
Cucumis pubescens	Cucurbitaceae
Echinochloa frumentacea	Poaceae
Hedera helix	Araliaceae
Momordica dioica	Cucurbitaceae
Oxalis corniculata	Oxalidaceae
Rubus ellipticus	Rosaceae
Rumex hastatus	Polygonaceae
Solanum nigrum	Solanaceae
Thevetia neriifolia	Apocynaceae
Urtica dioca	Urticaceae

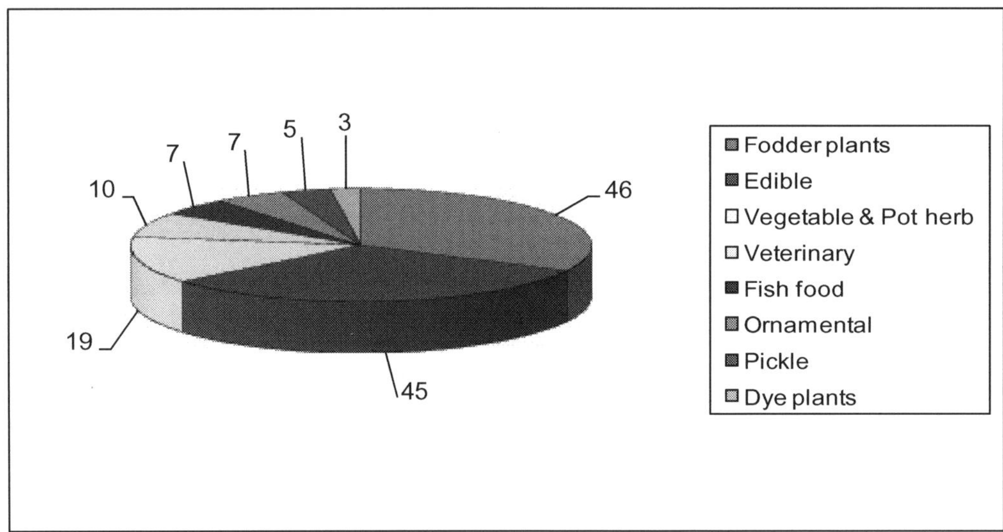

Figure 19   Pie diagram showing relative number of wetland plants used for various ethnobotanical purposes in District Bilaspur.

Table 48   Nutritional Components of Some Wetland Edibles of District Bilaspur (On Dry Weight Basis)

Plant	Car-boh-ydrates(%)	Protiens (%)	Fats (%)	Ash (%)	Fi-bres (%)	Starch (%)	Minerals (mg/ 100gm)	Vitamins (I.U.mg/ 100gm)
Amaranthus paniculatus (Aerial parts)	55-60	-	-	-	-	-	-	-
Cannabis sativa (Seeds)	-	21.51	30.4	4.60	18.64	-	-	A, 434; B,40

Plant	Car-boh-ydrates(%)	Protiens (%)	Fats (%)	Ash (%)	Fi-bres (%)	Starch (%)	Minerals (mg/100gm)	Vitamins (I.U.mg/100gm)
Chenopodium album (Seeds)	-	15.4-16.8	5.8-8.1	-	18.4-21.5	-	-	-
Chenopodium ambrosiodes (Fruits)	-	-	2.5	-	-	-	Mg Phosphate 0.5-1.0% Ammonia	C,1.02 mg/g
(Fresh sprouts)	-	-	-	-	-	-	-	-
Colocasia antiquorum (Corm)	-	-	-	-	-	25	-	0.001% Vitamin C
Convolvulus arvensis (Seeds)	-	-	4.7	-	-	-	-	-
Cucumis pubescence (Fruits)	22.22	16.68	-	10.48	24.44	-	Phosphors P²O⁵ 0.61%	-
Cynodon dactylon (Aerial parts)	10.47	-	28.2	11.8	-	-	-	C, 66-96
Fragaria vesca (Fruits)	87-88	-	14.3	0.6-0.7	-	-	-	-
Nasturtium officinale (Leaves)	-	-	-	-	-	-	Iron 2 mg/100g	Vitamins A,B,C,E

(*Source:* *Anonymous, 1948-76a; Arora & Pandey, 1996; Bennet* et al.*, 1991; Dwivedi, 1993; Harborne & Baxter, 2001; Rana, 2004; Sharma & Rana, 2005).*

For supplementing fodder requirements, 46 wetland species belonging to 40 genera under 19 families are valued by the local populace (Table 49). Of these, species marked with asterisk are greatly favoured for their high nutritious value (Table 49). Predominantly, the forage diversity belongs to Poaceae (14 species), followed by Cyperaceae (6), Fabaceae (6), Amaranthaceae, Asteraceae, Polygonaceae, Urticaceae (2 each) and the rest of the families such as Acanthaceae, Apiaceae, Apocynaceae, Araliaceae, Bignoniaceae, Commelinaceae, Convolvulaceae, Euphorbiaceae, Leeaceae, Moraceae, Scrophulariaceae and Solanaceae are poorly represented by one species each (Figure. 19; Table 49). As a matter of fact, commonly used genera of the wetland areas are those of *Cyperus* (5 species: Cyperaceae), *Cymbopogon* (2 species: Poaceae) and *Polygonum* (2 species: Polygonaceae) with 23 fodder species, *Apluda mutica, Bromus catharticus, Centella asiatica, Coix lachryma–jobi, Cymbopogon martinii, Cyperus compressus, Cyperus flabelliformis, Cyperus iria, Cyperus rotundus, Dichanthium annulatum, Digitaria griffithii, Echinochloa frumentacea, Eleusine indica, Gomphrena celosioides, Heteropogon contortus, Lathyrus aphaca, Lindernia ciliata, Medicago denticulata, Pennisetum lanatum, Poa supina, Sorghum halepense, Trifolium resupinatum, Trigonella pubescence* being new additions to the list of 400 species documented for our country (Anonymous, 1994). Undeniably, their importance as invaluable genetic stocks cannot be underestimated and hold considerable scope for popularization.

Table 49 Fodder Wetland Plants of District Bilaspur

Name	Family
Amaranthus viridis	Amaranthaceae
*Apluda mutica**	Poaceae
*Bromus catharticus**	Poaceae
*Centella asiatica**	Apiaceae
*Coix lachryma – jobi**	Poaceae
Convolvulus arvensis	Convolvulaceae
Commelina diffusa	Commelinaceae
Cymbopogon citratus	Poaceae
*Cymbopogon martinii**	Poaceae
Cynodon dactylon	Poaceae
*Cyperus compressus**	Cyperaceae
Cyperus distans	Cyperaceae
*Cyperus flabelliformis**	Cyperaceae
*Cyperus iria**	Cyperaceae
*Cyperus rotundus**	Cyperaceae
Debregeasia hypoleuca	Urticaceae
*Dichanthium annulatum**	Poaceae
*Digitaria griffithii**	Poaceae
*Echinochloa frumentacea**	Poaceae
*Eleusine indica**	Poaceae
Eriophorum comosum	Cyperaceae
Ficus roxburghii	Moraceae
*Gomphrena celosioides**	Amaranthaceae
Hedera helix	Araliaceae
*Heteropogon contortus**	Poaceae
*Lathyrus aphaca**	Fabaceae
Leea crispa	Leeaceae
Lepidagathis cuspidata	Acanthaceae
Lespedeza sericea	Fabaceae
*Lindernia ciliata**	Scrophulariaceae
*Medicago denticulata**	Fabaceae
Nerium indicum	Apocynaceae
Oroxylum indicum	Bignoniaceae
*Pennisetum lanatum**	Poaceae
Physalis longifolia	Solanaceae
*Poa supina**	Poaceae
Polygonum minus	Polygonaceae
Polygonum pulchrum	Polygonaceae
Ricinus communis	Euphorbiaceae
Sonchus arvensis	Asteraceae
*Sorghum halepense**	Poaceae
Tridax procumbens	Asteraceae
*Trifolium resupinatum**	Fabaceae
*Trigonella pubescence**	Fabaceae
Urtica dioica	Urticaceae
Vicia sativa	Fabaceae

* High fodder value

Of the estimated total wetland plant resources of the region, cordage requirements are largely fulfilled from 6 plants species belonging to 6 genera under 5 families (Table 50). For this purpose, commonly employed species are: *Abelmoschus crinitus, Canna indica, Cannabis sativa, Eleusine indica, Eriophorum comosum, Heteropogon contortus*. In usages, the family Poaceae predominates and rest of families like Cannabinaceae, Cannaceae, Cyperaceae and Malvaceae are represented by one species each. For fuel wood purposes, 3 plants, namely *Debregeasia hypoleuca, Ficus hispida, Ficus roxburghii* belonging to 2 genera, 3 species and under 2 families Urticaceae and Moraceae are being employed (Table 54). For dyeing, three species *Calamintha umbrosum, Impatiens balsamina, Phyllanthus urinaria* belonging to Lamiaceae, Balsaminaceae, Euphorbiaceae, respectively are employed (Figure. 19; Table 52). However, thatching requirements are largely met with from 5 species belonging to 5 genera under 2 families with predominance of Poaceae followed by Cyperaceae. In this regard, species commonly employed are: *Apluda mutica, Cyperus rotundus, Heteropogon contortus, Phragmites karka, Saccharum spontanium* (Table 51). Ornamentally, seven plant species (*Anisomeles indica, Canna indica, Coix lachryma – jobi, Cynodon dactylon, Gloriosa superba, Phragmites karka, Pteris cretica*) belonging to five families with predominance of Poaceae (3 species), followed by Cannaceae, Lamiaceae, Liliaceae, Pteridaceae are being used (Figure., 19; Table 53). Evidently, dependance on a great diversity of plant species for their sustenance ensures a year-round supply, and at the same time it reflects their conservational wisdom as well the practise ensures less pressure on the surrounding wetland plant diversity.

Table 50   Wetland Plants Used for Cordages

Plants	Family
Abelmoschus crinitus	Malvaceae
Canna indica	Cannaceae
Cannabis sativa	Cannabinaceae
Eleusine indica	Poaceae
Eriophorum comosum	Cyperaceae
Heteropogon contortus	Poaceae

Table 51   Wetland Plants used for Thatching Purposes

Plants	Family
Apluda mutica	Poaceae
Cyperus rotundus	Cyperaceae
Heteropogon contortus	Poaceae
Phragmites karka	Poaceae
Saccharum spontanium	Poaceae

Table 52   Dye Yielding Wetland Plants of District Bilaspur

Name	Family
Calmintha umbrosum	Lamiaceae
Impatiens balsamina	Balsaminaceae
Phyllanthus urinaria	Euphorbiaceae

Table 53 Ornamental Wetland Plants of District Bilaspur

Name	Family
Anisomeles indica	Lamiaceae
Canna indica	Cannaceae
Coix lachryma - jobi	Poaceae
Cynodon dactylon	Poaceae
Gloriosa superba	Liliaceae
Phragmites karka	Poaceae
Pteris cretica	Pteridaceae

Table 54 Fuel Wood Wetland Plants of District Bilaspur

Name	Family
Debregeasia hypoleuca	Urticaceae
Ficus hispida	Moraceae
Ficus roxburghii	Moraceae

For data of interest on agricultural related activities, cottage and rural industries, plants such as *Lannea coromandelica* (agricultural implements), *Eriophorum comosum, Heteropogon contortus, Saccharum munja* (basketry, matting), *Eleusine indica, Heteropogon contortus, Ipomoea carnea, Medicago denticulata, Paspalum distichum, Saccharum munja* (soil binder), *Ceratophyllum demersum, Crotolaria alata, Equisetum arvense, Equisetum debile, Hydrilla verticillata, Lantana camara, Medicago denticulata* (green manure), *Cymbopogon citratus, Lantana camara* (insect repellent) and *Chenopodium ambrosiodes* (snake repellent) are valued considerably (Table 55). Besides, plant usages for other miscellaneous activities/purposes like colouring hands (*Impatiens balsamina*), fish poison (*Conyza bonariensis, Cyathocline purpurea, Polygonum barbatum*), hallucination (*Cannabis sativa*), paints & varnish (*Cannabis sativa*), writing pen (*Saccharum munja*), poison (*Abrus precatorius*), soap (*Cannabis sativa, Convolvulus arvensis*), touch therapy (*Aerva sanguinolenta*), vermillion (*Selaginella chrysocaulos*) and worshipping (*Barleria cristata, Mirabilis jalapa, Reinwardtia indica*), (Table 55) are unique and have not been reported earlier in published ethnobotanical literature.

Table 55 Wetland Plants Used for Material Culture and other Miscellaneous Purposes

Use	Plant	Genera/Species
Agricultural implements	*Lannea coromandelica*	1/1
Baskets & mats	*Eriophorum comosum, Heteropogon contortus, Saccharum munja*	3/3
Colouring hands	*Impatiens balsamina*	1/1
Fish poison	*Conyza bonariensis, Cyathocline purpurea, Polygonum barbatum*	3/3
Green manure	*Ceratophyllum demersum, Crotolaria alata, Equisetum arvense, Equisetum debile, Hydrilla verticillata, Lantana camara, Medicago denticulata*	6/7
Hallucination	*Cannabis sativa*	1/1
Healing	*Achyranthes aspera*	1/1
Insect repellent	*Cymbopogon citratus, Lantana camara*	2/2
Milk increaser	*Gomphrena celosioides, Hedychium spicatum*	2/2

Use	Plant	Genera/Species
Paints & varnish	*Cannabis sativa*	1/1
Pen	*Saccharum munja*	1/1
Poison	*Abrus precatorius*	1/1
Snake repellent	*Chenopodium ambrosiodes*	1/1
Soap	*Cannabis sativa, Convolvulus arvensis*	2/2
Soil binder	*Eleusine indica, Heteropogon contortus, Ipomoea carnea, Medicago denticulata, Paspalum distichum, Saccharum munja*	6/6
Touch therapy	*Aerva sanguinolenta*	1/1
Vermillion	*Selaginella chrysocaulos*	1/1
Worshipping	*Barleria cristata, Mirabilis jalapa, Reinwardtia indica*	3/3

Based on analysis of data presented in Table 57, it is evident that several multipurpose species like *Amaranthus viridis* (Amaranthaceae, 4 uses), *Canna indica* (Cannaceae, 4 uses), *Heteropogon contortus* (Poaceae, 4 uses), *Ipomoea cairica* (Convolvulaceae, 4 uses), *Lantana camara* (Verbenaceae, 4 uses), *Abelmoschus crinitus* (Malvaceae, 3 uses), *Ampelopteris prolifera* (Thelypteridaceae, 3 uses), *Cannabis sativa* (Cannabinaceae, 3 uses), *Cassia occidentalis* (Fabaceae, 3 uses), *Centella asiatica* (Apiaceae, 3 uses), *Ceratophyllum demersum* (Ceratophyllaceae, 3 uses), *Coix lachryma – jobi* (Poaceae, 3 uses), *Cymbopogon citratus* (Poaceae, 3 uses), *Debregeasia hypoleuca* (Urticaceae, 3 uses), *Dioscorea belophylla* (Dioscoreaceae, 3 uses), *Dioscorea bulbifera* (Dioscoreaceae, 3 uses), *Echinochloa frumentacea* (Poaceae, 3 uses), *Eleusine indica* (Poaceae, 3 uses), *Ficus roxburghii* (Moraceae, 3 uses), *Gloriosa superba* (Liliaceae, 3 uses), *Hedera helix* (Araliaceae, 3 uses), *Mentha longifolia* (Lamiaceae, 3 uses), *Mirabilis jalapa* (Nyctaginaceae, 3 uses), *Nasturtium officinale* (Brassicaceae, 3 uses), *Potamogeton crispus* (Naiadaceae, 3 uses), *Rumex hastatus* (Polygonaceae, 3 uses), *Saccharum munja* (Poaceae, 3 uses), *Urtica dioica* (Urticaceae, 3 uses), etc., are well represented in the wetland areas of district Bilaspur. Notably, the resultant information is of great significance for achieving the goal of economic benefits for the local communities inhabiting fringe areas of wetlands of the district, and offers a challenging task for the agronomists.

For livestock ailments, the locals are well acquainted with the use of herbal veterinary medicine (Table 56). As many as 10 species under 10 genera belonging to 9 families are used for curing the prevalent veterinary ailments. These are: sexual weakness (*Abelmoschus crinitus*), tongue and mouth sores (*Apluda mutica*), antidote for poisonous fodder (*Canna indica*), wounds and sores (*Chenopodium album, Curcuma longa*), stomach disorders (*Cissampelos pareira*), fractured limbs (*Cryptolepis buchanani*), weakness (*Girardinia heterophylla*) and enhancing lactation (*Gomphrena celosioides, Hedychium spicatum*). As a matter of fact, clinical validation of these claims, however, be needed before their use by animal husbandry practitioners.

Table 56 Ethnoveterinary Important Wetland Plants of District Bilaspur

Name	Ailments	Family
Abelmoschus crinitus	Sexual weakness	Malvaceae
Apluda mutica	Tongue & mouth sores	Poaceae
Canna indica	Antidote for poisonous fodder	Cannaceae
Chenopodium album	Wounds & sores	Chenopodiaceae
Cissampelos pareira	Stomach disorders	Menispermaceae
Cryptolepis buchanani	Fractured limbs	Asclepiadaceae
Curcuma longa	Wounds	Zingiberaceae

Girardinia heterophylla	Weakness	Urticaceae
Gomphrena celosioides	Enhancing lactation	Amaranthaceae
Hedychium spicatum	Enhancing lactation	Zingiberaceae

Table 57 Wetland Plants of District Bilaspur Classified According to Use

1-Dye, 2-Edible, 3-Fish food, 4-Fodder, 5-Household Articles, 6-Magico-religious Belief, 7-Medicinal, 8-Ornamental, 9-Pot herb, 10-Sacred, 11-Scouring Teeth, 12-Miscellaneous (Biofertilizer, Detoxifying agent, Expel worms, Fermentation, Fish Farming, Fish poison, Flavour Curies, Green Manure, Heena substitute, Insect Repellant, Kill Maggots, Kill worms, Meal Plates, Milk Increaser, Perfumery, Purgative, Skin Troubles).

Species	Use											
	1	2	3	4	5	6	7	8	9	10	11	12
Abelmoschus crinitus		X			X		X					
Abrus precatorius							X					X
Abutilon indicum							X					
Achyranthes aspera							X				X	
Acorus calamus							X					
Adiantum capillus-veneris							X					X
Adiantum incisum							X					
Aerva sanguinolenta						X						X
Aeschynomene aspera		X										X
Ageratum conyzoides							X					
Ajuga bracteosa							X					
Amaranthus gangeticus		X					X					
Amaranthus paniculatus		X					X					
Amaranthus tricolor var. *gangeticus*		X					X					
Amaranthus viridis		X		X			X		X			
Ampelopteris prolifera		X					X	X				
Andrographis paniculata							X					
Anisomeles indica						X	X					
Apluda mutica				X			X					
Argemone mexicana							X					
Artemisia indica							X					
Artemisia scoparia							X					
Bacopa monnieri							X					X
Barleria cristata						X	X					
Begonia picta		X					X					
Bidens pilosa							X				X	
Blyxa auberti							X					
Boehmeria platyphylla							X					
Bromus catharticus				X								
Bryophyllum calycinum							X					
Calamintha umbrosum	X						X					
Canna indica					X		X	X				X
Cannabis sativa					X		X					X
Capsella bursa-pastoris		X					X					
Cardiospermum halicacobum							X					
Cassia absus							X					

Species	Use											
	1	2	3	4	5	6	7	8	9	10	11	12
Cassia occidentalis		X					X					X
Centella asiatica				X			X					X
Ceratophyllum demersum			X				X					X
Cheilanthes bicolor							X					
Chenopodium album		X					X					
Chenopodium ambrosiodes		X					X					
Chenopodium murale		X										
Christella dentata							X					
Cissampelos pareira							X					X
Clematis gouriana							X					
Coix lachryma - jobi							X	X				X
Colocasia antiquorum		X										
Commelina diffusa		X		X								
Commelina paludosa		X							X			
Convolvulus arvensis							X					X
Conyza bonariensis							X					X
Conyza stricta							X					
Costus speciosus							X					
Crotolaria alata							X					X
Crotolaria mysorensis							X					
Cryptolepis buchanani							X					
Cucumis pubescens		X										
Curcuma longa							X					
Cyathocline purpurea							X					X
Cymbopogon citratus				X			X					X
Cymbopogon martinii				X			X					
Cynodon dactylon							X	X				
Cyperus compressus				X								
Cyperus distans												X
Cyperus flabelliformis				X								
Cyperus iria				X			X					
Cyperus rotundus							X					X
Datura stramonium						X	X					
Debregeasia hypoleuca				X			X					X
Dichanthium annulatum				X								
Dicliptera roxburghiana							X					
Digitaria griffithii				X								
Dioscorea belophylla		X					X					X
Dioscorea bulbifera		X			X							X
Echinochloa frumentacea		X		X			X					
Eleusine indica				X	X		X					
Emilia sonchifolia						X	X					
Equisetum arvense							X					
Equisetum debile							X					X
Eriophorum comosum					X							
Eupatorium adenophorum					X		X					
Euphorbia geniculata							X					
Euphorbia helioscopia							X					
Euphorbia hirta							X					

Species	1	2	3	4	5	6	7	8	9	10	11	12
Euphorbia parviflora							X					
Ficus hispida		X					X					
Ficus roxburghii		X			X							X
Fragaria indica		X					X					
Fragaria nubicola		X					X					
Fumaria indica				X			X					
Galium aparine				X			X					
Geranium nepalense							X					
Girardinia heterophylla					X		X					
Gloriosa superba							X	X				X
Gnaphalium pensylvanicum							X					
Gomphrena celosioides				X								
Hedera helix		X		X			X					
Hedychium spicatum				X			X					
Heteropogon contortus				X	X		X					X
Hydrilla verticillata			X									X
Impatiens balsamina	X											X
Ipomoea cairica			X		X		X					X
Ipomoea carnea							X					X
Ipomoea muricata							X			X		
Ipomoea nil							X					
Ipomoea pestigridis							X					
Justicia simplex							X					
Lactuca dissecta							X					
Lannea coromandelica					X		X					
Lantana camara					X		X				X	X
Lathyrus aphaca				X			X					
Leea crispa				X			X					
Lepidagathis cuspidata				X			X					
Lespedeza sericea				X								
Lindernia ciliata				X								
Macrotyloma unifloruma							X					
Marchantia palmata							X					
Marsilea minuta		X					X					
Martynia annua							X					X
Medicago denticulata				X								
Melothria heterophylla		X					X					
Mentha longifolia		X					X					X
Mentha piperita							X					
Micromeria biflora							X					X
Mirabilis jalapa		X					X			X		
Momordica dioica		X					X					
Mucuna pruriens							X					
Najas graminea			X									X
Najas indica			X									X
Nasturtium officinale		X					X					X
Nerium indicum							X					X
Ocimum basilicum							X			X		
Oroxylum indicum							X					X

Species	Use											
	1	2	3	4	5	6	7	8	9	10	11	12
Oxalis corniculata		X					X					
Parthenium hysterophorus							X					
Paspalum distichum				X								X
Pennisetum lanatum				X								X
Phragmites karka								X				X
Phyllanthus urinaria	X						X					
Physalis longifolia				X								
Physalis minima		X					X					
Plectranthus coetsa							X					
Plumbago zeylanica							X					
Poa supina				X								
Pogostemon plectranthoides							X					
Polygala arvensis							X					
Polygonum barbatum				X			X					
Polygonum barbatum sub sp. *gracile*				X								X
Polygonum donii							X					
Polygonum glabrum				X			X					
Polygonum lapathifolium				X			X					
Polygonum minus				X			X					
Polygonum plebejjum							X					
Polygonum pulchrum				X			X					
Polygonum serrulatum							X					
Potamogeton crispus			X				X					X
Potamogeton pectinatus			X									X
Pteris cretica		X						X				
Pupalia lappacea							X					
Reinwardtia indica							X			X		
Rhynchoglossum obliquum							X					
Ricinus communis				X			X					
Roylea cinerea							X					
Rubus ellipticus		X					X					
Ruellia patula							X					
Rumex hastatus		X		X			X					
Rumex nepalensis		X					X					
Rungia pectinata							X					
Saccharum munja						X	X					X
Saccharum spontanium							X					X
Salix oxycarpa							X					
Saussurea heteromala							X					
Selaginella chrysocaulos												X
Setaria tomentosa		X										X
Sida acuta							X					
Silene conoidea							X					
Solanum nigrum		X					X					
Solanum xanthocarpum		X					X					
Sonchus arvensis							X					
Sonchus asper							X					
Sorghum halepense		X		X								
Spilanthes acmella var. *oleracea*		X					X					

Species	Use											
	1	2	3	4	5	6	7	8	9	10	11	12
Spilanthes paniculata							X					X
Stellaria media		X			X		X					
Strobilanthes atropurpurens							X					
Strobilanthes dalhousianus							X					
Taraxacum officinale							X					X
Thalictrum reniforme							X					
Thevetia neriifolia		X										
Trichodesma indicum							X					
Tridax procumbens				X			X					
Trifolium resupinatum				X			X					
Trigonella pubescence				X								
Uraria picta							X					
Urena lobata							X					
Urtica dioica		X		X			X					
Vernonia cinerea							X					
Veronica anagalis-aquatica							X					
Vicia sativa				X								X
Viola pilosa							X					
Withania somnifera							X					
Xanthium strumarium							X					

Comparative results of the total importance values (TIV) calculated for the first time for all the wetland resources of the region revealed that out of a total of 203 presently collected species (Table 58), plants namely, *Ficus roxburghii* (Moraceae) – 45%; *Amaranthus viridis* (Amaranthaceae), *Coix lachryma – jobi* (Poaceae), *Heteropogon contortus* (Poaceae) – 40%; *Cynodon dactylon* (Poaceae) – 35%; *Canna indica* (Cannaceae), *Cassia occidentalis* (Fabaceae), *Centella asiatica* (Apiaceae), *Cymbopogon citratus* (Poaceae), *Ficus hispida* (Moraceae), *Echinochloa frumentacea* (Poaceae), *Impatiens balsamina* (Balsaminaceae), *Rumex hastatus* (Polygonaceae), *Urtica dioica* (Urticaceae) - 30%; *Abelmoschus crinitus* (Malvaceae), *Achyranthes aspera* (Amaranthaceae), *Convolvulus arvensis* (Convolvulaceae), *Debregeasia hypoleuca* (Urticaceae), *Dioscorea belophylla* (Dioscoreaceae), *Eleusine indica* (Poaceae), *Hedera helix* (Araliaceae), *Ipomoea cairica* (Convolvulaceae), *Ipomoea carnea* (Convolvulaceae), *Nasturtium officinale* (Brassicaceae), *Oxalis corniculata* (Oxalidaceae), *Solanum nigrum* (Solanaceae), *Cyperus rotundus* (Cyperaceae) - 25%; are found to have high economic potential values, and have good contribution in rural economy of populace inhabiting fringe areas of wetlands of district Bilaspur. Hopefully, their impotance as invaluable genetic stocks cannot be underestimated and can be profitably made use of for the welfare of mankind.

Table 58   Total Importance Values (TIV) of Ethnobotanically Important Wetland Plants of District Bilaspur

[1-Dye, 2-Edible, 3- Fuel, 4- Fodder, 5-Oil, 6- Medicine, 7-Tannin/Gum, 8-Timber, 9-Fibre, 10- Miscellaneous (Biofertilizer, Detoxifying Agent, Expelling Worms, Fermentation, Fish Farming, Fish Poison, Flavouring Curries, Green Manure, Heena Substitute, Insect Repellent, Killing Maggots, Killing Worms, Meal Plates, Milk Enhancer, Perfumery, Purgative, Skin Troubles, Soil Binder)]

Species	Dye	Edible	Fuel	Fodder	Oil	Medicine	Tannin/Gum	Timber	Fibre	Miscellaneous	TIV %
Ficus roxburghii	-	++	++	+	-	+	-	-	+	++	45
Amaranthus viridis	-	++	-	+	-	++	++	-	-	+	40
Coix lachryma - jobi	-	-	-	++	-	++	++	-	-	++	40
Heteropogon contortus	-	-	-	++	-	++	-	-	++	++	40
Cynodon dactylon	-	-	-	++	-	++	++	-	-	+	35
Canna indica	-	-	-	-	-	++	++	-	-	++	30
Cassia occidentalis	-	++	-	-	-	++	-	-	-	++	30
Centella asiatica	-	-	-	++	-	++	-	-	-	++	30
Cymbopogon citratus	-	-	-	++	-	++	-	-	-	++	30
Echinochloa frumentacea	-	+	-	++	-	+	-	-	-	++	30
Ficus hispida	-	++	++	-	-	++	-	-	-	-	30
Impatiens balsamina	++	-	-	-	-	++	-	-	-	++	30
Rumex hastatus	-	++	-	++	-	++	-	-	-	-	30
Urtica dioica	-	++	-	++	-	+	-	-	-	+	30
Abelmoschus crinitus	-	+	-	+	-	++	-	-	+	-	25
Achyranthes aspera	-	+	-	-	-	++	-	-	-	++	25
Convolvulus arvensis	-	-	-	+	-	++	-	-	-	++	25
Cyperus rotundus	-	-	-	++	-	++	-	-	-	+	25
Debregeasia hypoleuca	-	-	++	+	-	++	-	-	-	-	25
Dioscorea belophylla	-	++	-	-	-	++	-	-	-	+	25
Eleusine indica	-	-	-	++	-	++	-	-	-	+	25
Hedera helix	-	++	-	+	-	++	-	-	-	-	25
Ipomoea cairica	-	-	++	-	-	++	-	-	-	+	25
Ipomoea carnea	-	-	-	-	-	++	+	-	-	++	25
Nasturtium officinale	-	++	-	-	-	++	-	-	-	+	25
Oxalis corniculata	-	++	-	-	-	++	-	-	-	+	25
Solanum nigrum	-	++	-	-	-	++	-	-	-	+	25
Abrus precatorius	-	-	-	-	-	++	-	-	-	++	20
Amaranthus gangeticus	-	++	-	-	-	++	-	-	-	-	20
Amaranthus tricolor var. *gangeticus*	-	++	-	-	-	++	-	-	-	-	20
Ampelopteris prolifera	-	+	-	-	-	++	+	-	-	-	20
Apluda mutica	-	-	-	++	-	++	-	-	-	-	20
Bacopa monnieri	-	-	-	-	-	++	-	-	-	++	20
Boehmeria platyphylla	-	-	-	-	-	++	-	-	-	++	20
Calamintha umbrosum	++	-	-	-	-	++	-	-	-	-	20
Cannabis sativa	-	-	-	-	-	+	-	-	+	++	20
Capsella bursa-pastoris	-	++	-	-	-	++	-	-	-	-	20
Chenopodium album	-	++	-	-	-	++	-	-	-	-	20
Chenopodium ambrosiodes	-	++	-	-	-	++	-	-	-	-	20

Species	Dye	Edible	Fuel	Fodder	Oil	Medicine	Tannin/Gum	Timber	Fibre	Miscellaneous	TIV %
Cissampelos pareira	-	-	-	-	-	++	-	-	-	++	20
Conyza bonariensis	-	-	-	-	-	++	-	-	-	++	20
Cyathocline purpurea	-	-	-	-	-	++	-	-	-	++	20
Dioscorea bulbifera	-	++	-	-	-	-	-	-	-	++	20
Eriophorum comosum	-	-	-	++	-	-	-	-	-	++	20
Fragaria indica	-	++	-	-	-	++	-	-	-	-	20
Fumaria indica	-	-	-	++	-	++	-	-	-	-	20
Galium aparine	-	-	-	++	-	++	-	-	-	-	20
Gloriosa superba	-	-	-	-	-	++	-	-	-	++	20
Hedychium spicatum	-	-	-	++	-	++	-	-	-	-	20
Lannea coromandelica	-	-	-	-	-	++	-	++	-	-	20
Lantana camara	-	-	-	-	+	++	-	-	-	+	20
Lathyrus aphaca	-	-	-	++	-	++	-	-	-	-	20
Marsilea minuta	-	++	-	-	-	++	-	-	-	-	20
Martynia annua	-	-	-	-	-	++	-	-	-	++	20
Melothria heterophylla	-	++	-	-	-	++	-	-	-	-	20
Mentha longifolia	-	-	-	-	-	++	-	-	-	++	20
Mentha piperita	-	-	-	-	-	++	-	-	-	++	20
Momordica dioica	-	++	-	-	-	++	-	-	-	-	20
Nerium indicum	-	-	-	-	-	++	-	-	-	++	20
Oroxylum indicum	-	-	+	-	-	++	-	-	-	+	20
Phyllanthus urinaria	+	-	-	-	-	++	-	-	-	+	20
Potamogeton crispus	-	-	-	-	-	++	-	-	-	++	20
Ricinus communis	-	-	-	+	+	++	-	-	-	-	20
Rubus ellipticus	-	++	-	-	-	++	-	-	-	-	20
Rumex nepalensis	-	++	-	-	-	++	-	-	-	-	20
Sonchus asper	-	+	-	+	-	++	-	-	-	-	20
Spilanthes paniculata	-	-	-	-	-	++	-	-	-	++	20
Tridax procumbens	-	-	-	++	-	++	-	-	-	-	20
Amaranthus paniculatus	-	++	-	-	-	+	-	-	-	-	15
Commelina diffusa	-	+	-	++	-	-	-	-	-	-	15
Commelina paludosa	-	-	-	++	-	-	+	-	-	-	15
Crotolaria alata	-	-	-	-	-	+	-	-	-	++	15
Cymbopogon martinii	-	-	-	++	-	+	-	-	-	-	15
Euphorbia hirta	-	-	-	-	-	+	++	-	-	-	15
Fragaria nubicola	-	+	-	-	-	++	-	-	-	-	15
Girardinia heterophylla	-	-	-	-	+	++	-	-	-	-	15
Leea crispa	-	-	-	+	-	++	-	-	-	-	15
Lepidagathis cuspidata	-	-	-	+	-	++	-	-	-	-	15
Micromeria biflora	-	-	-	-	-	++	-	-	-	+	15
Paspalum distichum	-	-	-	++	-	-	-	-	-	+	15
Pennisetum lanatum	-	-	-	++	-	-	-	-	-	+	15
Physalis minima	-	+	-	-	-	++	-	-	-	-	15
Plumbago zeylanica	-	-	-	-	-	++	-	-	-	+	15
Silene conoidea	-	-	-	-	-	++	-	-	-	+	15
Solanum xanthocarpum	-	+	-	-	-	++	-	-	-	-	15
Sonchus arvensis	-	-	-	+	-	++	-	-	-	-	15
Sorghum halepense	-	+	-	++	-	-	-	-	-	-	15
Stellaria media	-	-	-	-	+	++	-	-	-	-	15
Taraxacum officinale	-	-	-	-	-	+	-	-	-	++	15

Species	Dye	Edible	Fuel	Fodder	Oil	Medicine	Tannin/Gum	Timber	Fibre	Miscellaneous	TIV %
Vicia sativa	-	+	-	++	-	-	-	-	-	-	15
Abutilon indicum	-	-	-	-	-	++	-	-	-	-	10
Acorus calamus	-	-	-	-	-	++	-	-	-	-	10
Adiantum capillus-veneris	-	-	-	-	-	+	-	-	-	+	10
Adiantum incisum	-	-	-	-	-	++	-	-	-	-	10
Aeschynomene aspera	-	+	-	-	-	-	-	-	-	+	10
Ageratum conyzoides	-	-	-	-	-	++	-	-	-	-	10
Ajuga bracteosa	-	-	-	-	-	++	-	-	-	-	10
Andrographis paniculata	-	-	-	-	-	++	-	-	-	-	10
Anisomeles indica	-	-	-	-	-	+	-	-	-	+	10
Argemone mexicana	-	-	-	-	-	++	-	-	-	-	10
Artemisia indica	-	-	-	-	-	++	-	-	-	-	10
Artemisia scoparia	-	-	-	-	-	++	-	-	-	-	10
Barleria cristata	-	-	-	-	-	+	-	-	-	+	10
Begonia picta	-	-	-	-	-	+	-	-	-	+	10
Bidens pilosa	-	-	-	-	-	++	-	-	-	-	10
Blyxa auberti	-	-	-	-	-	++	-	-	-	-	10
Bromus catharticus	-	-	-	+	-	-	-	-	-	+	10
Bryophyllum calycinum	-	-	-	-	-	++	-	-	-	-	10
Cardiospermum halicacobum	-	-	-	-	-	++	-	-	-	-	10
Cassia absus	-	-	-	-	-	+	-	-	-	+	10
Ceratophyllum demersum	-	-	-	-	-	+	-	-	-	+	10
Cheilanthes bicolor	-	-	-	-	-	++	-	-	-	-	10
Christella dentata	-	-	-	-	-	+	-	-	-	+	10
Clematis gouriana	-	-	-	-	-	++	-	-	-	-	10
Conyza stricta	-	-	-	-	-	++	-	-	-	-	10
Costus speciosus	-	-	-	-	-	++	-	-	-	-	10
Crotolaria mysorensis	-	-	-	-	-	+	-	-	-	+	10
Cryptolepis buchanani	-	-	-	-	-	++	-	-	-	-	10
Cucumis pubescens	-	++	-	-	-	-	-	-	-	-	10
Curcuma longa	-	-	-	-	-	++	-	-	-	-	10
Cyperus distans	-	-	-	+	-	-	-	-	-	+	10
Cyperus flabelliformis	-	-	-	++	-	-	-	-	-	-	10
Cyperus iria	-	-	-	+	-	+	-	-	-	-	10
Datura stramonium	-	-	-	-	-	++	-	-	-	-	10
Dichanthium annulatum	-	-	-	++	-	-	-	-	-	-	10
Dicliptera roxburghiana	-	-	-	-	-	++	-	-	-	-	10
Digitaria griffithii	-	-	-	++	-	-	-	-	-	-	10
Emilia sonchifolia	-	-	-	-	-	++	-	-	-	-	10
Equisetum arvense	-	-	-	-	-	++	-	-	-	-	10
Equisetum debile	-	-	-	-	-	+	-	-	-	+	10
Eupatorium adenophorum	-	-	-	-	-	+	-	-	-	+	10
Euphorbia geniculata	-	-	-	-	-	++	-	-	-	-	10
Euphorbia helioscopia	-	-	-	-	-	+	-	-	-	+	10
Euphorbia parviflora	-	-	-	-	-	++	-	-	-	-	10
Geranium nepalense	-	-	-	-	-	++	-	-	-	-	10
Gnaphalium pensylvanicum	-	-	-	-	-	++	-	-	-	-	10

Species	Dye	Edible	Fuel	Fodder	Oil	Medicine	Tannin/Gum	Timber	Fibre	Miscellaneous	TIV %
Ipomoea muricata	-	-	-	-	-	++	-	-	-	-	10
Ipomoea nil	-	-	-	-	-	++	-	-	-	-	10
Justicia simplex	-	-	-	-	-	++	-	-	-	-	10
Lactuca dissecta	-	-	-	-	-	++	-	-	-	-	10
Lespedeza sericea	-	-	-	++	-	-	-	-	-	+	10
Macrotyloma unifloruma	-	-	-	-	-	++	-	-	-	-	10
Marchantia palmata	-	-	-	-	-	++	-	-	-	-	10
Medicago denticulata	-	-	-	++	-	-	-	-	-	-	10
Mucuna pruriens	-	-	-	-	-	++	-	-	-	-	10
Ocimum basilicum	-	-	-	-	-	++	-	-	-	-	10
Parthenium hysterophorus	-	-	-	-	-	++	-	-	-	-	10
Phragmites karka	-	-	-	-	-	-	+	-	-	+	10
Plectranthus coetsa	-	-	-	-	-	++	-	-	-	-	10
Poa supina	-	-	-	++	-	-	-	-	-	-	10
Pogostemon plectranthoides	-	-	-	-	-	++	-	-	-	-	10
Polygala arvensis	-	-	-	-	-	++	-	-	-	-	10
Polygonum barbatum	-	-	-	-	-	++	-	-	-	-	10
Polygonum barbatum sub sp. *gracile*	-	-	-	+	-	+	-	-	-	-	10
Polygonum minus	-	-	-	+	-	+	-	-	-	-	10
Polygonum plebejjum	-	-	-	-	-	++	-	-	-	-	10
Potamogeton pectinatus	-	-	-	-	-	-	-	-	-	++	10
Pteris cretica	-	+	-	-	-	-	+	-	-	-	10
Pupalia lappacea	-	-	-	-	-	++	-	-	-	-	10
Reinwardtia indica	-	-	-	-	-	++	-	-	-	-	10
Rhynchoglossum obliquum	-	-	-	-	-	++	-	-	-	-	10
Roylea cinerea	-	-	-	-	-	++	-	-	-	-	10
Ruellia patula	-	-	-	-	-	++	-	-	-	-	10
Rungia pectinata	-	-	-	-	-	++	-	-	-	-	10
Saccharum munja	-	-	-	-	-	-	-	-	-	++	10
Saccharum spontanium	-	-	-	-	-	-	-	-	-	++	10
Salix oxycarpa	-	-	-	-	-	++	-	-	-	-	10
Saussurea heteromala	-	-	-	-	-	++	-	-	-	-	10
Selaginella chrysocaulos	-	-	-	-	-	-	-	-	-	++	10
Sida acuta	-	-	-	-	-	++	-	-	-	-	10
Spilanthes acmella var. *oleracea*	-	-	-	-	-	++	-	-	-	-	10
Strobilanthes atropurpurens	-	-	-	-	-	++	-	-	-	-	10
Thalictrum reniforme	-	-	-	-	-	++	-	-	-	-	10
Thevetia neriifolia	-	++	-	-	-	-	-	-	-	-	10
Trichodesma indicum	-	-	-	-	-	++	-	-	-	-	10
Trifolium resupinatum	-	-	-	++	-	-	-	-	-	-	10
Trigonella pubescence	-	-	-	++	-	-	-	-	-	-	10
Urena lobata	-	-	-	-	-	++	-	-	-	-	10
Vernonia cinerea	-	-	-	-	-	++	-	-	-	-	10

Species	Dye	Edible	Fuel	Fodder	Oil	Medicine	Tannin/Gum	Timber	Fibre	Miscellaneous	TIV %
Veronica anagalis-aquatica	-	-	-	-	-	++	-	-	-	-	10
Viola pilosa	-	-	-	-	-	++	-	-	-	-	10
Withania somnifera	-	-	-	-	-	++	-	-	-	-	10
Xanthium strumarium	-	-	-	-	-	++	-	-	-	-	10
Aerva sanguinolenta	-	-	-	-	-	-	-	-	-	+	5
Chenopodium murale	-	+	-	-	-	-	-	-	-	-	5
Colocasia antiquorum	-	+	-	-	-	-	-	-	-	-	5
Cyperus compressus	-	-	-	+	-	-	-	-	-	-	5
Gomphrena celosioides	-	-	-	+	-	-	-	-	-	-	5
Hydrilla verticillata	-	-	-	-	-	-	-	-	-	+	5
Ipomoea pestigridis	-	-	-	-	-	+	-	-	-	-	5
Lindernia ciliata	-	-	-	+	-	-	-	-	-	-	5
Mirabilis jalapa	-	-	-	-	-	+	-	-	-	-	5
Najas graminea	-	-	-	-	-	-	-	-	-	+	5
Najas indica	-	-	-	-	-	-	-	-	-	+	5
Physalis longifolia	-	-	-	+	-	-	-	-	-	-	5
Polygonum donii	-	-	-	-	-	+	-	-	-	-	5
Polygonum glabrum	-	-	-	-	-	+	-	-	-	-	5
Polygonum lapathifolium	-	-	-	-	-	+	-	-	-	-	5
Polygonum pulchrum	-	-	-	-	-	+	-	-	-	-	5
Polygonum serrulatum	-	-	-	-	-	+	-	-	-	-	5
Setaria tomentosa	-	-	-	-	-	-	-	-	-	+	5
Strobilanthes dalhousianus	-	-	-	-	-	+	-	-	-	-	5
Uraria picta	-	-	-	-	-	+	-	-	-	-	5

(+ : Normal Use; ++ : Maximal Use)

So far as the observed density and availability of ethnobotanically important wetland plant resources of the region is concerened, 40 species have been found to be very abundant, 93 abundant and 70 rare (*Achyranthes aspera, Adiantum capillus-veneris, Adiantum incisum, Apluda mutica, Argemone mexicana, Calamintha umbrosum, Canna indica, Cannabis sativa, Cassia occidentalis, Centella asiatica, Cynodon dactylon, Cyperus rotundus, Dicliptera roxburghiana, Eriophorum comosum, Eupatorium adenophorum, Galium aparine, Heteropogon contortus, Hydrilla verticillata, Ipomoea carnea, Lantana camara, Macrotyloma unifloruma, Marchantia palmata, Mentha longifolia, Mentha piperita, Nasturtium officinale, Oxalis corniculata, Parthenium hysterophorus, Plumbago zeylanica, Pteris cretica, Reinwardtia indica, Ricinus communis, Rubus ellipticus, Rumex nepalensis, Salix oxycarpa, Taraxacum officinale, Thalictrum reniforme, Vernonia cinerea, Veronica anagalis-aquatica, Viola pilosa, Xanthium strumarium*; 93 abundant species are *Abelmoschus crinitus, Acorus calamus, Aerva sanguinolenta, Ageratum conyzoides, Ajuga bracteosa, Amaranthus viridis, Ampelopteris prolifera, Anisomeles indica, Artemisia indica, Artemisia scoparia, Barleria cristata, Bidens pilosa, Boehmeria platyphylla, Bryophyllum calycinum, Capsella bursa-pastoris, Cassia absus, Ceratophyllum demersum, Cheilanthes bicolor, Chenopodium album, Christella dentata, Cissampelos pareira, Commelina diffusa, Commelina paludosa, Costus speciosus, Cryptolepis buchanani, Cucumis pubescens, Cymbopogon citratus, Cymbopogon martinii, Cyperus compressus, Datura stramonium, Debregeasia hypoleuca, Dichanthium annulatum, Digi-*

taria griffithii, Echinochloa frumentacea, Eleusine indica, Equisetum arvense, Equisetum debile, Euphorbia geniculata, Euphorbia hirta, Ficus hispida, Ficus roxburghii, Fumaria indica, Geranium nepalense, Hedera helix, Hedychium spicatum, Impatiens balsamina, Ipomoea muricata, Lactuca dissecta, Lathyrus aphaca, Leea crispa, Lindernia ciliata, Marsilea minuta, Medicago denticulata, Melothria heterophylla, Mirabilis jalapa, Mucuna pruriens, Najas graminea, Najas indica, Nerium indicum, Oroxylum indicum, Paspalum distichum, Pennisetum lanatum, Phragmites karka, Phyllanthus urinaria, Plectranthus coetsa, Poa supina, Polygonum barbatum, Polygonum barbatum sub sp. *gracile, Polygonum plebejjum, Potamogeton crispus, Potamogeton pectinatus, Pupalia lappacea, Rhynchoglossum obliquum, Roylea cinerea, Rumex hastatus, Saccharum munja, Saccharum spontanium, Saussurea heteromala, Selaginella chrysocaulos, Silene conoidea, Solanum nigrum, Solanum xanthocarpum, Sonchus arvensis, Spilanthes acmella* var. *oleracea, Spilanthes paniculat, Strobilanthes dalhousianus, Trichodesma indicum, Trifolium resupinatum, Trigonella pubescence, Uraria picta, Urena lobata, Urtica dioica, Vicia sativa;* 70 rare species are *Abrus precatorius, Abutilon indicum, Aeschynomene aspera, Amaranthus gangeticus, Amaranthus paniculatus, Amaranthus tricolor* var. *gangeticus, Andrographis paniculata, Bacopa monnieri, Begonia picta, Blyxa auberti, Bromus catharticus, Cardiospermum halicacobum, Chenopodium ambrosiodes, Chenopodium murale, Clematis gouriana, Coix lachryma – jobi, Colocasia antiquorum, Convolvulus arvensis, Conyza bonariensis, Conyza stricta, Crotolaria alata, Crotolaria mysorensis, Curcuma longa, Cyathocline purpurea, Cyperus distans, Cyperus flabelliformis, Cyperus iria, Dioscorea belophylla, Dioscorea bulbifera, Emilia sonchifolia, Euphorbia helioscopia, Euphorbia parviflora, Fragaria indica, Fragaria nubicola, Girardinia heterophylla, Gloriosa superba, Gnaphalium pensylvanicum, Gomphrena celosioides, Ipomoea cairica, Ipomoea nil, Ipomoea pestigridis, Justicia simplex, Lannea coromandelica, Lepidagathis cuspidata, Lespedeza sericea, Martynia annua, Micromeria biflora, Momordica dioica, Ocimum basilicum, Physalis longifolia, Physalis minima, Pogostemon plectranthoides, Polygala arvensis, Polygonum donii, Polygonum glabrum, Polygonum lapathifolium, Polygonum minus, Polygonum pulchrum, Polygonum serrulatum, Ruellia patula, Rungia pectinata, Setaria tomentosa, Sida acuta, Sonchus asper, Sorghum halepense, Stellaria media, Strobilanthes atropurpurens, Thevetia neriifolia, Tridax procumbens, Withania somnifera* (Table 59). Among these, species like *Ipomoea carica, Dioscorea belophylla* and *Veronica anagalis-aquatica* have been listed as rare and endangered flowering plant species (Rao, 2006); *Datura stramonium, Equisetum arvense* as threatened medicinal plant resources in W. Himalaya (Kaul, 1996); *Andrographis paniculata, Bacopa monnieri, Centella asiatica, Plumbago zeylanica* as endangered medicinal plants (Rao & Rajasekharan, 2002); *Acorus calamus, Gloriosa superba, Hedychium spicata* as IUCN status endangered medicinal plants having various threat levels of 2, 1, 5; 4, 2, 1, 3, 5 and 4, 5, 2, respectively; *Abrus precatorius, Achyranthes aspera, Ageratum conyzoides, Chenopodium ambrosoides, Cissampelos pariera, Euphorbia hirta, Fumaria indica, Oxalis corniculata, Polygonum glabrum, Solanum nigrum, Taraxacum officinale, Vernonia cineria, Xanthium strumarium* as species which may be exported subject to licenscing and ceiling limit (Subramani et al., 2005) whereas those like *Acorus calamus, Costus speciosus, Gloriosa superba* and *Withania somnifera* are listed in negative list of export by Government of India (Ahuja, 2001). In order to ward off the danger of extinction of the above listed species, Government regulation on wild collections of endangered species is necessary as well as there is need to assess the cause of their diversity loss. Moreover, policies regarding their conservation and cultivation should also be encouraged and strictly enforced with participation of local people, village panchayats, NGOs and forest department.

Table 59 Observed Density and Availability (ODA) of Ethnobotanically Important Wetland Plants of District Bilaspur (H.P.)

Botanical Name	Flowering and Fruiting Season	Observed Density and Availability
Abelmoschus crinitus Wall.	July-September	++
Abrus precatorius L.	July-August	+
Abutilon indicum L. Sweet	July-October	+
Achyranthes aspera L.	May-October	+++
Acorus calamus L.	June-July	++
Adiantum capillus-veneris L. f.	May-October	+++
Adiantum incisum Forsk.	June-December	+++
Aerva sanguinolenta (L.) Blume	July-October	++
Aeschynomene aspera L.	October-December	+
Ageratum conyzoides L.	August-October	++
Ajuga bracteosa Wall. *ex* Benth.	January-March	++
Amaranthus gangeticus L. Syst.	July-September	+
Amaranthus paniculatus L.	September-November	+
Amaranthus tricolor (L.) var. *gangeticus* (L.) Fiori	June-November	+
Amaranthus viridis L.	May-June	++
Ampelopteris prolifera (Retz.) Copeland.	August-October	++
Andrographis paniculata (Burm. F.) Wall. *ex* Nees	October-December	+
Anisomeles indica (L.) O. Kuntze	September-December	++
Apluda mutica Auct.	September-November	+++
Argemone mexicana L.	April-May	+++
Artemisia indica Waldst. & Kit.	August-October	++
Artemisia scoparia Waldst. & Kit.	August-September	++
Bacopa monnieri (L.) Pennell	August-October	+
Barleria cristata L.	July-October	++
Begonia picta Sm.	August-September	+
Bidens pilosa L.	June-September	++
Blyxa auberti Rich.	May-October	+
Boehmeria platyphylla Don	July-September	++
Bromus catharticus Vahl	July-October	+
Bryophyllum calycinum Salisb.	July-November	++
Calamintha umbrosum (M.B.) C. Koch	April-August	+++
Canna indica L.	July-September	+++
Cannabis sativa L.	June-September	+++
Capsella bursa-pastoris (L.) Medik.	April-October	++
Cardiospermum halicacobum L.	August-March	+
Cassia absus L.	July-September	++
Cassia occidentalis L.	May-September	+++
Centella asiatica L.	April-May	+++
Ceratophyllum demersum L.	October-February	++
Cheilanthes bicolor (Roxb.) Fraser-Jenkins.	July-October	++
Chenopodium album L.	November-April	++
Chenopodium ambrosiodes L.	June-September	+
Chenopodium murale L.	January-April	+
Christella dentata (Forsk.) Brownsey & Jermy.	May-October	++
Cissampelos pareira L.	April-August	++
Clematis gouriana Roxb.	June- September	+
Coix lachryma – jobi L.	November-February	+
Colocasia antiquorum Schott	June- October	+

Botanical Name	Flowering and Fruiting Season	Observed Density and Availability
Commelina diffusa Burm. F.	July-September	++
Commelina paludosa Burm.	June-August	++
Convolvulus arvensis L.	December-March	+
Conyza bonariensis L.	June-october	+
Conyza stricta Willd.	September-October	+
Costus speciosus Sm.	July-August	++
Crotolaria alata Buch.-Ham.	July-November	+
Crotolaria mysorensis Roth	July-October	+
Cryptolepis buchanani Roem. & Schult.	April-November	++
Cucumis pubescens Willd.	May-August	++
Curcuma longa Wall.	April-June	+
Cyathocline purpurea (Don) Kuntze	April-May	+
Cymbopogon citratus Stapf	June -August	++
Cymbopogon martinii Stapf	September-February	++
Cynodon dactylon (L.) Pers.	August-September	+++
Cyperus compressus L.	August-December	++
Cyperus distans L.f.	August-December	+
Cyperus flabelliformis Rottb.	August-December	+
Cyperus iria L.	July-October	+
Cyperus rotundus L.	July-October	+++
Datura stramonium L.	June-October	++
Debregeasia hypoleuca Wedd.	January-April	++
Dichanthium annulatum Hack.	September-December	++
Dicliptera roxburghiana Nees	October-December	+++
Digitaria griffithii (Hook.f.) Henn.	October-December	++
Dioscorea belophylla Voigt. *ex* Haines.	September-November	+
Dioscorea bulbifera L.	September-February	+
Echinochloa frumentacea Link	July-February	++
Eleusine indica Gaertn.	August-November	++
Emilia sonchifolia (L.) DC.	December-April	+
Equisetum arvense L.	July-October	++
Equisetum debile Roxb.	August-December	++
Eriophorum comosum Wall.	July-September	+++
Eupatorium adenophorum Spreng.	February-May	+++
Euphorbia geniculata Ort. *ex* Boiss.	August-November	++
Euphorbia helioscopia L.	March-August	+
Euphorbia hirta L.	March-August	++
Euphorbia parviflora L.	May-June	+
Ficus hispida L.f.	March-May	++
Ficus roxburghii Wall.	April-August	++
Fragaria indica Andr.	June-August	+
Fragaria nubicola Lindl.	May-October	+
Fumaria indica (Hausskn.) Pugsley.	February-March	++
Galium aparine L.	July-August	+++
Geranium nepalense Sweet	June- September	++
Girardinia heterophylla Decne.	September-October	+
Gloriosa superba L.	July-September	+
Gnaphalium pensylvanicum Willd.	December-May	+
Gomphrena celosioides Mart.	January-April	+
Hedera helix Clarke	June-November	++
Hedychium spicatum Buch.-Ham. *ex* Sm.	July-September	++

Botanical Name	Flowering and Fruiting Season	Observed Density and Availability
Heteropogon contortus (L.) Beauv. *ex* Roem. & Schult.	September-December	+++
Hydrilla verticillata (L.f.) Royle	January-April	+++
Impatiens balsamina L.	July-September	++
Ipomoea cairica L.	May-July	+
Ipomoea carnea Facq.	July- September	+++
Ipomoea muricata Jacq.	June-August	++
Ipomoea nil (L.) Roth	June-September	+
Ipomoea pestigridis L.	September-December	+
Justicia simplex Don	June-August	+
Lactuca dissecta Don	May-September	++
Lannea coromandelica (Houtt.) Merr.	March-June	+
Lantana camara L.	April-June	+++
Lathyrus aphaca L.	April-May	++
Leea crispa Willd.	August-September	++
Lepidagathis cuspidata Nees	December-June	+
Lespedeza sericea Miq.	November-February	+
Lindernia ciliata Colsm.	August-February	++
Macrotyloma unifloruma (Lam.) Verdc.	July-september	+++
Marchantia palmata Nees	January-April	+++
Marsilea minuta L.	November-April	++
Martynia annua L.	August-November	+
Medicago denticulata Willd.	February-April	++
Melothria heterophylla Cogn.	May-October	++
Mentha longifolia L.	November-February	+++
Mentha piperita L.	July-September	+++
Micromeria biflora (Buch.-Ham.) Benth.	January-July	+
Mirabilis jalapa L.	June-September	++
Momordica dioica Roxb. *ex* Willd.	July-September	+
Mucuna pruriens DC.	July-September	++
Najas graminea Dd.	August-October	++
Najas indica (Willd.) Cham.	August-September	++
Nasturtium officinale R. Br.	April-September	+++
Nerium indicum Mill.	April-June	++
Ocimum basilicum L.	June-September	+
Oroxylum indicum Vent.	June-August	++
Oxalis corniculata L.	April-November	+++
Parthenium hysterophorus L.	August-September	+++
Paspalum distichum L.	August-October	++
Pennisetum lanatum Klotzsch	June-October	++
Phragmites karka Roxb.	May-June	++
Phyllanthus urinaria L.	July-September	++
Physalis longifolia Nutt.	July-October	+
Physalis minima L.	June-August	+
Plectranthus coetsa Buch.-Ham. *ex* D. Don	October-November	++
Plumbago zeylanica L.	October-March	+++
Poa supina Schrad.	August-September	++
Pogostemon plectranthoides Desf.	February-April	+
Polygala arvensis Willd.	July-December	+
Polygonum barbatum L.	April-July	++
Polygonum barbatum L. sub sp. *gracile* Dansar.	April-July	++
Polygonum donii Meisn.	April-July	+

Botanical Name	Flowering and Fruiting Season	Observed Density and Availability
Polygonum glabrum Willd.	January-September	+
Polygonum lapathifolium L.	November-February	+
Polygonum minus Huds.	January-December	+
Polygonum plebejjum Br. Prodr.	October-April	++
Polygonum pulchrum Blume	October-March	+
Polygonum serrulatum Lag.	June-October	+
Potamogeton crispus L.	December-June	++
Potamogeton pectinatus L.	October-January	++
Pteris cretica L.	May-October	+++
Pupalia lappacea Juss.	July-September	++
Reinwardtia indica Dumort.	February-April	+++
Rhynchoglossum obliquum Blume	July-August	++
Ricinus communis L.	March-May	+++
Roylea cinerea Baill.	May-October	++
Rubus ellipticus Sm.	January-May	+++
Ruellia patula Jacq.	January-April	+
Rumex hastatus Don	November-April	++
Rumex nepalensis Spreng.	May-October	+++
Rungia pectinata (L.) Nees	January-February	+
Saccharum munja L.	January-April	++
Saccharum spontanium L.	November-December	++
Salix oxycarpa Anderss.	April-June	+++
Saussurea heteromala (Don) Hand.-Mazz.	April-June	++
Selaginella chrysocaulos Hook. *et* Grev.	August-October	++
Setaria tomentosa Kunth	January-May	+
Sida acuta Burm. f.	September-December	+
Silene conoidea L.	March-May	++
Solanum nigrum L.	January-July	++
Solanum xanthocarpum Schrad. & Wendl.	March-April	++
Sonchus arvensis L.	May-June	++
Sonchus asper Hill	March-April	+
Sorghum halepense Wall.	July-October	+
Spilanthes acmella L.var. *oleracea* Jacq.	October-March	++
Spilanthes paniculata Wall *ex* DC.	October-March	++
Stellaria media L.	March-April	+
Strobilanthes atropurpurens Nees	October-February	+
Strobilanthes dalhousianus Clarke	July-October	++
Taraxacum officinale Wigg	March-November	+++
Thalictrum reniforme Wall.	June-August	+++
Thevetia neriifolia Juss. *ex* Steud.	July-September	+
Trichodesma indicum Br.	September-December	++
Tridax procumbens L.	July-August	+
Trifolium resupinatum L.	May-June	++
Trigonella pubescence Edgew.	March-May	++
Uraria picta Desf.	July-November	++
Urena lobata L.	July-September	++
Urtica dioica L.	June-October	++
Vernonia cinerea (L.) Less.	October-March	+++
Veronica anagalis-aquatica L.	November-January	+++
Vicia sativa L.	January-February	++
Viola pilosa Blume	February-April	+++

Botanical Name	Flowering and Fruiting Season	Observed Density and Availability
Withania somnifera Dunal	September-October	+
Xanthium strumarium L.	September-November	+++

(+ Rare extent; ++ Considerable extent; +++ Abundant)

Table 60   Ethnobotanically Important Wetland Plants of District Bilaspur with Additional Uses

Botanical Name	Family
Abelmoschus crinitus	Malvaceae
Abrus precatorius	Fabaceae
Abutilon indicum	Malvaceae
Achyranthes aspera	Amaranthaceae
Acorus calamus	Araceae
Adiantum capillus-veneris	Adiantaceae
Adiantum incisum	Adiantaceae
Aeschynomene aspera	Fabaceae
Ageratum conyzoides	Asteraceae
Ajuga bracteosa	Lamiaceae
Amaranthus gangeticus	Amaranthaceae
Amaranthus tricolor var. *gangeticus*	Amaranthaceae
Amaranthus viridis	Amaranthaceae
Ampelopteris prolifera	Thelypteridaceae
Andrographis paniculata	Acanthaceae
Anisomeles indica	Lamiaceae
Artemisia indica	Asteraceae
Artemisia scoparia	Asteraceae
Bacopa monnieri	Scrophulariaceae
Barleria cristata	Acanthaceae
Begonia picta	Begoniaceae
Bidens pilosa	Asteraceae
Blyxa auberti	Hydrocharidaceae
Boehmeria platyphylla	Urticaceae
Bromus catharticus	Poaceae
Bryophyllum calycinum	Crassulaceae
Calamintha umbrosum	Lamiaceae
Cannabis sativa	Cannabinaceae
Capsella bursa-pastoris	Brassicaceae
Cardiospermum halicacobum	Sapindaceae
Cassia absus	Fabaceae
Cassia occidentalis	Fabaceae
Centella asiatica	Apiaceae
Ceratophyllum demersum	Ceratophyllaceae
Cheilanthes bicolor	Sinopteridaceae
Chenopodium album	Chenopodiaceae
Chenopodium murale	Chenopodiaceae
Christella dentata	Thelypteridaceae
Clematis gouriana	Ranunculaceae
Coix lachryma – jobi	Poaceae

Botanical Name	Family
Colocasia antiquorum	Araceae
Commelina diffusa	Commelinaceae
Commelina paludosa	Commelinaceae
Conyza bonariensis	Asteraceae
Costus speciosus	Zingiberaceae
Crotolaria alata	Fabaceae
Cryptolepis buchanani	Asclepiadaceae
Cucumis pubescens	Cucurbitaceae
Cyathocline purpurea	Asteraceae
Cymbopogon citratus	Poaceae
Cymbopogon martinii	Poaceae
Cynodon dactylon	Poaceae
Cyperus distans	Cyperaceae
Cyperus flabelliformis	Cyperaceae
Cyperus iria	Cyperaceae
Cyperus rotundus	Cyperaceae
Datura stramonium	Solanaceae
Debregeasia hypoleuca	Urticaceae
Dichanthium annulatum	Poaceae
Dicliptera roxburghiana	Acanthaceae
Dioscorea belophylla	Dioscoreaceae
Dioscorea bulbifera	Dioscoreaceae
Echinochloa frumentacea	Poaceae
Eleusine indica	Poaceae
Emilia sonchifolia	Asteraceae
Equisetum arvense	Equisetaceae
Equisetum debile	Equisetaceae
Eriophorum comosum	Cyperaceae
Eupatorium adenophorum	Asteraceae
Euphorbia geniculata	Euphorbiaceae
Euphorbia helioscopia	Euphorbiaceae
Euphorbia hirta	Euphorbiaceae
Euphorbia parviflora	Euphorbiaceae
Ficus hispida	Moraceae
Ficus roxburghii	Moraceae
Fragaria indica	Rosaceae
Fumaria indica	Fumariaceae
Galium aparine	Rubiaceae
Geranium nepalense	Geraniaceae
Girardinia heterophylla	Urticaceae
Gloriosa superba	Liliaceae
Gomphrena celosioides	Amaranthaceae
Hedera helix	Araliaceae
Heteropogon contortus	Poaceae
Hydrilla verticillata	Hydrocharidaceae
Impatiens balsamina	Balsaminaceae
Ipomoea carnea	Convolvulaceae
Ipomoea muricata	Convolvulaceae
Ipomoea pestigridis	Convolvulaceae
Justicia simplex	Acanthaceae
Lactuca dissecta	Asteraceae
Lannea coromandelica	Anacardiaceae

Botanical Name	Family
Lantana camara	Verbenaceae
Lathyrus aphaca	Fabaceae
Leea crispa	Leeaceae
Lepidagathis cuspidata	Acanthaceae
Lespedeza sericea	Fabaceae
Lindernia ciliata	Scrophulariaceae
Marchantia palmata	Marchantiaceae
Martynia annua	Martyniaceae
Melothria heterophylla	Cucurbitaceae
Mentha longifolia	Lamiaceae
Mirabilis jalapa	Nyctaginaceae
Momordica dioica	Cucurbitaceae
Najas graminea	Naiadaceae
Najas indica	Naiadaceae
Nasturtium officinale	Brassicaceae
Nerium indicum	Apocynaceae
Ocimum basilicum	Lamiaceae
Oroxylum indicum	Bignoniaceae
Oxalis corniculata	Oxalidaceae
Parthenium hysterophorus	Asteraceae
Phragmites karka	Poaceae
Phyllanthus urinaria	Euphorbiaceae
Physalis longifolia	Solanaceae
Plectranthus coetsa	Lamiaceae
Plumbago zeylanica	Plumbaginaceae
Polygala arvensis	Polygalaceae
Polygonum barbatum	Polygonaceae
Polygonum barbatum sub sp. *gracile*	Polygonaceae
Polygonum donii	Polygonaceae
Polygonum glabrum	Polygonaceae
Polygonum lapathifolium	Polygonaceae
Polygonum serrulatum	Polygonaceae
Potamogeton crispus	Naiadaceae
Potamogeton pectinatus	Naiadaceae
Pteris cretica	Pteridaceae
Pupalia lappacea	Amaranthaceae
Reinwardtia indica	Linaceae
Rhynchoglossum obliquum	Gesneriaceae
Ricinus communis	Euphorbiaceae
Roylea cinerea	Lamiaceae
Rubus ellipticus	Rosaceae
Ruellia patula	Acanthaceae
Rumex hastatus	Polygonaceae
Rumex nepalensis	Polygonaceae
Saccharum munja	Poaceae
Saccharum spontanium	Poaceae
Salix oxycarpa	Salicaceae
Saussurea heteromala	Asteraceae
Selaginella chrysocaulos	Selaginellaceae
Setaria tomentosa	Poaceae
Silene conoidea	Caryophyllaceae
Solanum nigrum	Solanaceae

Botanical Name	Family
Solanum xanthocarpum	Solanaceae
Sonchus asper	Asteraceae
Spilanthes acmella var. *oleracea*	Asteraceae
Spilanthes paniculata	Asteraceae
Stellaria media	Caryophyllaceae
Strobilanthes atropurpurens	Acanthaceae
Taraxacum officinale	Asteraceae
Thalictrum reniforme	Ranunculaceae
Thevetia neriifolia	Apocynaceae
Trichodesma indicum	Boraginaceae
Tridax procumbens	Asteraceae
Trifolium resupinatum	Fabaceae
Trigonella pubescence	Fabaceae
Urena lobata	Malvaceae
Urtica dioica	Urticaceae
Vernonia cinerea	Asteraceae
Vicia sativa	Fabaceae
Viola pilosa	Violaceae
Withania somnifera	Solanaceae

Table 61   Wetland Plants Reported for the First Time from District Bilaspur

Plants	Family
Amaranthus paniculatus	Amaranthaceae
Amaranthus tricolor var. *gangeticus*	Amaranthaceae
Andrographis paniculata	Acanthaceae
Artemisia indica	Asteraceae
Blyxa auberti	Hydrocharidaceae
Bromus catharticus	Poaceae
Calamintha umbrosum	Lamiaceae
Colocasia antiquorum	Araceae
Commelina diffusa	Commelinaceae
Conyza bonariensis	Asteraceae
Cryptolepis buchanani	Asclepiadaceae
Cucumis pubescens	Cucurbitaceae
Cymbopogon citratus	Poaceae
Cyperus distans	Cyperaceae
Cyperus flabelliformis	Cyperaceae
Digitaria griffithii	Poaceae
Eupatorium adenophorum	Asteraceae
Gnaphalium pensylvaticum	Asteraceae
Ipomoea muricata	Convolvulaceae
Justicia simplex	Acanthaceae
Macrotyloma unifloruma	Fabaceae
Mentha piperita	Lamiaceae
Najas graminea	Naiadaceae
Najas indica	Naiadaceae
Physalis longifolia	Solanaceae
Polygonum barbatum	Polygonaceae
Polygonum barbatum sub sp. *gracile*	Polygonaceae

Plants	Family
Polygonum donii	Polygonaceae
Polygonum minus	Polygonaceae
Polygonum pulchrum	Polygonaceae
Reinwardtia indica	Linaceae
Ruellia patula	Acanthaceae
Saccharum munja	Poaceae
Setaria tomentosa	Poaceae
Sida acuta	Malvaceae
Spilanthes acmella var. *oleracea*	Asteraceae
Spilanthes paniculata	Asteraceae
Strobilanthes dalhousianus	Acanthaceae
Trifolium resupinatum	Fabaceae
Vicia sativa	Fabaceae
Viola pilosa	Violaceae

Table 62   Ethnobotanically Important Wetland Plants Not Reported in Literature

Plants	Family
Blyxa auberti	Hydrocharidaceae
Gomphrena celosioides	Amaranthaceae
Najas indica	Naiadaceae
Pupalia lappacea	Amaranthaceae
Trigonella pubescence	Fabaceae

On critical screening of the presently generated data and its comparison with earlier ethnobotanical findings presented in different treatises, books and research papers revealed (Ambasta, 1986; Singh & Maheshwari, 1983; Singh, 1993; Singh & Prakash, 1994; Siwakoti & Siwakoti, 2000; Sood et al., 2001) that the ethnobotanical usages of 5 wetland plant species, e.g., *Blyxa auberti, Gomphrena celosiodes, Najas indica, Pupalia lappacea* and *Trigonella pubescence* are recorded for the first time (Table 62); and new usages for 163 already reported wild plants of various Indian tribes (Table 60) *Abelmoschus crinitus, Abrus precatorius, Abutilon indicum, Achyranthes aspera, Acorus calamus, Adiantum capillus-veneris, Adiantum incisum, Aeschynomene aspera, Ageratum conyzoides, Ajuga bracteosa, Amaranthus gangeticus, Amaranthus tricolor* var. *gangeticus, Amaranthus viridis, Ampelopteris prolifera, Andrographis paniculata, Anisomeles indica, Artemisia indica, Artemisia scoparia, Bacopa monnieri, Barleria cristata, Begonia picta, Bidens pilosa, Boehmeria platyphylla, Bromus catharticus, Bryophyllum calycinum, Calamintha umbrosum, Canna indica, Cannabis sativa, Capsella bursa-pastoris, Cardiospermum halicacobum, Cassia absus, Cassia occidentalis, Centella asiatica, Ceratophyllum demersum, Cheilanthes bicolor, Chenopodium album, Chenopodium murale, Christella dentata, Cissampelos pareira, Clematis gouriana, Coix lachryma – jobi, Colocasia antiquorum, Commelina diffusa, Commelina paludosa, Conyza bonariensis, Costus speciosus, Crotolaria alata, Cryptolepis buchanani, Cucumis pubescens, Curcuma longa, Cyathocline purpurea, Cymbopogon citratus, Cymbopogon martinii, Cynodon dactylon, Cyperus distans, Cyperus flabelliformis, Cyperus iria, Cyperus rotundus, Datura stramonium, Debregeasia hypoleuca, Dichanthium annulatum, Dicliptera roxburghiana, Dioscorea belophylla, Dioscorea bulbifera, Echinochloa frumentacea, Eleusine indica, Emilia sonchifolia, Equisetum arvense, Equisetum debile, Eriophorum comosum, Eupatorium adenophorum, Euphorbia geniculata,*

Euphorbia helioscopia, Euphorbia hirta, Euphorbia parviflora, Ficus hispida, Ficus roxburghii, Fragaria indica, Fumaria indica, Galium aparine, Geranium nepalense, Girardinia heterophylla, Gloriosa superba, Hedera helix, Heteropogon contortus, Hydrilla verticillata, Impatiens balsamina, Ipomoea carnea, Ipomoea muricata, Ipomoea pestigridis, Justicia simplex, Lactuca dissecta, Lannea coromandelica, Lantana camara, Lathyrus aphaca, Leea crispa, Lepidagathis cuspidata, Lespedeza sericea, Lindernia ciliata, Marchantia palmata, Martynia annua, Melothria heterophylla, Mentha longifolia, Mirabilis jalapa, Momordica dioica, Najas graminea, Nasturtium officinale, Nerium indicum, Ocimum basilicum, Oroxylum indicum, Oxalis corniculata, Parthenium hysterophorus, Phragmites karka, Phyllanthus urinaria, Physalis longifolia, Plectranthus coetsa, Plumbago zeylanica, Polygala arvensis, Polygonum barbatum, Polygonum barbatum sub sp. *gracile, Polygonum donii, Polygonum glabrum, Polygonum lapathifolium, Polygonum serrulatum, Potamogeton crispus, Potamogeton pectinatus, Pteris cretica, Reinwardtia indica, Rhynchoglossum obliquum, Ricinus communis, Roylea cinerea, Rubus ellipticus, Ruellia patula, Rumex hastatus, Rumex nepalensis, Saccharum munja, Saccharum spontanium, Salix oxycarpa, Saussurea heteromala, Selaginella chrysocaulos, Setaria tomentosa, Silene conoidea, Solanum nigrum, Solanum xanthocarpum, Sonchus asper, Spilanthes acmella* var. *oleracea, Spilanthes paniculata, Stellaria media, Strobilanthes atropurpurens, Taraxacum officinale, Thalictrum reniforme, Thevetia neriifolia, Trichodesma indicum, Tridax procumbens, Trifolium resupinatum, Urena lobata, Urtica dioica, Vernonia cinerea, Vicia sativa, Viola pilosa, Withania somnifera* form significant contributions of the present study. Not only that, as many as 41 plant species form new floristic records for the region (Table LXI) as these have not been enlisted in any of the earlier flora published of the state of Himachal Pradesh (Chowdhery & Wadhwa, 1984; Polunin & Stainton, 1984; Stainton, 1988).

Altogether, a total of 92 species in 75 genera under 40 families have been recorded as aliens from the wetlands of district Bilaspur (H.P.) (Table 63), representing 0.54% of the Indian flora. Of these, six species *Canna indica, Chenopodium album, Cucumis pubescens, Cymbopogon martinii, Mentha piperita* and *Mirabilis jalapa* occur wild as well as cultivated in the region (Table 8). Statistically, the ratio of exotics to native species is 0.82:1 in the region (Tables 63, 64). From habit-wise analysis, it is indicated that herbs with 81 species (88.04%) predominate followed by shrubs 5 (5.43%), undershrubs 3 (3.26%), tree, climber and liana 1 species (1.08%) each (Figure. 20). Among these, dicotyledonous taxa with 66 species grouped under 54 genera and 28 families constitute maximum (71.74%) proportion of wetland plant resources of the study area (Table 65), followed by monocots (20 species; Table 66) distributed in 15 genera under 7 families such as Poaceae (*Bromus catharticus, Cymbopogon martinii, Cynodon dactylon, Heteropogon contortus, Poa supina, Setaria tomentosa, Sorghum halepense*), Naiadaceae (*Najas graminea, Najas indica, Potamogeton crispus, Potamogeton pectinatus*), Hydrocharitaceae (*Blyxa auberti, Hydrilla verticillata*), Cyperaceae (*Cyperus compressus, Cyperus distans, Cyperus rotundus*), Commelinaceae (*Commelina diffusa, Commelina paludosa*), Araceae (*Acorus calamus*), Cannaceae (*Canna indica*) (Table 66) and Pteridophytes (6 species) represented by 6 genera under 5 families (Tables 67, 68), namely Thelypteridaceae (*Ampelopteris prolifera, Christella dentata*), Adiantaceae (*Adiantum capillus-veneris*), Sinopteridaceae (*Cheilanthes bicolor*), Equisetaceae (*Equisetum debile*) and Pteridaceae (*Pteris cretica*). In the analysis of families, largest number of alien plant species belongs to Asteraceae (13), Fabaceae (8), Euphorbiaceae (5), Convolvulaceae (5), Polygonaceae (5), Acanthaceae (3), Amaranthaceae (3), etc., amongst dicots (Table 69) and Poaceae (7) Naiadaceae (4) and Cyperaceae (3) amongst monocots (Table 66). Relatively, the top ten alien families contribute 56 species with proportion of 60.86% (Tables 69, 70). Seven families: Brassicaceae, Caryophyllaceae, Cucurbitaceae, Chenopodiaceae, Hydrocharitaceae, Commelinaceae and Thelypteridaceae with two species each, and the remaining twenty three families, viz., Urticaceae, Apocynaceae, Araliaceae, Begoniaceae, Crassulaceae, Gesneriaceae, Lamiaceae, Linaceae, Martyniaceae, Nyctaginaceae, Oxalidaceae, Papaveraceae, Rosaceae, Rubiaceae, Scrophulariaceae, Solanaceae, Verbenaceae, Araceae, Cannaceae, Adiantaceae, Sinopteridaceae, Equisetaceae and Pteridaceae are represented by one species each (Tables 68 - 70). Importantly, the genera with highest number of alien species are *Ipomoea, Polygonum* (4 each), *Amaranthus, Cyperus, Euphorbia,* (3 each), *Chenopodium, Commelina, Conyza, Crotolaria, Najas, Potamogeton* and *Strobilanthes* (2 each) (Tables 71, 72). Furthermore, it is evident from Tables 73 - 78 that most of alien wetland species have been introduced intentionally for various purposes. Among these 72 (78.26%) wetland plants have been introduced predominantly for medicinal purposes (Table 75), followed by fodder 21 (22.82%; Table 73), edible 18 (19.56%; Table 48), i.e., *Amaranthus paniculatus, Amaranthus tricolor* var. *gangeticus, Amaranthus viridis, Ampelopteris prolifera, Capsella bursa-pastoris, Cassia occidentalis, Chenopodium album, Chenopodium ambrosiodes, Commelina diffusa, Cucumis pubescens, Fragaria nubicola, Hedera helix, Melothria heterophylla, Nasturtium officinale, Oxalis corniculata, Rumex nepalensis, Sorghum halepense, Thevetia neriifolia;* fish food 6 (6.52%; Table 78), i.e. *Hydrilla verticillata, Ipomoea cairica, Najas graminea, Najas indica, Potamogeton crispus, Potamogeton pectinatus;* crop plants 6 (6.52%; Table 8), ornamental and ethnoveterinary 3 (3.26%; Tables 76, 77) each. Sustainable exploitation of all these species may help in socio-economic amelioration of the locals in the region.

Table 63   Exotic Wetland Plants of District Bilaspur

Botanical Name	Family	Origin
Adiantum capillus-veneris L. f.	Adiantaceae	Europe, North America, Central and South America, Eurasia
Amaranthus viridis L.	Amaranthaceae	Caribbean Territories, South America,
Amaranthus paniculatus L.	Amaranthaceae	Central America, Eastern America
Amaranthus tricolor (L.) var. *gangeticus* (L.) Fiori	Amaranthaceae	Tropical Asia, Africa and America, China
Ampelopteris prolifera (Retz.) Copeland.	Thelypteridaceae	Australia, Old World Tropics
Argemone mexicana L.	Papaveraceae	Mexico and Caribbean
Artemisia scoparia Waldst. & Kit.	Asteraceae	Euresia, Africa, Asia Temperate, Pakistan
Begonia picta Sm.	Begoniaceae	Yunnan province of China
Bidens pilosa L.	Asteraceae	Java, South Africa
Blyxa auberti Rich.	Hydrocharitaceae	Australia
Boehmeria platyphylla Don	Urticaceae	Malay Islands, China, Japan, Africa
Bromus catharticus Vahl	Poaceae	South America, California
Bryophyllum calycinum Salisb.	Crassulaceae	Madagascar
Canna indica L.	Cannaceae	Central and Southern America
Capsella bursa-pastoris (L.) Medik.	Brassicaceae	Middle East
Cassia occidentalis L.	Fabaceae	Cosmopolitan in the Tropics, America
Cheilanthes bicolor (Roxb.) Fraser-Jenkins.	Sinopteridaceae	U.S.A., California
Chenopodium album L.	Chenopodiaceae	South America, Central America, Mexico
Chenopodium ambrosiodes L.	Chenopodiaceae	Central America
Christella dentata (Forsk.) Brownsey & Jermy.	Thelypteridaceae	Phillippines Island
Commelina diffusa Burm. f.	Commelinaceae	Tropics and Subtropics regions
Commelina paludosa Burm.	Commelinaceae	U. S. A.

Botanical Name	Family	Origin
Convolvulus arvensis L.	Convolvulaceae	Africa- Algeria, Egypt, Libya, Moroco, Tunisia; Asia-Temperate- Afganistan, Armenia, Azerbaijan, China, Cyprus, Egypt, Iran, Israel, Jordan, Kazakhstan, Kyrgyzstan, Lebanon, Mongolia, Syria, Tajikistan, Turkey, Turkmenistan, Uzbekistan; Asia-Tropical-Nepal; Pakistan; Europe-Albania, Austria, Belarus, Belgium, Bulgaria, Czechoslovakia, Denmark, Estonia, Finland, Former Yugoslavia, France, Germany, Greece, Hungary, Ireland, Italy, Latvia, Lithuania, Moldova, Netherlands, Norway, Poland, Romania, Russian Federation, Spain, Sweden, Switzerland, Ukraine, United Kingdom
Conyza bonariensis L.	Asteraceae	North America and Central America, Canada
Conyza stricta Willd.	Asteraceae	Eastern Africa
Crotolaria alata Buch.-Ham.	Fabaceae	Java
Crotolaria mysorensis Roth	Fabaceae	Tropical region
Cucumis pubescens Willd.	Cucurbitaceae	Asia
Cyathocline purpurea (Don) Kuntze	Asteraceae	Java
Cymbopogon martinii Stapf	Poaceae	America
Cynodon dactylon (L.) Pers.	Poaceae	Tropical Africa
Cyperus compressus L.	Cyperaceae	Asia
Cyperus distans L.f.	Cyperaceae	Asia
Cyperus rotundus L.	Cyperaceae	America, Africa
Equisetum debile Roxb.	Equisetaceae	Europe, North Africa
Eupatorium adenophorum Spreng.	Asteraceae	Mexico, America
Euphorbia geniculata Ort. *ex* Boiss.	Euphorbiaceae	Tropical – Australia, Pacific Islands
Euphorbia helioscopia L.	Euphorbiaceae	Afghanistan, Japan
Euphorbia parviflora L.	Euphorbiaceae	Tropics of both the Hemispheres, Australia, Pacific Islands
Fragaria nubicola Lindl.	Rosaceae	N. America, Eurasia
Galium aparine L.	Rubiaceae	Europe, North America, Australia, Alaska, Bulgaria
Gnaphalium pensylvanicum Willd.	Asteraceae	Egypt
Gomphrena celosioides Mart.	Amaranthaceae	America
Hedera helix Clarke	Araliaceae	Europe, Japan
Heteropogon contortus (L.) Beauv. *ex* Roem. & Schult.	Poaceae	Mediterranean region, Carribbean Territories, N. America, Oceania
Hydrilla verticillata (L.f.) Royle	Hydrocharitaceae	Asia, Africa, Australia
Ipomoea cairica L.	Convolvulaceae	Tropical Africa and Asia; Mediternannean region, Taiwan, Malaysia
Ipomoea carnea Facq.	Convolvulaceae	China, Java, South America
Ipomoea muricata Jacq.	Convolvulaceae	Japan
Ipomoea nil (L.) Roth	Convolvulaceae	Tropical and Sub-Tropical regions
Justicia simplex Don	Acanthaceae	Abyssinia, Maylaya, Loochoo Isles
Lactuca dissecta Don	Asteraceae	Afganistan and Baluchistan
Lantana camara L.	Verbenaceae	Caribbean Territories, Central U.S., Hawaii, Australia.

Botanical Name	Family	Origin
Lathyrus aphaca L.	Fabaceae	Continental U.S., Caribbean Territories
Martynia annua L.	Martyniaceae	Mexico and Central America
Medicago denticulata Willd.	Fabaceae	Europe, Asia, North Africa
Melothria heterophylla Cogn.	Cucurbitaceae	Central Asia
Mentha piperita L.	Lamiaceae	Europe
Mirabilis jalapa L.	Nyctaginaceae	Mexico
Najas graminea Dd.	Naiadaceae	Asia
Najas indica (Willd.) Cham.	Naiadaceae	Phillippine Islands
Nasturtium officinale R. Br.	Brassicaceae	Latin America
Oxalis corniculata L.	Oxalidaceae	Caribbean Territories
Phyllanthus urinaria L.	Euphorbiaceae	Panatropical, Africa, Tropical America
Physalis longifolia Nutt.	Solanaceae	U. S. A. , Callifornia
Poa supina Schrad.	Poaceae	Central Europe, South Australia,
Polygonum lapathifolium L.	Polygonaceae	W. and N. Asia, Europe, Africa and America
Polygonum minus Huds.	Polygonaceae	Europe, Temperate and Tropical Asia, Europe
Polygonum pulchrum Blume	Polygonaceae	Java, Philippines, Tropical and South Africa
Polygonum serrulatum Lag.	Polygonaceae	Australia, West Asia, South Europe, Africa, America
Potamogeton crispus L.	Naiadaceae	N. and S. Temperate and SubTropical regions
Potamogeton pectinatus L.	Naiadaceae	Europe, Asia
Pteris cretica L.	Pteridaceae	America, Japan, South Korea, Taiwan, Hawaii
Reinwardtia indica Dumort.	Linaceae	Euresia
Rhynchoglossum obliquum Blume	Gesneriaceae	Throughout the Malayan Archipelago
Ricinus communis L.	Euphorbiaceae	North Africa
Rumex nepalensis Spreng.	Polygonaceae	Ethopia
Saussurea heteromala (Don) Hand.-Mazz.	Asteraceae	Temperate and Arctic regions of Asia, Europe, North America
Setaria tomentosa Kunth	Poaceae	Temperate and Tropical regions
Silene conoidea L.	Caryophyllaceae	Continental US
Sonchus asper Hill	Asteraceae	Caribbean Territories, U. S. A., Oceania
Sorghum halepense Wall.	Poaceae	Tropical and Subtropical regions, The Mediterranean Regions, U.S.A. Michigan, States for Johnsongrass, Northern Europe, Asia
Spilanthes paniculata Wall. *ex* DC.	Asteraceae	Tropical America
Stellaria media L.	Caryophyllaceae	Croatia, Armenia
Strobilanthes atropurpurens Nees	Acanthaceae	America, Japan
Strobilanthes dalhousianus Clarke	Acanthaceae	New Mexico, U.K.
Taraxacum officinale Wigg	Asteraceae	W. Texas, Mexico
Thevetia neriifolia Juss. *ex* Steud.	Apocynaceae	Mexico and Central America
Urtica dioica L.	Urticaceae	Europe, Southern Africa, Andes, Australia
Vernonia cinerea (L.) Less.	Asteraceae	Indo-Malaysia region; Tropical Africa, Australia, New Zealand
Veronica anagalis-aquatica L.	Scrophulariaceae	Europe, North Asia, South Africa, North America
Vicia sativa L.	Fabaceae	China, Mediterranean Region

Table 64 Indigenous Wetland Plants of District Bilaspur

Botanical Name	Family	Origin
Abelmoschus crinitus Wall.	Malvaceae	W. Africa, South Asia, China, India
Abrus precatorius L.	Fabaceae	India,
Abutilon indicum L. Sweet	Malvaceae	India, South-East Asia, Old World Tropics
Achyranthes aspera L.	Amaranthaceae	Mediterranean, N. Africa, Europe, India, Pakistan, Asia
Acorus calamus L.	Araceae	Europe, India, Myanmar, Sri Lanka
Adiantum incisum Forsk.	Adiantaceae	Mexico-Ecuador, Galapagos, Parshoshan, Germany, Scotland, Iran, India, Africa
Aerva sanguinolenta (L.) Blume	Amaranthaceae	India, Asia tropical, Pakistan
Aeschynomene aspera L.	Fabaceae	Asia, India, Bangladesh, China, Bhutan, Combodia,
Ageratum conyzoides L.	Asteraceae	Tropical America; India, South America
Ajuga bracteosa Wall. *ex* Benth.	Lamiaceae	India, Afganistan, China, Japan, Myanmar, Nepal
Amaranthus gangeticus L. Syst.	Amaranthaceae	Tropical India, Bhutan, Afganistan, Tropical Asia, China
Andrographis paniculata (Burm. f.) Wall. *ex* Nees	Acanthaceae	East Indies; China, India, Sri-Lanka
Anisomeles indica (L.) O. Kuntze	Lamiaceae	Tropical and Subtropical India, Sri Lanka, Malaysia, Southern China, Philippines
Apluda mutica Auct.	Poaceae	Eastern Tropical Asia, Malaya, Australia, India, Pakistan
Artemisia indica Waldst. & Kit.	Asteraceae	Temperate Europe, India and Asia, Siam, Java
Bacopa monnieri (L.) Pennell	Scrophulariaceae	Southern Asia, India
Barleria cristata L.	Acanthaceae	India
Calamintha umbrosum (M.B.) C. Koch	Lamiaceae	Afganistan, India, Caucasus, China, Japan, Java
Cannabis sativa L.	Cannabinaceae	Central Asia and Himalaya, Caucasus, China, Iran, India, Middle east
Cardiospermum halicacobum L.	Sapindaceae	East Indies, India, Nepal, Africa
Cassia absus L.	Fabaceae	India; Tropical America
Centella asiatica L.	Apiaceae	India
Ceratophyllum demersum L.	Ceratophyllaceae	India
Chenopodium murale L.	Chenopodiaceae	India
Cissampelos pareira L.	Menispermaceae	Tropical India, Southeast Asia, East Africa, American Tropics
Clematis gouriana Roxb.	Ranunculaceae	Indo-Malaysia; Philippines; Java, India, Malaysia, China
Coix lachryma - jobi L.	Poaceae	China; India; Malaysia; Myanmar Nepal; Pakistan; Philippines; Sri Lanka
Colocasia antiquorum Schott	Araceae	India, China, Tropical East Asia
Costus speciosus Sm.	Zingiberaceae	India
Cryptolepis buchanani Roem. & Schult.	Asclepiadaceae	India, Bhutan, Pakistan, Myanmar, Sri Lanka
Curcuma longa Wall.	Zingiberaceae	India, Southern Asia
Cymbopogon citratus Stapf	Poaceae	India, Sri Lanka, Malaysia
Cyperus flabelliformis Rottb.	Cyperaceae	Madagascar, Asia, Africa, Australia, India
Cyperus iria L.	Cyperaceae	Asia, East Africa, India

Botanical Name	Family	Origin
Datura stramonium L.	Solanaceae	Caspian region of Europe, North America and Indian subcontinent
Debregeasia hypoleuca Wedd.	Urticaceae	W. Asia to the W. Himalayas
Dichanthium annulatum Hack.	Poaceae	India
Dicliptera roxburghiana Nees	Acanthaceae	India
Digitaria griffithii (Hook. f.) Henn.	Poaceae	India
Dioscorea belophylla Voigt. *ex* Haines.	Dioscoreaceae	Hindustani Center, India, Pakistan
Dioscorea bulbifera L.	Dioscoreaceae	S. W. Asia
Echinochloa frumentacea Link	Poaceae	Europe; India
Eleusine indica Gaertn.	Poaceae	Europe; India; Sri Lanka
Emilia sonchifolia (L.) DC.	Asteraceae	India; Sri Lanka
Equisetum arvense L.	Equisetaceae	Europe, North Africa, Northern Asia, America, Britain, North America, India
Eriophorum comosum Wall.	Cyperaceae	Asia, India, Pakistan, Nepal
Euphorbia hirta L.	Euphorbiaceae	Tropical - India, Australia
Ficus hispida L. f.	Moraceae	India, Asia
Ficus roxburghii Wall.	Moraceae	China; Bhutan; India; Nepal; Malaysia; Myanmar; Pakistan; Thailand; Vietnam
Fragaria indica Andr.	Rosaceae	India
Fumaria indica (Hausskn.) Pugsley.	Fumariaceae	Europe; Temperate – Asia, Afganistan, Persia, India
Geranium nepalense Sweet	Geraniaceae	India; Nepal; Ceylon; China; Japan
Girardinia heterophylla Decne.	Urticaceae	Travancore (India), Sri Lanka, Java
Gloriosa superba L.	Liliaceae	Tropical India; Tropical and Southern Africa, Madgascar, Indonesia, Malaysia
Hedychium spicatum Buch.- Ham. *ex* Sm.	Zingiberaceae	Tropical Asia, Himalaya, China, Java, The Greater Galan, East Indies, India, Bhutan, Nepal
Impatiens balsamina L.	Balsaminaceae	India; Myanmar, China
Ipomoea pestigridis L.	Convolvulaceae	Tropical Africa, Asia, India
Lannea coromandelica (Houtt.) Merr.	Anacardiaceae	Bangladesh; Bhutan; Cambodia; China; India – Andaman and Nicobar, Arunachal Pradesh, Assam, Bihar, Gujarat, Himachal Pradesh, Jammu and Kashmir, Karnatka, Kerala, Madhya Pradesh, Maharashtra, Meghalaya, Orissa, Rajasthan, Sikkim, Tamil Nadu, Tripura, Uttar Pradesh, West Bengal; Laos; Myanmar; Nepal; Pakistan; Sri Lanka; Thailand; Vietnam
Leea crispa Willd.	Leeaceae	The Concan, India, Myanmar, Bhutan
Lepidagathis cuspidata Nees	Acanthaceae	S. W. of Himalayas, India, Pakistan
Lespedeza sericea Miq.	Fabaceae	Australia; Bhutan; China; Hong Kong; India; Indonesia; Japan; Korea; Mayanmar; Nepal; Pakistan; Papua New Guinea; Philippines; Taiwan
Lindernia ciliata Colsm.	Scrophulariaceae	South-east Asia, China, Indian subcontinent, Tropical Australia, India, Burma, Taiwan
Macrotyloma unifloruma (Lam.) Verdc.	Fabaceae	South Asia, Africa, Asia, India, Bhutan
Marchantia palmata Nees	Marchantiaceae	India

Botanical Name	Family	Origin
Marsilea minuta L.	Marsileaceae	Europe, South Asia, Mexico, Florida, Eurasia
Mentha longifolia L.	Lamiaceae	Africa – Algeria, Ethopia, Kenya, Lesotho, Morocco, Namibia, South Africa, Swaziland, Tunisia, Zimbabwe; Asia-Temperate – Afganistan, Cyprus, Egypt, Iran, Iraq, Israel, Lebanon, Syria, Turkey, Armenia, Azerbaijan, Georgia, Russian Fedration (Ciscaucasia, Dagestan, Western Siberia); Asia-Tropical- India, Nepal, Pakistan; Europe – Sweden, Austria, Belgium, Czechoslovakia, Germany, Hungary, Poland, Switzerland, Belarus, Estonia, Latvia, Lithuania, Moldova, Russian Fedration (Europian part), Ukraine, Albania, Bulgaria, Former Yugoslavia, Portugal, Spain, Greece, Italy, Romania, France,
Micromeria biflora (Buch.-Ham.) Benth.	Lamiaceae	India-Himalaya
Momordica dioica Roxb. *ex* Willd.	Cucurbitaceae	India
Mucuna pruriens DC.	Fabaceae	India
Nerium indicum Mill.	Apocynaceae	Africa- Algeria, Ethiopia Libya, Morocco, Tunisia; Asia- Temperate – Afganistan, China, Cyprus, Iran, Iraq, Israel, Jordan, Lebanon, Oman, United Arab Emirates, Syria, Turkey; Asia-Tropical – India – Jammu and Kashmir, Punjab, Nepal, Pakistan; Europe – Albania, FormerYugoslavia, Greece, Italy, Malta; France; Portugal, Spain
Ocimum basilicum L.	Lamiaceae	Tropical Asia, Africa, India
Oroxylum indicum Vent.	Bignoniaceae	India
Parthenium hysterophorus L.	Asteraceae	Tropical and Subtropical America, India
Paspalum distichum L.	Poaceae	N.W. India, Malabar, Malacca, Andaman Island
Pennisetum lanatum Klotzsch	Poaceae	India, Tibet
Phragmites karka Roxb.	Poaceae	South Asia, North America, India, Australia
Physalis minima L.	Solanaceae	India; Baluchistan; Ceylon; Afganistan; Tropical – Africa; Australia
Plectranthus coetsa Buch.-Ham. *ex* D. Don	Lamiaceae	Tropical America, Africa, South east Asia, India, Eurasia, Malaya
Plumbago zeylanica L.	Plumbaginaceae	Southern India, Malaysia
Pogostemon plectranthoides Desf.	Lamiaceae	India
Polygala arvensis Willd.	Polygalaceae	Tropical Asia, Australia, India, Nicobar Islands.
Polygonum barbatum L.	Polygonaceae	India
Polygonum barbatum L. sub sp. *gracile* Dansar.	Polygonaceae	Munipure, Java, China and Japan

Botanical Name	Family	Origin
Polygonum donii Meisn.	Polygonaceae	Java, China, Japan, India
Polygonum glabrum Willd.	Polygonaceae	India, Sri-Lanka, China, Australia, Africa, America
Polygonum plebejjum Br. Prodr.	Polygonaceae	Tropical India, Bhutan, Afganistan
Pupalia lappacea Juss.	Amaranthaceae	Tropical Asia and Africa, India, Indonesia, Bangladesh
Roylea cinerea Baill.	Lamiaceae	India, Nepal
Rubus ellipticus Sm.	Rosaceae	S.E. Asia, India, Himalaya
Ruellia patula Jacq.	Acanthaceae	Ethiopia, Southern United Status, Eastern Himalayas (India, Bhutan, Nepal)
Rumex hastatus Don	Polygonaceae	Asia, India
Rungia pectinata (L.) Nees	Acanthaceae	India; Sri Lanka
Saccharum munja L.	Poaceae	China, India
Saccharum spontanium L.	Poaceae	Tropical Asia, Europe, Turkmenistan, Iran, Southeast Asia, New Guinea, India, Eastern Africa, Southern
Salix oxycarpa Anderss.	Salicaceae	India, Europe
Selaginella chrysocaulos Hook. *et* Grev.	Selaginellaceae	S.W. China, N.Myanmar, N.E. India, Himalaya
Sida acuta Burm.f.	Malvaceae	India
Solanum nigrum L.	Solanaceae	India, Asia, Africa
Solanum xanthocarpum Schrad. & Wendl.	Solanaceae	India
Sonchus arvensis L.	Asteraceae	India
Spilanthes acmella L.var. *oleracea* Jacq.	Asteraceae	South Asia, India
Thalictrum reniforme Wall.	Ranunculaceae	India
Trichodesma indicum Br.	Boraginaceae	India, Myanmar, Philippines, Mauritius
Tridax procumbens L.	Asteraceae	Tropical America, India
Trigonella pubescence Edgew.	Fabaceae	Kunawar
Trifolium resupinatum L.	Fabaceae	Central and South Europe, The Mediterranean, South west Asia, U. S. A.
Uraria picta Desf.	Fabaceae	Africa, South Asia, Malaysia, Phillippines, Africa, India, China
Urena lobata L.	Malvaceae	East Indies, Australia
Viola pilosa Blume	Violaceae	Bhutan, China, India, Myanmar, Nepal, Pakistan, North Europe, Asia, Japan
Withania somnifera Dunal	Solanaceae	Africa – Algeria, Angola, Cape Verde, Egypt, Ethiopia, Kenya, Liberia, Morocco, Nigeria, Somalia, South Africa, Sudan, Tanzania, Tunisia, Uganda, Zambia, Zimbabwe; Asia Temperate – Afganistan, Arabia, Iran, Iraq, Israel, Jordan, Lebanon, Syria, Turkey; Asia-Tropical – India, Pakistan, Sri Lanka; Spain Europe-Greece, Italy,
Xanthium strumarium L.	Asteraceae	America, India, Sri Lanka

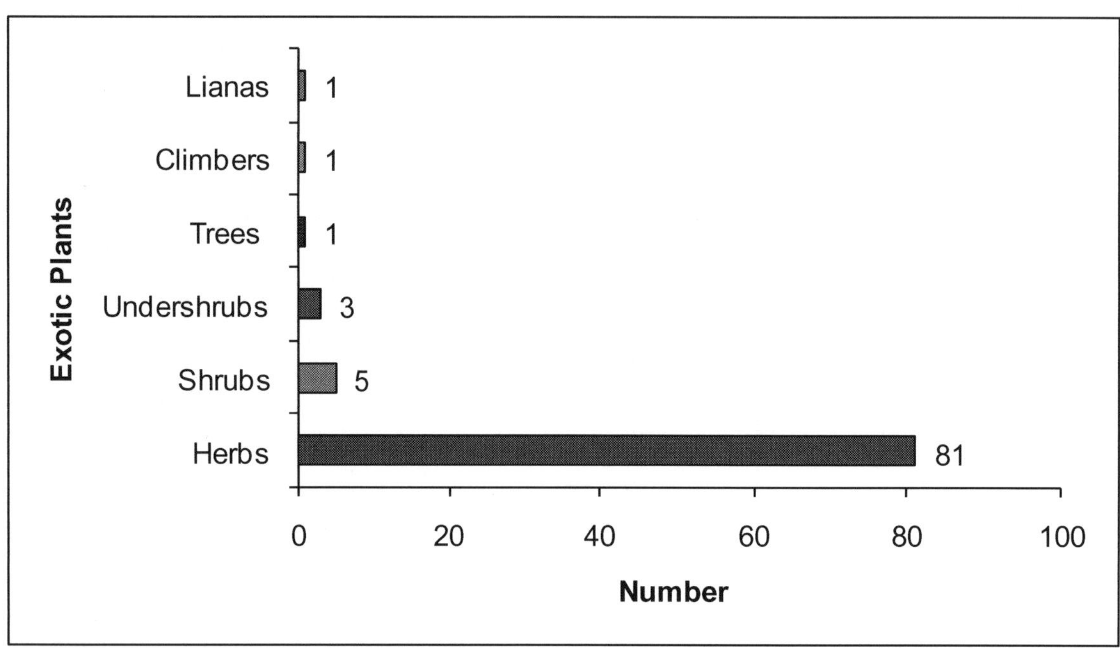

Figure 20 Ethnobotanically important wetland exotic herbs, shrubs, undershrubs, trees and climbers of District Bilaspur.

Table 65   Exotic Wetland Dicots of District Bilaspur

Botanical Name	Family	Origin
Amaranthus viridis L.	Amaranthaceae	Caribbean Territories, South America,
Amaranthus paniculatus L.	Amaranthaceae	Central America, Eastern America
Argemone mexicana L.	Papaveraceae	Mexico, Caribbean
Artemisia scoparia Waldst. & Kit.	Asteraceae	Euresia, Africa, Asia Temperate, Pakistan
Begonia picta Sm.	Begoniaceae	Yunnan province of China
Bidens pilosa L.	Asteraceae	Java, South Africa
Boehmeria platyphylla Don	Urticaceae	Malay, Islands, China, Japan, Africa
Bryophyllum calycinum Salisb.	Crassulaceae	Madagascar
Capsella bursa-pastoris (L.) Medik.	Brassicaceae	Middle East
Cassia occidentalis L.	Fabaceae	Cosmopolitan in the Tropics, America
Chenopodium album L.	Chenopodiaceae	South America, Central America, Mexico
Chenopodium ambrosiodes L.	Chenopodiaceae	Central America
Convolvulus arvensis L.	Convolvulaceae	Africa- Algeria, Egypt, Libya, Moroco, Tunisia; Asia-Temperate- Afganistan, Armenia, Azerbaijan, China, Cyprus, Egypt, Iran, Israel, Jordan, Kazakhstan, Kyrgyzstan, Lebanon, Mongolia, Syria, Tajikistan, Turkey, Turkmenistan, Uzbekistan; Asia-Tropical- Nepal; Pakistan; Eurpope-Albania, Austria, Belarus, Belgium, Bulgaria, Czechoslovakia, Denmark, Estonia, Finland, Former Yugoslavia, France, Germany, Greece, Hungary, Ireland, Italy, Latvia, Lithuania, Moldova, Netherlands, Norway, Poland, Romania, Russian Federation, Spain, Sweden, Switzerland, Ukraine, United Kingdom
Conyza bonariensis L.	Asteraceae	North America and Central America, Canada

Botanical Name	Family	Origin
Conyza stricta Willd.	Asteraceae	Eastern Africa
Crotolaria alata Buch.-Ham.	Fabaceae	Java
Crotolaria mysorensis Roth	Fabaceae	Tropical region
Cucumis pubescens Willd.	Cucurbitaceae	Asia
Cyathocline purpurea (Don) Kuntze	Asteraceae	Java
Eupatorium adenophorum Spreng.	Asteraceae	Mexico, America
Euphorbia geniculata Ort. *ex* Boiss.	Euphorbiaceae	Tropical – Australia, Pacific Islands
Euphorbia helioscopia L.	Euphorbiaceae	Afghanistan, Japan
Euphorbia parviflora L.	Euphorbiaceae	Tropics of both the Hemispheres, Australia, Pacific Islands
Fragaria nubicola Lindl.	Rosaceae	N. America, Eurasia
Galium aparine L.	Rubiaceae	Europe, North America, Australia, Alaska, Bulgaria
Gnaphalium pensylvanicum Willd.	Asteraceae	Egypt
Gomphrena celosioides Mart.	Amaranthaceae	America
Hedera helix Clarke	Araliaceae	Europe, Japan
Ipomoea cairica L.	Convolvulaceae	Tropical Africa and Asia; Mediternannean region, Taiwan, Malaysia
Ipomoea carnea Facq.	Convolvulaceae	China, Java, South America
Ipomoea muricata Jacq.	Convolvulaceae	Japan
Ipomoea nil (L.) Roth	Convolvulaceae	Tropical and Sub-Tropical regions
Justicia simplex Don	Acanthaceae	Abyssinia, Malaya, Loochoo Isles
Lactuca dissecta Don	Asteraceae	Afganistan and Baluchistan
Lantana camara L.	Verbenaceae	Caribbean Teritories, Central U.S., Hawaii, Australia.
Lathyrus aphaca L.	Fabaceae	Continental U.S., Caribbean Territories
Martynia annua L.	Martyniaceae	Mexico and Central America
Medicago denticulata Willd.	Fabaceae	Europe, Asia, North Africa
Melothria heterophylla Cogn.	Cucurbitaceae	Central Asia
Mentha piperita L.	Lamiaceae	Europe
Mirabilis jalapa L.	Nyctaginaceae	Mexico
Nasturtium officinale R. Br.	Brassicaceae	Latin America
Oxalis corniculata L.	Oxalidaceae	Caribbean Territories
Phyllanthus urinaria L.	Euphorbiaceae	Panatropical, Africa, Tropical America
Physalis longifolia Nutt.	Solanaceae	U. S. A. , California
Polygonum lapathifolium L.	Polygonaceae	W. and N. Asia, Europe, Africa and America
Polygonum pulchrum Blume	Polygonaceae	Java, Philippines, Tropical and South Africa
Polygonum serrulatum Lag.	Polygonaceae	Australia, West Asia, South Europe, Africa, America
Reinwardtia indica Dumort.	Linaceae	Euresia
Rhynchoglossum obliquum Blume	Gesneriaceae	Throughout the Malayan Archipelago
Ricinus communis L.	Euphorbiaceae	North Africa
Rumex nepalensis Spreng.	Polygonaceae	Ethopia
Saussurea heteromala (Don) Hand.-Mazz.	Asteraceae	Temperate and Arctic regions of Asia, Europe, North America
Silene conoidea L.	Caryophyllaceae	Continental US
Sonchus asper Hill	Asteraceae	Caribbean Territories, USA, Oceania
Spilanthes paniculata Wall. *ex* DC.	Asteraceae	Tropical America
Stellaria media L.	Caryophyllaceae	Croatia, Armenia
Strobilanthes atropurpurens Nees	Acanthaceae	America, Japan
Strobilanthes dalhousianus Clarke	Acanthaceae	New Mexico, U.K.
Taraxacum officinale Wigg	Asteraceae	W. Texas and Mexico

Botanical Name	Family	Origin
Thevetia neriifolia Juss. *ex* Steud.	Apocynaceae	Mexico and Central America
Trifolium resupinatum L.	Fabaceae	Central and South Europe, The Mediterranean, U. S. A.
Urtica dioica L.	Urticaceae	Europe, Southern Africa, Andes, Australia
Vernonia cinerea (L.) Less.	Asteraceae	Indo-Malaysia region; Tropical Africa, Australia, New Zealand
Veronica anagalis-aquatica L.	Scrophulariaceae	Europe, North Asia, South Africa, North America
Vicia sativa L.	Fabaceae	China, Mediterranean Region

Table 66 Exotic Wetland Moncots of District Bilaspur

Botanical Name	Family	Origin
Acorus calamus L.	Araceae	Europe, Myanmar, Sri Lanka
Blyxa auberti Rich.	Hydrocharitaceae	Australia
Bromus catharticus Vahl	Poaceae	South America. California
Canna indica L.	Cannaceae	Central and Southern America
Commelina diffusa Burm. f.	Commelinaceae	Tropics and Subtropics regions
Commelina paludosa Burm.	Commelinaceae	U.S.A.
Cymbopogon martinii Stapf	Poaceae	America
Cynodon dactylon (L.) Pers.	Poaceae	Tropical Africa
Cyperus compressus L.	Cyperaceae	Asia
Cyperus distans L.f.	Cyperaceae	Asia
Cyperus rotundus L.	Cyperaceae	America, Africa
Heteropogon contortus (L.) Beauv. *ex* Roem. & Schult.	Poaceae	Mediterranean region, Carribbean Territories, N. America, Oceania
Hydrilla verticillata (L.f.) Royle	Hydrocharitaceae	Asia, Africa, Australia
Najas graminea Dd.	Naiadaceae	Asia
Najas indica (Willd.) Cham.	Naiadaceae	Phillippine Islands
Poa supina Schrad.	Poaceae	Central Europe, South Australia,
Potamogeton crispus L.	Naiadaceae	N. and S. Temperate and Subtropical regions
Potamogeton pectinatus L.	Naiadaceae	Europe, Asia
Setaria tomentosa Kunth	Poaceae	Temperate and Tropical regions
Sorghum halepense Wall.	Poaceae	Tropical and Subtropical regions, The Mediterranean Regions, U.S.A. Michigan, States for Johnsongrass, Northern Europe, Asia

Table 67 Taxonomical Components of Ethnobotanical Wetland Exotics of District Bilaspur

Plant Group	No. of Famlies	No. of Genera	No. of Species
Dicotyledons	28	54	66
Monocotyledons	07	15	20
Pteridophytes	05	06	06
Total	40	75	92

Table 68   Exotic Wetland Pteridophytes of District Bilaspur

Botanical Name	Family	Origin
Adiantum capillus-veneris L. f.	Adiantaceae	Europe, North America, Central and South America, Eurasia
Ampelopteris prolifera (Retz.) Copeland.	Thelypteridaceae	Australia, Old World Tropics
Cheilanthes bicolor (Roxb.) Fraser-Jenkins.	Sinopteridaceae	U.S.A., California
Christella dentata (Forsk.) Brownsey & Jermy.	Thelypteridaceae	Phillippens Island
Equisetum debile Roxb.	Equisetaceae	Europe, North Africa
Pteris cretica L.	Pteridaceae	America, Japan, South Korea, Taiwan, Hawaii

Table 69   Exotic Dicot Families with Corresponding Species in Wetland Areas of District Bilaspur

Family	Botanical Name	Genus/Species
Asteraceae	*Artemisia scoparia, Bidens pilosa, Conyza bonariensis, Conyza stricta, Cyathocline purpurea, Eupatorium adenophorum, Gnaphalium pensylvanicum, Lactuca dissecta, Saussurea heteromala, Sonchus asper, Spilanthes paniculata, Taraxacum officinale, Vernonia cinerea*	12/13
Fabaceae	*Cassia occidentalis, Crotolaria alata, Crotolaria mysorensis, Lathyrus aphaca, Medicago denticulata, Trifolium resupinatum, Uraria picta, Vicia sativa*	7/8
Euphorbiaceae	*Euphorbia geniculata, Euphorbia helioscopia, Euphorbia parviflora, Phyllanthus urinaria, Ricinus communis*	3/5
Convolvulaceae	*Convolvulus arvensis, Ipomoea cairica, Ipomoea carnea, Ipomoea muricata, Ipomoea nil*	2/5
Polygonaceae	*Polygonum lapathifolium, Polygonum minus, Polygonum pulchrum, Polygonum serrulatum, Rumex nepalensis*	2/5
Acanthaceae	*Justicia simplex, Strobilanthes atropurpurens, Strobilanthes dalhousianus*	2/3
Amaranthaceae	*Amaranthus paniculatus, Amaranthus viridis, Gomphrena celosioides*	2/3
Brassicaceae	*Capsella bursa-pastoris, Nasturtium officinale*	2/2
Caryophyllaceae	*Silene conoidea, Stellaria media*	2/2
Cucurbitaceae	*Cucumis pubescens, Melothria heterophylla*	2/2
Chenopodiaceae	*Chenopodium album, Chenopodium ambrosiodes*	1/2
Urticaceae	*Urtica dioica*	1/1
Apocynaceae	*Thevetia neriifolia*	1/1
Araliaceae	*Hedera helix*	1/1
Begoniaceae	*Begonia picta*	1/1
Crassulaceae	*Bryophyllum calycinum*	1/1
Gesneriaceae	*Rhynchoglossum obliquum*	1/1
Lamiaceae	*Mentha piperita*	1/1
Linaceae	*Reinwardtia indica*	1/1
Martyniaceae	*Martynia annua*	1/1
Nyctaginaceae	*Mirabilis jalapa*	1/1
Oxalidaceae	*Oxalis corniculata*	1/1
Papaveraceae	*Argemone mexicana*	1/1
Rosaceae	*Fragaria nubicola*	1/1

Family	Botanical Name	Genus/Species
Rubiaceae	*Galium aparine*	1/1
Scrophulariaceae	*Veronica anagalis- aquatica*	1/1
Solanaceae	*Physalis longifolia*	1/1
Verbenaceae	*Lantana camara*	1/1

Table 70   Exotic Monocot Families with Corresponding Species in Wetland Areas of District Bilaspur

Family	Botanical Name	Genus/Species
Poaceae	*Bromus catharticus, Cymbopogon martinii, Cynodon dactylon, Heteropogon contortus, Poa supina, Setaria tomentosa, Sorghum halepense*	7/7
Naiadaceae	*Najas graminea, Najas indica, Potamogeton crispus, Potamogeton pectinatus*	2/4
Hydrocharitaceae	*Blyxa auberti, Hydrilla verticillata*	2/2
Cyperaceae	*Cyperus compressus, Cyperus distans, Cyperus rotundus*	1/3
Commelinaceae	*Commelina diffusa, Commelina paludosa*	1/2
Araceae	*Acorus calamus*	1/1
Cannaceae	*Canna indica*	1/1

Table 71   Predominant Wetland Exotic Genera of District Bilaspur

Botanical Name	No. of Species
Ipomoea	4
Polygonum	4
Amaranthus	3
Cyperus	3
Euphorbia	3
Chenopodium	2
Commelina	2
Conyza	2
Crotolaria	2
Najas	2
Potamogeton	2
Strobilanthes	2
Adiantum	1
Ampelopteris	1
Argemone	1
Artemisia	1
Begonia	1
Bidens	1
Blyxa	1
Boehmeria	1
Bromus	1
Bryophyllum	1
Canna	1
Capsella	1
Cassia	1

Botanical Name	No. of Species
Cheilanthes	1
Christella	1
Convolvulus	1
Cucumis	1
Cyathocline	1
Cymbopogon	1
Cynodon	1
Equisetum	1
Eupatorium	1
Fragaria	1
Galium	1
Gnaphalium	1
Gomphrena	1
Hedera	1
Heteropogon	1
Hydrilla	1
Justicia	1
Lactuca	1
Lantana	1
Lathyrus	1
Martynia	1
Medicago	1
Melothria	1
Mentha	1
Mirabilis	1
Nasturtium	1
Oxalis	1
Phyllanthus	1
Physalis	1
Poa	1
Pteris	1
Reinwardtia	1
Rhynchoglossum	1
Ricinus	1
Rumex	1
Saussurea	1
Setaria	1
Silene	1
Sonchus	1
Sorghum	1
Spilanthes	1
Stellaria	1
Taraxacum	1
Thevetia	1
Urtica	1
Vernonia	1
Veronica	1
Vicia	1

Table 72   Predominant Exotic Wetland Monocot Genera of District Bilaspur

Genus	Family	Species
Cyperus	Cyperaceae	3
Commelina	Commelinaceae	2
Najas	Naiadaceae	2
Potamogeton	Naiadaceae	2
Blyxa	Hydrocharitaceae	1
Bromus	Poaceae	1
Canna	Cannaceae	1
Cymbopogon	Poaceae	1
Cynodon	Poaceae	1
Heteropogon	Poaceae	1
Hydrilla	Hydrocharitaceae	1

Table 73   Exotic Fodder Wetland Plants of District Bilaspur

Botanical Name	Family
Amaranthus viridis	Amaranthaceae
Bromus catharticus	Poaceae
Convolvulus arvensis	Convolvulaceae
Cymbopogon martinii	Poaceae
Cynodon dactylon	Poaceae
Cyperus compressus	Cyperaceae
Cyperus distans	Cyperaceae
Cyperus rotundus	Cyperaceae
Gomphrena celosioides	Amaranthaceae
Hedera helix	Araliaceae
Heteropogon contortus	Poaceae
Lathyrus aphaca	Fabaceae
Medicago denticulata	Fabaceae
Physalis longifolia	Solanaceae
Poa supina	Poaceae
Polygonum minus	Polygonaceae
Polygonum pulchrum	Polygonaceae
Ricinus communis	Euphorbiaceae
Sorghum halepense	Poaceae
Urtica dioica	Urticaceae
Vicia sativa	Fabaceae

Table 74   Exotic Edible Wetland Plants of District Bilaspur

Botanical Name	Family
Amaranthus paniculatus	Amaranthaceae
Amaranthus tricolor var. *gangeticus*	Amaranthaceae
Amaranthus viridis	Amaranthaceae
Ampelopteris prolifera	Thelypteridaceae
Capsella bursa-pastoris	Brassicaceae
Cassia occidentalis	Fabaceae
Chenopodium album	Chenopodiaceae
Chenopodium ambrosiodes	Chenopodiaceae

Botanical Name	Family
Commelina diffusa	Commelinaceae
Cucumis pubescens	Cucurbitaceae
Fragaria nubicola	Rosaceae
Hedera helix	Araliaceae
Melothria heterophylla	Cucurbitaceae
Nasturtium officinale	Brassicaceae
Oxalis corniculata	Oxalidaceae
Rumex nepalensis	Polygonaceae
Sorghum halepense	Poaceae
Thevetia neriifolia	Apocynaceae

Table 75   Exotic Medicinal Wetland Plants of District Bilaspur

Botanical Name	Family
Amaranthus paniculatus	Amaranthaceae
Amaranthus tricolor var. *gangeticus*	Amaranthaceae
Amaranthus viridis	Amaranthaceae
Ampelopteris prolifera	Thelypteridaceae
Argemone mexicana	Papaveraceae
Artemisia scoparia	Asteraceae
Begonia picta	Begoniaceae
Bidens pilosa	Asteraceae
Blyxa auberti	Hydrocharitaceae
Boehmeria platyphylla	Urticaceae
Bryophyllum calycinum	Crassulaceae
Canna indica	Cannaceae
Capsella bursa-pastoris	Brassicaceae
Cassia occidentalis	Fabaceae
Cheilanthes bicolor	Sinopteridaceae
Chenopodium album	Chenopodiaceae
Chenopodium ambrosiodes	Chenopodiaceae
Christella dentata	Thelypteridaceae
Convolvulus arvensis	Convolvulaceae
Conyza bonariensis	Asteraceae
Conyza stricta	Asteraceae
Crotolaria alata	Fabaceae
Crotolaria mysorensis	Fabaceae
Cyathocline purpurea	Asteraceae
Cymbopogon martinii	Poaceae
Cynodon dactylon	Poaceae
Equisetum debile	Equisetaceae
Eupatorium adenophorum	Asteraceae
Euphorbia geniculata	Euphorbiaceae
Euphorbia helioscopia	Euphorbiaceae
Euphorbia parviflora	Euphorbiaceae
Fragaria nubicola	Rosaceae
Galium aparine	Rubiaceae
Gnaphalium pensylvanicum	Asteraceae
Gomphrena celosioides	Amaranthaceae
Hedera helix	Araliaceae
Heteropogon contortus	Poaceae

Botanical Name	Family
Ipomoea cairica	Convolvulaceae
Ipomoea carnea	Convolvulaceae
Ipomoea muricata	Convolvulaceae
Ipomoea nil	Convolvulaceae
Justicia simplex	Acanthaceae
Lactuca dissecta	Asteraceae
Lantana camara	Verbenaceae
Lathyrus aphaca	Fabaceae
Martynia annua	Martyniaceae
Melothria heterophylla	Cucurbitaceae
Mentha piperita	Lamiaceae
Mirabilis jalapa	Nyctaginaceae
Nasturtium officinale	Brassicaceae
Oxalis corniculata	Oxalidaceae
Phyllanthus urinaria	Euphorbiaceae
Physalis longifolia	Solanaceae
Polygonum lapathifolium	Polygonaceae
Polygonum minus	Polygonaceae
Polygonum pulchrum	Polygonaceae
Polygonum serrulatum	Polygonaceae
Potamogeton crispus	Naiadaceae
Reinwardtia indica	Linaceae
Rhynchoglossum obliquum	Gesneriaceae
Ricinus communis	Euphorbiaceae
Rumex nepalensis	Polygonaceae
Saussurea heteromala	Asteraceae
Sonchus asper	Asteraceae
Spilanthes paniculata	Asteraceae
Stellaria media	Caryophyllaceae
Strobilanthes atropurpurens	Acanthaceae
Strobilanthes dalhousianus	Acanthaceae
Taraxacum officinale	Asteraceae
Urtica dioica	Urticaceae
Vernonia cinerea	Asteraceae
Veronica anagalis-aquatica	Scrophulariaceae

Table 76   Exotic Ornamental Wetland Plants of District Bilaspur

Name	Family
Canna indica	Cannaceae
Cynodon dactylon	Poaceae
Pteris cretica	Pteridaceae

Table 77   Exotic Ethnoveterinary Wetland Plants of District Bilaspur

Name	Family
Canna indica	Cannaceae
Chenopodium album	Chenopodiaceae
Gomphrena celosioides	Amaranthaceae

Table 78   Exotic Fish Food Wetland Plants of District Bilaspur

Name	Family
Hydrilla verticillata	Hydrocharitaceae
Ipomoea cairica	Convolvulaceae
Najas graminea	Naiadaceae
Najas indica	Naiadaceae
Potamogeton crispus	Naiadaceae
Potamogeton pectinatus	Naiadaceae

Table 79   Native Status of Exotic Wetland Plants of District Bilaspur (H. P.)

Continent/Country	Species	Genera	Family
Asia (Afganistan, Bangladesh, Bhutan, China, Japan, Myanmar, Nepal, Pakistan, Sri Lanka, Thailand etc.)	12	11	11
Central North and South American Continent (Brazil, Caribbean territories, Chile, Mexico, Venezuela, United States etc.)	34	32	22
African Continent (Algeria, Egypt, Ethiopia, Ghana, Kenya, South Africa, Tunisia, Zimbabwe etc.)	15	15	11
Europe (Austria, Belgium, Bulgaria, France, Germany, Italy, Romania, Spain, Ukraine etc.)	4	4	4
Common Origin (Asia, America, Africa, Europe)	24	19	11
Australian	3	3	3

Table 80   Origin and Habit of Wetland Exotic Plants of District Bilaspur

Categories	Sub-Categories	% age of Species
Origin	Asian	13.04%
	American (Central, South & North)	36.95%
	African	16.30%
	Europian	04.34%
	Australian	03.26%
	Common Origin	26.04%
Habit	Herbs	88.54%
	Shrubs	05.20%
	Undershrubs	03.12%
	Trees	01.04%
	Climbers	01.04%
	Lianas	01.04%

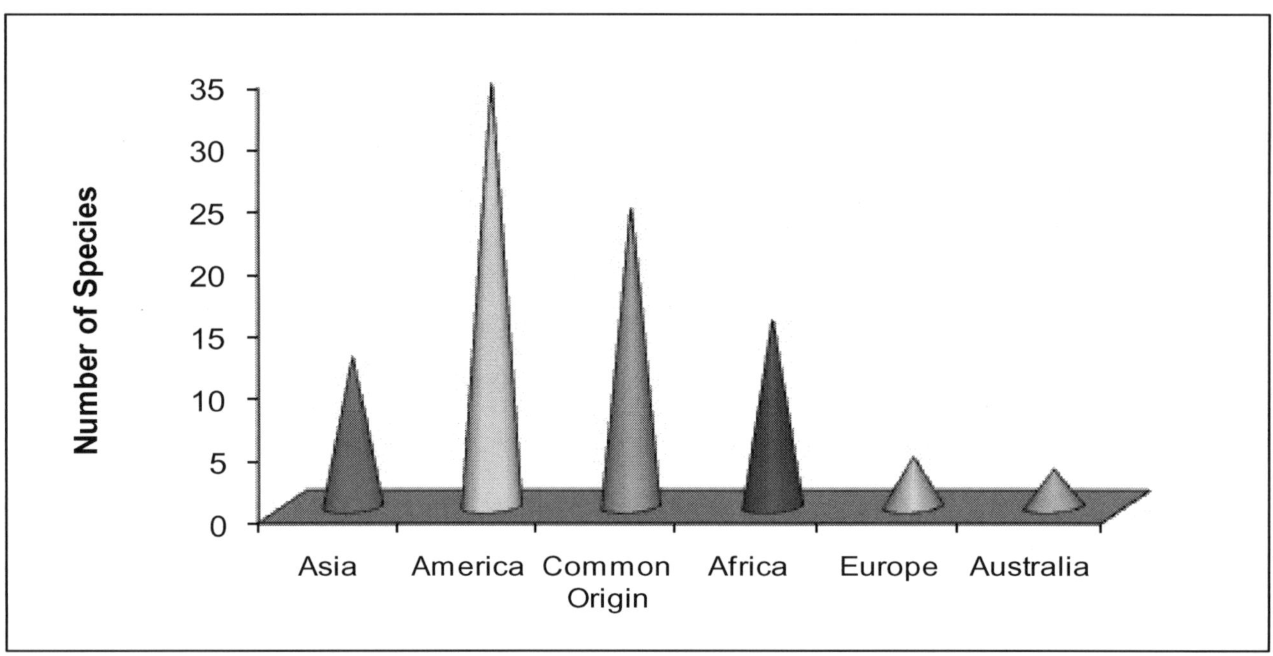

Figure 21 Histogram showing native status of ethnobotanically important wetland exotic plants of District Bilaspur.

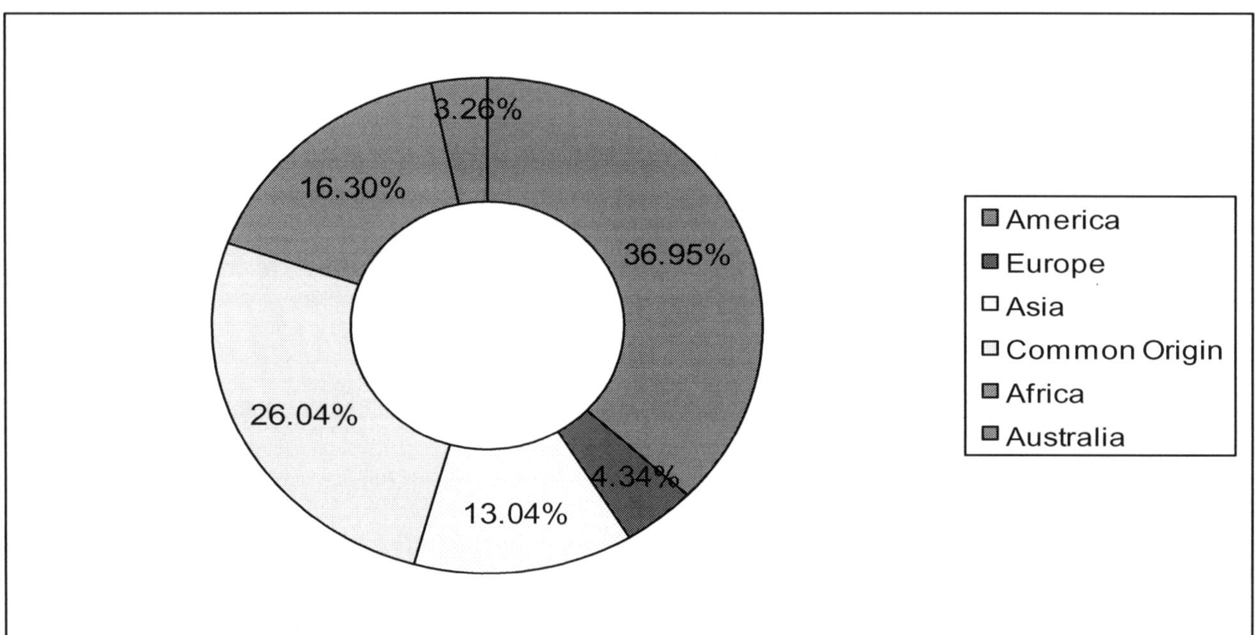

Figure 22 Doughnut showing native status of ethnobotanically important wetland exotic plants of District Bilaspur.

With regard to the nativity status of wetland exotics of the study area (Figures. 21, 22; Tables 79, 80), it is evident that highest number of alien plant species (34) belonging to 32 genera under 22 families with 36.95% overall corresponding representation trace their origin to Central, North and South America (Brazil, Caribbean Territories, Chile, Mexico, Venezuela, United States), followed by Common Origin (Asia, America, Africa, Europe) 26.04%, African Continent (Algeria, Egypt, ethiopia, Ghana, Kenya, South Africa, Tunisia, Zimbabwe, etc.) 16.30%, Asia (Afganistan, Bangladesh, Bhutan, China, Japan, Myanmar, Nepal, Pakistan, Sri Lanka, Thailand, etc.) 13.04%, Europe (Austria, Belgium, Bulgaria, France, Germany, Italy, Romania, Spain, Ukraine, etc.) 4.34% and Australian Continent with 3.26%. However, according to Nayar (1977), the American elements contribute 55% to the alien flora of the whole of India while the European elements are represented by only 15%. Various factors like increasing global trade and trade result in the translocation of species either deliberately or by accident and consequently account for homogenisation of earth's biodata (Drake et al., 1989; Kowarik, 2003; Perrings et al., 2005). From Indian perspective, Chatterjee (1947) opined that introduction and naturalization of foreign plants dates back to the early Aryans who invaded India from countries in the north-west Eurasia.

On comparing upon the wetland exotics recorded presently with catalogue of invasive alien flora of India listing 173 species under 117 genera of 44 families (Reddy, 2008), as many as of 24 invasive aliens have been recorded from the region with predominance of common origin elements (Asia, America, Africa, Europe) and Asteraceae (Figures. 23, 24; Tables 81, 82). These are: *Ageratum conyzoides, Argemone mexicana, Bidens pilosa, Cassia absus, Cassia occidentalis, Cyperus iria, Emilia sonchifolia, Gnaphalium pensylvanicum, Gomphrena celosioides, Impatiens balsamina, Ipomoea carnea, Ipomoea pestigridis, Lantana camara, Martynia annua, Mirabilis jalapa, Oxalis corniculata, Parthenium hysterophorus, Physalis longifolia, Saccharum spontanium, Sida acuta, Sonchus asper, Tridax procumbens, Urena lobata, Xanthium strumarium.* Furthermore, of the 33 species recorded as invaders of wetlands in the above catalogue (Reddy, 2008), five species, viz., *Cyperus iria, Gnaphalium pensylvanicum, Ipomoea carnea, Saccharum spontanium* and *Sonchus asper* are represented in the wetlands of the study area. Undeniably, the information generated presently on alien wetlands flora of the region are likely to provide vital clues for elucidation of the probable causes and consequences of the phenomenon of invasion, and likely to strengthen the policy efforts for effective predictive systems, monitoring tools and their management practices (see also Khuroo et al., 2006).

Table 82   Invasive Wetland Plants of District Bilaspur

Botanical Name	Family	Origin
Ageratum conyzoides L.	Asteraceae	Tropical America; India, South America
Argemone mexicana L.	Papaveraceae	Mexico and Caribbean
Bidens pilosa L.	Asteraceae	Java, South Africa
Cassia absus L.	Fabaceae	India; Tropical America
Cassia occidentalis L.	Fabaceae	Cosmopolitan in the Tropics, America
Cyperus iria L.	Cyperaceae	Asia, East Africa, India
Emilia sonchifolia (L.) DC.	Asteraceae	India; Sri Lanka
Gnaphalium pensylvanicum Willd.	Asteraceae	Egypt
Gomphrena celosioides Mart.	Amaranthaceae	America
Impatiens balsamina L.	Balsaminaceae	India; Myanmar, China
Ipomoea carnea Facq.	Convolvulaceae	China, Java, South America
Ipomoea pestigridis L.	Convolvulaceae	Tropical Africa, Asia, India
Lantana camara L.	Verbenaceae	Caribbean Teritories, Central U. S. A., Hawaii, Australia.
Martynia annua L.	Martyniaceae	Mexico and Central America
Mirabilis jalapa L.	Nyctaginaceae	Mexico

Botanical Name	Family	Origin
Oxalis corniculata L.	Oxalidaceae	Caribbean Territories
Parthenium hysterophorus L.	Asteraceae	Tropical and Subtropical America, India
Physalis longifolia Nutt.	Solanaceae	U. S. A. , Callifornia
Saccharum spontanium L.	Poaceae	Tropical Asia, Turkmenistan, Iran, Southeast Asia, New Guinea, India, Eastern Africa, Southern Europe
Sida acuta Burm.f.	Malvaceae	India
Sonchus asper Hill	Asteraceae	Caribbean Territories, U. S. A., Oceania
Tridax procumbens L.	Asteraceae	Tropical America, India
Urena lobata L.	Malvaceae	East Indies, Australia
Xanthium strumarium L.	Asteraceae	America, India, Sri Lanka

Table 83   Native Status of Invasive Wetland Plants of District Bilaspur (H. P.)

Continent/Country	Species	Genera	Family
Asia (Afghanistan, Bangladesh, Bhutan, China, Japan, Myanmar, Nepal, Pakistan, Sri Lanka, Thailand, etc.)	3	3	3
Central North and South American Continent (Brazil, Caribbean territories, Chile, Mexico, Venezuela, United States, etc.)	8	8	8
African Continent (Algeria, Egypt, Ethiopia, Ghana, Kenya, South Africa, Tunisia, Zimbabwe, etc.)	2	2	1
Common Origin (Asia, America, Africa, Europe)	11	10	7

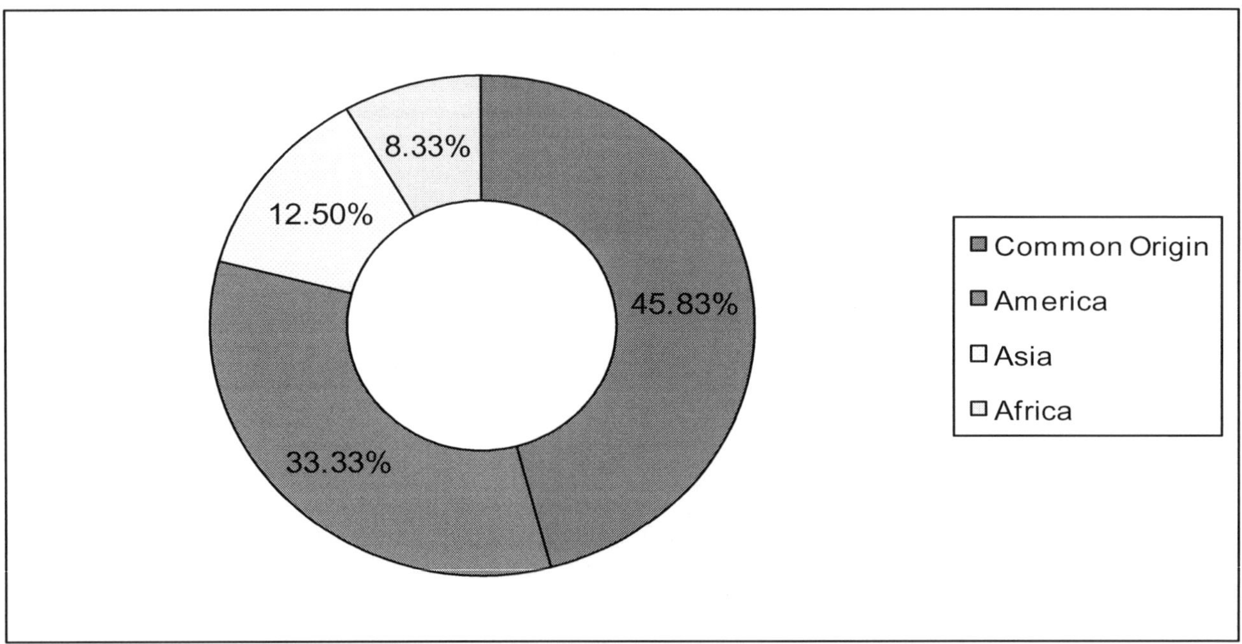

Figure 23   Doughnut showing native status of ethnobotanically important invasive wetland plants of District Bilaspur.

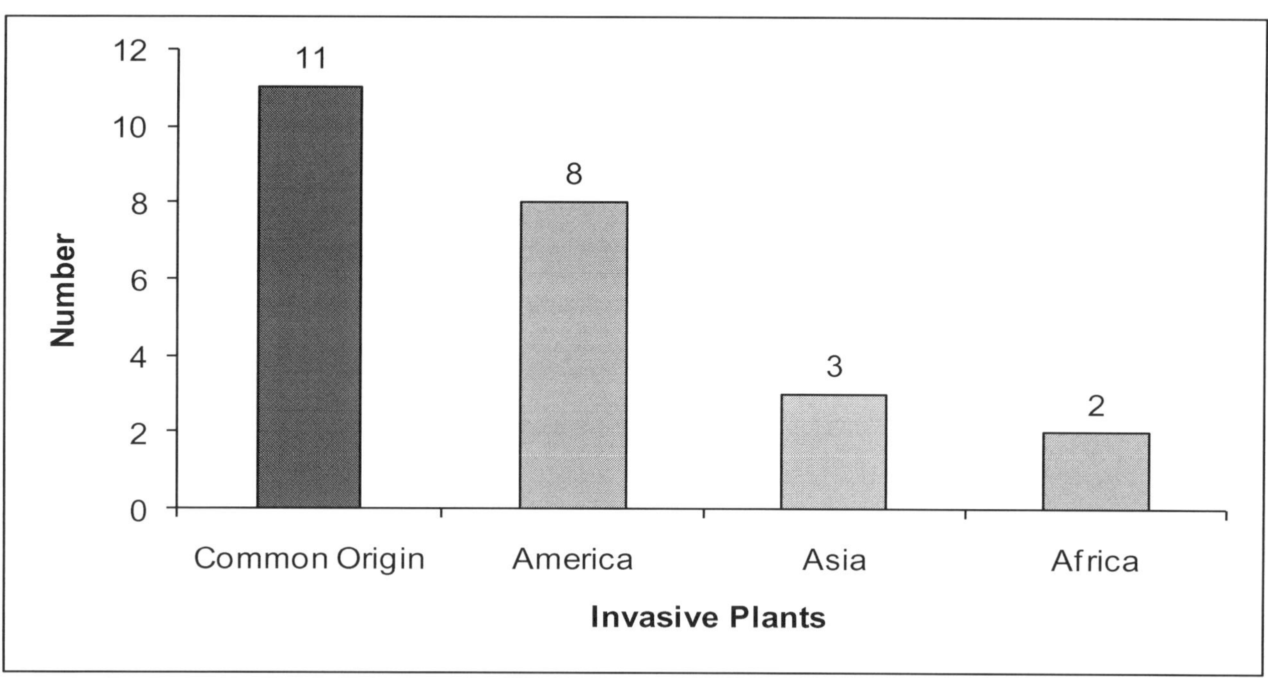

Figure 24   Histogram showing native status of ethnobotanically important invasive wetland plants of District Bilaspur.

In conclusion, it can be said that the wetland resources of the region play a significant role both in economic and social life of the local communities. However due to increasing unsustainable anthropogenic pressures in the region, the valuable native species and their habitats need to be protected and managed urgently by following a wealth of conservation oriented practices and adopting integrated management sustainable measures.

APPENDIX-1

Index to Wetland Plant Families of District Bilaspur (H.P.)

Family	Total Number of Genera and Species in a Family	Total Number of Species in a Genus
Acanthaceae	*Andrographis* (1), *Barleria* (1), *Dicliptera* (1), *Justicia* (1), *Lepidagathis* (1), *Ruellia* (1), *Rungia* (1), *Strobilanthes* (2)	8/9
Adiantaceae	*Adiantum* (2)	1/2
Amaranthaceae	*Achyranthes* (1), *Aerva* (1), *Amaranthus* (4), *Gomphrena* (1), *Pupalia* (1)	5/8
Anacardiaceae	*Lannea* (1)	1/1
Apiaceae	*Centella* (1)	1/1
Apocynaceae	*Nerium* (1), *Thevetia* (1)	2/2
Araceae	*Acorus* (1), *Colocasia* (1)	2/2
Araliaceae	*Hedera* (1)	1/1
Asclepiadaceae	*Cryptolepis* (1)	1/1
Asteraceae	*Ageratum* (1), *Artemisia* (2), *Bidens* (1), *Conyza* (2), *Cyathocline* (1), *Emilia* (1), *Eupatorium* (1), *Gnaphalium* (1), *Lactuca* (1), *Parthenium* (1), *Saussurea* (1), *Sonchus* (2), *Spilanthes* (2), *Taraxacum* (1), *Tridax* (1), *Vernonia* (1), *Xanthium* (1)	17/21
Balsaminaceae	*Impatiens* (1)	1/1
Begoniaceae	*Begonia* (1)	1/1
Bignoniaceae	*Oroxylum* (1)	1/1
Boraginaceae	*Trichodesma* (1)	1/1
Brassicacae	*Capsella* (1), *Nasturtium* (1)	2/2
Cannabinaceae	*Cannabis* (1)	1/1
Cannaceae	*Canna* (1)	1/1
Caryophyllaceae	*Silene* (1), *Stellaria* (1)	2/2
Ceratophyllaceae	*Ceratophyllum* (1)	1/1
Chenopodiaceae	*Chenopodium* (3)	1/3
Commelinaceae	*Commelina* (2)	1/2
Convolvulaceae	*Convolvulus* (1), *Ipomoea* (5)	2/6
Crassulaceae	*Bryophyllum* (1)	1/1
Cucurbitaceae	*Cucumis* (1), *Melothria* (1), *Momordica* (1)	3/3
Cyperaceae	*Cyperus* (5), *Eriophorum* (1)	2/6
Dioscoreaceae	*Dioscorea* (2)	1/1
Equisetaceae	*Equisetum* (2)	1/2
Euphorbiaceae	*Euphorbia* (4), *Phyllanthus* (1), *Ricinus* (1)	3/6
Fabaceae	*Abrus* (1), *Cassia* (2), *Crotolaria* (2), *Lathyrus* (1), *Lespedeza* (1), *Macrotyloma* (1), *Medicago* (1), *Mucuna* (1), *Trifolium* (1), *Trigonella* (1), *Uraria* (1), *Vicia* (1)	12/14
Fumariaceae	*Fumaria* (1)	1/1
Geraniaceae	*Geranium* (1)	1/1
Gesneriaceae	*Rhynchoglossum* (1)	1/1

Family	Total Number of Genera and Species in a Family	Total Number of Species in a Genus
Hydrocharitaceae	*Blyxa* (1), *Hydrilla* (1)	2/2
Lamiaceae	*Ajuga* (1), *Anisomeles* (1), *Calamintha* (1), *Mentha* (2), *Micromeria* (1), *Ocimum* (1), *Plectranthus* (2), *Pogostemon* (1), *Roylea* (1)	9/11
Leeaceae	*Leea* (1)	1/1
Liliaceae	*Gloriosa* (1)	1/1
Linaceae	*Reinwardtia* (1)	1/1
Malvaceae	*Abelmoschus* (1), *Abutilon* (1), *Sida* (1), *Urena* (1)	4/4
Marchantiaceae	*Marchantia* (1)	1/1
Marsileaceae	*Marsilea* (1)	1/1
Martyniaceae	*Martynia* (1)	1/1
Menispermaceae	*Cissampelos* (1)	1/1
Moraceae	*Ficus* (2)	1/2
Naiadaceae	*Najas* (2), *Potamogeton* (2)	2/4
Nyctaginaceae	*Mirabilis* (1)	1/1
Oxalidaceae	*Oxalis* (1)	1/1
Papaveraceae	*Argemone* (1)	1/1
Plumbaginaceae	*Plumbaginaceae* (1)	1/1
Poaceae	*Apluda* (1), *Bromus* (1), *Coix* (1), *Cymbopogon* (2), *Cynodon* (1), *Dichanthium* (1), *Digitaria* (1), *Echinochloa* (1), *Eleusine* (1), *Heteropogon* (1), *Paspalum* (1), *Pennisetum* (1), *Phragmites* (1), *Poa* (1), *Saccharum* (2), *Setaria* (1), *Sorghum* (1)	17/19
Polygalaceae	*Polygala* (1)	1/1
Polygonaceae	*Polygonum* (9), *Rumex* (2)	2/11
Pteridaceae	*Pteris* (1)	1/1
Ranunculaceae	*Clematis* (1), *Thalictrum* (1)	2/2
Rosaceae	*Fragaria* (2), *Rubus* (1)	2/3
Rubiaceae	*Galium* (1)	1/1
Salicaceae	*Salix* (1)	1/1
Sapindaceae	*Cardiospermum* (1)	1/1
Scrophulariaceae	*Bacopa* (1), *Lindernia* (1), *Veronica* (1)	3/3
Selaginellaceae	*Selaginella* (1)	1/1
Sinopteridaceae	*Cheilanthes* (1)	1/1
Solanaceae	*Datura* (1), *Physalis* (2), *Solanum* (2), *Withania* (1)	4/6
Thelypteridaceae	*Ampelopteris* (1), *Christella* (1)	2/2
Urticaceae	*Boehmeria* (1), *Debregeasia* (1), *Girardinia* (1), *Urtica* (1)	4/4
Verbenaceae	*Lantana* (1)	1/1
Violaceae	*Viola* (1)	1/1
Zingiberaceae	*Costus* (1), *Curcuma* (1), *Hedychium* (1)	3/3

APPENDIX-2

Index to Local Names of Ethnobotanically Important Wetland Resources of District Bilaspur (H.P.)

Local Name/s	Botanical Name
Akhae	*Rubus ellipticus* Sm.
Ambi	*Rumex hastatus* Don
Arand	*Ricinus communis* L.
Arlu	*Oroxylum indicum* Vent.
Ashvagandha	*Withania somnifera* Dunal
Baan	*Ruellia patula* Jacq.
Bachhu	*Polygala arvensis* Willd.
Bada Kaer, Roothi	*Vicia sativa* L.
Bada karkara	*Spilanthes acmella* L.var. *oleracea* Jacq.
Bada-dudhia	*Euphorbia helioscopia* L.
Bada-dudhia	*Euphorbia hirta* L.
Badi-aeluan	*Cassia occidentalis* L.
Badi-dredae	*Urena lobata* L.
Badi-kogsi	*Girardinia heterophylla* Decne.
Bagad	*Eriophorum comosum* Wall.
Bahupatre	*Pteris cretica* L.
Bajayanti mala	*Coix lachryma - jobi* L.
Baj-bhanga	*Chenopodium murale* L.
Bakerbel	*Clematis gouriana* Roxb.
Bakra	*Geranium nepalense* Sweet
Ban haldi	*Hedychium spicatum* Buch.-Ham. *ex* Sm.
Banapsha	*Viola pilosa* Blume
Bandarpuncha	*Heteropogon contortus* (L.) Beauv. *ex* Roem. & Schult.
Ban-Haldi	*Curcuma longa* Wall.
Barae	*Acorus calamus* L.
Barchita	*Sorghum halepense* Wall.
Bariara	*Sida acuta* Burm.f.
Baru	*Pennisetum lanatum* Klotzsch
Baru	*Phragmites karka* Roxb.
Basant	*Reinwardtia indica* Dumort.
Basinga	*Anisomeles indica* (L.) O. Kuntze
Bathu	*Chenopodium album* L.
Bathu	*Chenopodium ambrosiodes* L.
Beeiuns	*Salix oxycarpa* Anderss.
Beuli	*Abelmoschus crinitus* Wall.
Bhabhri	*Ocimum basilicum* L.
Bhaerda	*Pogostemon plectranthoides* Desf.
Bhains	*Melothria heterophylla* Cogn.

Local Name/s	Botanical Name
Bhang	*Cannabis sativa* L.
Bhatindu	*Cissampelos pareira* L.
Bhee-kaphal	*Fragaria nubicola* Lindl.
Bhursalae	*Sonchus asper* Hill
Binni	*Lespedeza sericea* Miq.
Bishkhulti	*Polygonum minus* Huds.
Brahmi	*Centella asiatica* L.
Chakshu	*Cassia absus* L.
Chaksu	*Thalictrum reniforme* Wall.
Charmaar	*Artemisia indica* Waldst. & Kit.
Cheeta	*Plumbago zeylanica* L.
Chikadu	*Parthenium hysterophorus* L.
Chini ghas	*Paspalum distichum* L.
Chinjadoo	*Xanthium strumarium* L.
Chink booti, Dudhli, Taern	*Cryptolepis buchanani* Roem. & Schult.
Chirua	*Poa supina* Schrad.
Chofki basar	*Apluda mutica* Auct.
Cholai	*Amaranthus viridis* L.
Chollayi	*Amaranthus paniculatus* L.
Chooly, Badi–Kandayi	*Argemone mexicana* L.
Chota karkara	*Spilanthes paniculata* Wall. *ex* DC.
Chu	*Najas indica* (Willd.) Cham.
Chuch	*Nasturtium officinale* R. Br.
Chura	*Commelina paludosa* Burm.
Churisaroj	*Artemisia scoparia* Waldst. & Kit.
Dabra	*Uraria picta* Desf.
Dabru	*Ficus hispida* L.f.
Dakari	*Hedera helix* Clarke
Dhanpatri	*Galium aparine* L.
Dhatura	*Datura stramonium* L.
Dhudu chana, Jangli chana	*Silene conoidea* L.
Doob	*Cynodon dactylon* (L.) Pers.
Doob	*Cyperus iria* L.
Drabu, Muthisag	*Polygonum plebejjum* Br. Prodr.
Dragal	*Dioscorea bulbifera* L.
Dryagal	*Mucuna pruriens* DC.
Dudhua-bel	*Convolvulus arvensis* L.
Dushkanda	*Rhynchoglossum obliquum* Blume
Fox tailed asperagus	*Equisetum debile* Roxb.
Fuka fucha	*Cardiospermum halicacobum* L.
Gandh-bhadra	*Cyathocline purpurea* (Don) Kuntze
Ganeechi	*Cyperus distans* L.f.
Gathban	*Strobilanthes dalhousianus* Clarke
Ghangolae	*Leea crispa* Willd.
Ghaudan	*Ipomoea muricata* Jacq.
Ghaudan	*Ipomoea nil* (L.) Roth
Ghuni	*Digitaria griffithii* (Hook.f.) Henn.
Gingu	*Lindernia ciliata* Colsm.
Gujar	*Lannea coromandelica* (Houtt.) Merr.
Heui	*Physalis minima* L.
Hodari	*Conyza stricta* Willd.
Jaladhru	*Blyxa auberti* Rich.

Local Name/s	Botanical Name
Jalaj, Shaival	*Ceratophyllum demersum* L.
Jalaz	*Potamogeton crispus* L.
Jalaz	*Potamogeton pectinatus* L.
Jalbhangru	*Rumex nepalensis* Spreng.
Jalchida	*Najas graminea* Dd.
Jalnema	*Bacopa monnieri* (L.) Pennell
Jangdu	*Bromus catharticus* Vahl
Jangli bhindi	*Abutilon indicum* L. Sweet
Jangli bhindi	*Solanum xanthocarpum* Schrad. &Wendl.
Jangli Kachalu	*Colocasia antiquorum* Schott
Jangli methi	*Trigonella pubescence* Edgew.
Jangli-tulsi	*Calamintha umbrosum* (M.B.) C. Koch
Jhunka	*Crotolaria alata* Buch.-Ham.
Juffa, Pitpapda	*Justicia simplex* Don
Jungli-pudina, Pudina	*Mentha longifolia* L.
Kadavi, Sunachari	*Taraxacum officinale* Wigg
Kadvya	*Roylea cinerea* Baill.
Kalibasuti	*Eupatorium adenophorum* Spreng.
Kalmegh	*Andrographis paniculata* (Burm. f.) Wall. *ex* Nees
Kaner	*Thevetia neriifolia* Juss. *ex* Steud.
Kaner, Ghanira	*Nerium indicum* Mill.
Kangu	*Gomphrena celosioides* Mart.
Kanv	*Martynia annua* L.
Kapala	*Commelina diffusa* Burm. f.
Kaseru	*Strobilanthes atropurpurens* Nees
Kathu	*Christella dentata* (Forsk.) Brownsey & Jermy.
Kathua	*Polygonum serrulatum* Lag.
Katur	*Polygonum barbatum* L.
Kedae ki challi	*Costus speciosus* Sm.
Khabad-dudhia	*Euphorbia geniculata* Ort. *ex* Boiss.
Khagori	*Setaria tomentosa* Kunth
Khal	*Lactuca dissecta* Don
Khattu	*Begonia picta* Sm.
Khokani	*Medicago denticulata* Willd.
Kirdu	*Dichanthium annulatum* Hack.
Kogsi	*Urtica dioica* L.
Kosha	*Saccharum spontanium* L.
Kulthi	*Macrotyloma unifloruma* (Lam.) Verdc.
Kyagnu	*Polygonum pulchrum* Blume
Lal Dudhi	*Euphorbia parviflora* L.
Lal sag	*Amaranthus gangeticus* L. Syst.
Lal Sag	*Amaranthus tricolor* (L.) var. *gangeticus* (L.) Fiori
Lamb	*Bidens pilosa* L.
Langhi	*Gloriosa superba* L.
Madir	*Echinochloa frumentacea* Link
Makoda Ghas	*Cymbopogon citratus* Stapf
Malora	*Oxalis corniculata* L.
Matakain	*Marchantia palmata* Nees
Mathaedi	*Saussurea heteromala* (Don) Hand.-Mazz.
Meetha karela	*Momordica dioica* Roxb. *ex* Willd.
Morni	*Barleria cristata* L.
Morpanki	*Adiantum incisum* Forsk.

Local Name/s	Botanical Name
Motha	*Cyperus rotundus* L.
Naal	*Saccharum munja* L.
Naktibuti, Jablota	*Ipomoea carnea* Facq.
Neelkanthi	*Ajuga bracteosa* Wall. *ex* Benth.
Padyala	*Stellaria media* L.
Paryoon, Chamrala	*Boehmeria platyphylla* Don
Patharchatta	*Bryophyllum calycinum* Salisb.
Patkayai, Kale kyanun	*Solanum nigrum* L.
Perseen	*Trifolium resupinatum* L.
Photial wagpadi	*Ipomoea pestigridis* L.
Photnu	*Cucumis pubescens* Willd.
Phuldali	*Rungia pectinata* (L.) Nees
Pipermint	*Mentha piperita* L.
Pitpapda	*Fumaria indica* (Hausskn.) Pugsley.
Pooin anvalah	*Phyllanthus urinaria* L.
Pooshathala	*Emilia sonchifolia* (L.) DC.
Popti	*Physalis longifolia* Nutt.
Pugru	*Tridax procumbens* L.
Puin Aakha	*Fragaria indica* Andr.
Pushanbanda	*Micromeria biflora* (Buch.-Ham.) Benth.
Puthkanda	*Lepidagathis cuspidata* Nees
Puthkanda, Lathjeera	*Achyranthes aspera* L.
Railway creeper	*Ipomoea cairica* L.
Rajputana	*Eleusine indica* Gaertn.
Raten	*Abrus precatorius* L.
Rohini	*Polygonum barbatum* L. sub sp. *gracile* Dansar.
Rohini	*Polygonum lapathifolium* L.
Rugosika, Sehet bund	*Equisetum arvense* L.
Rukhadi	*Trichodesma indicum* Br.
Rusha ghas, Palmarosa	*Cymbopogon martinii* Stapf
Sadevi	*Veronica anagalis-aquatica* L.
Sadhi	*Sonchus arvensis* L.
Safed Rohini	*Polygonum donii* Meisn.
Sahdaiya	*Vernonia cinerea* (L.) Less.
Saru	*Plectranthus coetsa* Buch.-Ham. *ex* D. Don
Saundi	*Dicliptera roxburghiana* Nees
Seksi	*Capsella bursa-pastoris* (L.) Medik.
Sena	*Ampelopteris prolifera* (Retz.) Copeland.
Sena	*Pupalia lappacea* Juss.
Senu	*Polygonum glabrum* Willd.
Shesherda	*Conyza bonariensis* L.
Shivkali	*Mirabilis jalapa* L.
Shopa	*Cheilanthes bicolor* (Roxb.) Fraser-Jenkins.
Shyaru	*Debregeasia hypoleuca* Wedd.
Sindoor	*Selaginella chrysocaulos* Hook. *et* Grev.
Sirsan	*Crotolaria mysorensis* Roth
Sola	*Cyperus flabelliformis* Rottb.
Solu	*Aeschynomene aspera* L.
Sudershan	*Canna indica* L.
Sudu	*Lathyrus aphaca* L.
Sufed-phulia	*Aerva sanguinolenta* (L.) Blume
Sukaru	*Gnaphalium pensylvanicum* Willd.

Local Name/s	Botanical Name
Tardi	*Dioscorea belophylla* Voigt. *ex* Haines.
Teurya	*Impatiens balsamina* L.
Thangi, Jala	*Hydrilla verticillata* (L.f.) Royle
Trayambalu	*Ficus roxburghii* Wall.
Tripatre	*Marsilea minuta* L.
Tuli ghas	*Cyperus compressus* L.
Ujadu	*Lantana camara* L.
Ukal Booty, Phulnu	*Ageratum conyzoides* L.
Vikrantaa	*Adiantum capillus-veneris* L. f.

APPENDIX-3

Index to Medicinal Uses of Ethnobotanically Important Wetland Resources of District Bilaspur (H.P.)

1.	**Abdominal Pain**	*Cymbopogon citratus*
2.	Anaemia	*Chenopodium murale, Commelina paludosa, Curcuma longa, Ficus roxburghii, Nasturtium officinale, Phyllanthus urinaria*
3.	Antidote	*Dioscorea bulbifera*
4.	Antiseptic	*Curcuma longa*
5.	Appetiser	*Oxalis corniculata*
6.	Asthma / Shortness of Breath	*Adiantum capillus-veneris, Datura stramonium, Phyllanthus urinaria, Polygala arvensis, Withania somnifera*
7.	Bacterial Infections	*Christella dentata*
8.	Blister, Burns & Boils	*Ageratum conyzoides, Ajuga bracteosa, Emilia sonchifolia, Fragaria indica, Mirabilis jalapa, Oxalis corniculata, Polygala arvensis, Ricinus communis, Sonchus arvensis, Veronica anagalis-aquatica*
9.	Blood Pressure	*Adiantum incisum*
10.	Body Pain & Massage	*Acorus calamus, Begonia picta, Emilia sonchifolia, Ipomoea pestigridis*
11.	Bone fracture	*Boehmeria platyphylla, Cassia occidentalis, Conyza stricta, Curcuma longa, Debregeasia hypoleuca, Equisetum debile, Oroxylum indicum, Urtica dioica*
12.	Bowel complaints	*Andrographis paniculata, Cyperus rotundus, Emilia sonchifolia, Euphorbia geniculata, Ipomoea muricata, Ipomoea nil, Polygonum plebejjum, Vernonia cinerea*
13.	Bronchitis & Chest infections	*Bacopa monnieri, Cyathocline purpurea, Polygala arvensis, Solanum xanthocarpum, Viola pilosa*
14.	Cancer	*Equisetum arvense*
15.	Children Ailments	*Artemisia indica*
16.	Cholera	*Euphorbia helioscopia, Euphorbia hirta, Mucuna pruriens*
17.	Colic pain	*Polygonum glabrum*
18.	Contraceptive	*Abrus precatorius, Gloriosa superba, Selaginella chrysocaulos*
19.	Convulsions	*Artemisia indica, Eleusine indica*
20.	Cough	*Abrus precatorius, Adiantum capillus-veneris, Bacopa monnieri, Barleria cristata, Bidens pilosa, Boehmeria platyphylla, Cannabis sativa, Cissampelos pareira, Galium aparine, Ipomoea carnea, Justicia simplex, Lathyrus aphaca, Ocimum basilicum, Polygonum donii, Rubus ellipticus, Sonchus arvensis, Strobilanthes atropurpurens, Viola pilosa*
21.	Cuts & Wounds	*Acorus calamus, Cassia occidentalis, Urtica dioica*
22.	Diabetes	*Ficus hispida, Silene conoidea*
23.	Dropsy	*Argemone mexicana, Achyranthes aspera*

24.	Dysentery / Diarrhoea	*Abelmoschus crinitus, Andrographis paniculata, Anisomeles indica, Cassia occidentalis, Cissampelos pareira, Costus speciosus, Ficus hispida, Lannea coromandelica, Lathyrus aphaca, Polygonum plebejjum, Spilanthes paniculata*
25.	Epilepsy	*Acorus calamus, Ampelopteris prolifera, Argemone mexicana, Martynia annua*
26.	Expel worms	*Nasturtium officinale*
27.	Eye lotion	*Ageratum conyzoides*
28.	Eye & Ear complaints	*Ageratum conyzoides, Artemisia scoparia, Begonia picta*
29.	Fever	*Abrus precatorius, Ajuga bracteosa, Andrographis paniculata, Cheilanthes bicolour, Cymbopogon martinii, Cyperus rotundus, Datura stramonium, Ipomoea nil, Oroxylum indicum, Physalis minima, Taraxacum officinale*
30.	Flatulence	*Andrographis paniculata, Cannabis sativa, Urena lobata*
31.	Gonorrhoea	*Equisetum debile, Gloriosa superba*
32.	Gout, Rheumatism & Muscular Pain	*Acorus calamus, Adiantum capillus-veneris, Ajuga bracteosa, Crotolaria alata, Cryptolepis buchanani, Cynodon dactylon, Hedera helix, Hedychium spicatum, Mentha piperita, Oroxylum indicum, Plumbago zeylanica, Ricinus communis, Spilanthes acmella*
33.	Hair growth	*Adiantum capillus-veneris, Ipomoea muricata, Ipomoea nil*
34.	Headache	*Acorus calamus, Ageratum conyzoides, Artemisia indica, Artemisia scoparia, Bacopa monnieri, Clematis gouriana, Gnaphalium pensylvanicum, Ipomoea nil, Mentha piperita, Nerium indicum, Oxalis corniculata, Plectranthus coetsa, Plumbago zeylanica, Ricinus communis,Rumex hastatus, Rungia pectinata, Salix oxycarpa, Saussurea heteromala, Stellaria media, Xanthium strumarium*
35.	Hysteria	*Acorus calamus*
36.	Immunity	*Convolvulus arvensis*
37.	Influenza	*Cymbopogon citratus, Plumbago zeylanica*
38.	Insect bite	*Dioscorea bulbifera*
39.	Insect repellent	*Lantana camara*
40.	Intestinal worms	*Selaginella chrysocaulos*
41.	Irritable ulcers	*Sonchus arvensis*
42.	Kidney's stone	*Adiantum capillus-veneri, Euphorbia hirta, Macrotyloma unifloruma, Mucuna pruriens*
43.	Leucorrhoea	*Dioscorea belophylla, Thalictrum reniforme*
44.	Liver Disorders	*Andrographis paniculata, Argemone mexicana, Costus speciosus, Cymbopogon citratus, Hedychium spicatum, Solanum nigrum*
45.	Measles	*Euphorbia parviflora*
46.	Memory Boosters	*Centella asiatica, Cynodon dactylon*
47.	Neuralgia	*Andrographis paniculata, Equisetum arvense*
48.	Nose bleeding	*Ageratum conyzoides*
49.	Paralysis	*Abrus precatorius, Spilanthes paniculata, Nerium indicum*
50.	Piles	*Cynodon dactylon, Mucuna pruriens, Solanum nigrum*
51.	Pneumonia	*Polygonum plebejjum*
52.	Rickets	*Cryptolepis buchanani*
53.	Ringworm	*Withania somnifera*

54.	**Skin diseases, Itching & Warts**	*Adiantum incisum, Andrographis paniculata, Ampelopteris prolifera, Argemone mexicana, Artemisia indica, Barleria cristata, Boehmeria platyphylla, Bryophyllum calycinum, Cassia occidentalis, Cardiospermum halicacobum, Centella asiatica, Ceratophyllum demersum, Chenopodium album, Cissampelos pareira, Costus speciosus, Curcuma longa, Cyperus rotundus, Eleusine indica, Euphorbia geniculata, Euphorbia helioscopia, Euphorbia hirta, Fumaria indica, Geranium nepalense, Impatiens balsamina, Lannea coromandelica, Martynia annua, Mucuna pruriens, Nerium indicum, Pogostemon plectranthoides, Rhynchoglossum obliquum, Ricinus communis, Roylea cinerea, Rumex hastatus, Rumex nepalensis, Trichodesma indicum, Tridax procumbens*
55.	Slipped naval	*Parthenium hysterophorus*
56.	Snake bite	*Polygala arvensis, Uraria picta, Withania somnifera*
57.	Sore & Ulcers	*Ajuga bracteosa, Apluda mutica, Bidens pilosa, Conyza bonariensis, Ipomoea carnea, Polygonum pulchrum*
58.	Spermatorrhoea	*Ajuga bracteosa, Ficus roxburghii, Melothria heterophylla*
59.	Spleen disorders	*Physalis minima*
60.	Stomach Pain	*Ajuga bracteosa, Canna indica, Cannabis sativa, Cassia occidentalis, Cissampelos pareira, Crotolaria mysorensis, Eupatorium adenophorum, Ficus roxburghii, Hedera helix, Melothria heterophylla, Mentha piperita, Ocimum basilicum, Pupalia lappacea, Taraxacum officinale, Trichodesma indicum, Tridax procumbens*
61.	Swellings	*Adiantum capillus-veneris, Cardiospermum halicacobum, Gloriosa superba, Lantana camara, Ricinus communis, Roylea cinerea, Spilanthes acmella, Withania somnifera*
62.	Throat & Mouth Infection	*Apluda mutica, Bidens pilosa, Emilia sonchifolia, Fragaria nubicola, Melothria heterophylla, Polygonum barbatum, Spilanthes paniculata*
63.	Toothache	*Abutilon indicum, Acorus calamus, Barleria cristata, Bidens pilosa, Ficus hispida, Ipomoea carnea, Lannea coromandelica, Lantana camara, Ocimum basilicum, Oroxylum indicum, Salix oxycarpa, Spilanthes paniculata, Thalictrum reniforme*
64.	Tuberculosis	*Coix lachryma - jobi*
65.	Urinary Disorders	*Canna indica, Costus speciosus, Galium aparine, Micromeria biflora, Mucuna pruriens, Plumbago zeylanica, Rungia pectinata*
66.	Vaginal disorder	*Viola pilosa*
67.	Vomiting	*Mentha longifolia*
68.	Weakness	*Ruellia patula, Saussurea heteromala*

APPENDIX-4

Wetland Plants / Plant Parts Used for Various Medicinal Purposes

Aerial plant parts	*Amaranthus paniculatus, Bidens pilosa, Capsella bursa-pastoris, Chenopodium ambrosiodes, Euphorbia geniculata, Nasturtium officinale, Polygonum barbatum* sub sp. *gracile, Sonchus asper, Viola pilosa, Xanthium strumarium*
Bark-Gum	*Lannea coromandelica*
Berries	*Withania somnifera*
Flowers	*Barleria cristata, Bidens pilosa, Canna indica, Datura stramonium, Impatiens balsamina, Mirabilis jalapa, Pogostemon plectranthoides, Reinwardtia indica, Spilanthes acmella* var. *oleracea, Spilanthes paniculata, Strobilanthes atropurpurens*
Fronds	*Adiantum capillus-veneris, Adiantum incisum, Cheilanthes bicolour, Christella dentata*
Fruits	*Coix lachryma – jobi, Cucumis pubescens, Datura stramonium, Ficus hispida, Ficus roxburghii, Fragaria indica, Hedychium spicatum, Lantana camara, Leea crispa, Martynia annua, Melothria heterophylla, Momordica dioica, Nerium indicum, Oroxylum indicum, Phyllanthus urinaria, Physalis minima, Rubus ellipticus, Solanum xanthocarpum, Thevetia neriifolia*
Latex	*Ipomoea carnea*
Leaves	*Abrus precatorius, Abutilon indicum, Achyranthes aspera, Aeschynomene aspera, Ageratum conyzoides, Ajuga bracteosa, Amaranthus gangeticus, Amaranthus paniculatus, Amaranthus tricolor* var. *gangeticus , Amaranthus viridis, Ampelopteris prolifera, Andrographis paniculata, Anisomeles indica, Artemisia scoparia, Bacopa monnieri, Barleria cristata, Bidens pilosa, Blyxa auberti, Boehmeria platyphylla, Bryophyllum calycinum, Canna indica, Cannabis sativa, Cardiospermum halicacobum, Cassia absus, Centella asiatica, Cassia occidentalis, Chenopodium album, Chenopodium murale, Christella dentata, Cissampelos pareira, Coix lachryma – jobi, Clematis gouriana, Colocassia antiquorum, Commelina diffusa, Commelina paludosa, Convolvulus arvensis, Cryptolepis buchanani, Cymbopogon citratus, Cynodon dactylon, Cyperus compressus, Cyperus distans, Datura stramonium, Debregeasia hypoleuca, Dioscorea belophylla, Emilia sonchifolia, Equisetum debile, Eupatorium adenophorum, Euphorbia geniculata, Euphorbia helioscopia, Euphorbia hirta, Ficus hispida, Ficus roxburghii, Fragaria nubicola, Girardinia heterophylla, Gnaphalium pensylvanicum, Hedera helix, Impatiens balsamina, Ipomoea carnea, Justicia simplex, Lannea coromandelica, Lantana camara, Lepidagathis cuspidata, Marsilea minuta, Martynia annua, Mentha longifolia, Mentha piperita, Mirabilis jalapa, Mucuna pruriens, Najas graminea, Nasturtium officinale, Nerium indicum, Ocimum basilicum, Oxalis corniculata, Phyllanthus urinaria, Physalis minima, Plectranthus coetsa, Plumbago zeylanica, Polygala arvensis, Pogostemon plectranthoides, Polygonum minus, Pupalia lappacea, Ricinus communis, Salix oxycarpa, Sida acuta, Sonchus asper, Spilanthes paniculata, Taraxacum officinale, Tridax procumbens, Urtica dioica, Vernonia cinerea, Withania somnifera*
Root Bark	*Nerium indicum, Salix oxycarpa*
Root Stock	*Acorus calamus*

Roots — *Abelmoschus crinitus, Abrus precatorius, Abutilon indicum, Acorus calamus, Aerva sanguinolenta, Amaranthus tricolor* var. *gangeticus, Ampelopteris prolifera, Andrographis paniculata, Argemone mexicana, Canna indica, Cassia occidentalis, Cheilanthes bicolour, Cissampelos pareira, Coix lachryma – jobi, Convolvulus arvensis, Cryptolepis buchanani, Cymbopogon citratus, Cyperus rotundus, Datura stramonium, Debregeasia hypoleuca, Dioscorea bulbifera, Emilia sonchifolia, Impatiens balsamina, Ipomoea cairica, Ipomoea pestigridis, Mirabilis jalapa, Momordica dioica, Mucuna pruriens, Oroxylum indicum, Parthenium hysterophorus, Phyllanthus urinaria, Plumbago zeylanica, Polygala arvensis, Polygonum plebejjum, Reinwardtia indica, Ricinus communis, Rubus ellipticus, Ruellia patula, Rumex nepalensis, Saccharum munja, Sida acuta, Sonchus arvensis, Sonchus asper, Taraxacum officinale, Thalictrum reniforme, Uraria picta, Vernonia cinerea, Viola pilosa, Withania somnifera*

Seeds — *Canna indica, Cannabis sativa, Cardiospermum halicacobum, Cassia absus, Cassia occidentalis, Coix lachryma – jobi, Commelina diffusa, Costus speciosus, Crotolaria mysorensis, Datura stramonium, Debregeasia hypoleuca, Dioscorea bulbifera, Emilia sonchifolia, Impatiens balsamina, Ipomoea muricata, Ipomoea nil, Ipomoea pestigridis, Lathyrus aphaca, Macrotyloma unifloruma, Mirabilis jalapa, Mucuna pruriens, Nasturtium officinale, Oroxylum indicum, Phyllanthus urinaria, Ricinus communis, Saccharum munja, Sida acuta, Sonchus arvensis, Sorghum halepense, Urena lobata, Xanthium strumarium*

Shoots — *Rumex hastatus, Urtica dioica*

Spores — *Selaginella chrysocaulos*

Sporocarp — *Marsilea minuta*

Stem — *Abelmoschus crinitus, Abrus precatorius, Acorus calamus, Aerva sanguinolenta, Amaranthus paniculatus, Amaranthus tricolor* var. *gangeticus, Begonia picta, Canna indica, Cannabis sativa, Cardiospermum halicacobum, Cassia occidentalis, Cheilanthes bicolour, Coix lachryma – jobi, Colocasia antiquorum, Costus speciosus, Curcuma longa, Cymbopogon citratus, Cyperus iria, Cyperus rotundus, Dioscorea belophylla, Dioscorea bulbifera, Eriophorum comosum, Hedychium spicatum, Ipomoea cairica, Lantana camara, Oroxylum indicum, Ricinus communis, Saccharum munja, Withania somnifera*

Stem Bark — *Cryptolepis buchanani, Nerium indicum*

Stem Latex — *Ficus hispida*

Tuberous Roots — *Gloriosa superba*

Whole Plant — *Abelmoschus crinitus, Adiantum capillus-veneris, Adiantum incisum, Ageratum conyzoides, Amaranthus gangeticus, Amaranthus viridis, Andrographis paniculata, Artemisia scoparia, Bacopa monnieri, Begonia picta, Boehmeria platyphylla, Bromus catharticus, Calamintha umbrosum, Cardiospermum halicacobum, Centella asiatica, Ceratophyllum demersum, Chenopodium album, Cissampelos pareira, Convolvulus arvensis, Conyza bonariensis, Conyza stricta, Crotolaria alata, Cyathocline purpurea, Cymbopogon citratus, Cymbopogon martinii, Cynodon dactylon, Cyperus flabelliformis, Cyperus iria, Dichanthium annulatum, Dicliptera roxburghiana, Digitaria griffithii, Echinochloa frumentacea, Eleusine indica, Equisetum arvense, Equisetum debile, Euphorbia hirta, Euphorbia parviflora, Fragaria indica, Fumaria indica, Galium aparine, Geranium nepalense, Girardinia heterophylla, Gomphrena celosioides, Heteropogon contortus, Hydrilla verticillata, Ipomoea carnea, Ipomoea nil, Justicia simplex, Lathyrus aphaca, Lespedeza sericea, Lindernia ciliata, Marchantia palmata, Medicago denticulata, Mentha longifolia, Mentha piperita, Micromeria biflora, Najas graminea, Najas indica, Oxalis corniculata, Parthenium hysterophorus, Paspalum distichum, Pennisetum lanatum, Phragmites karka, Physalis longifolia, Plumbago zeylanica, Poa supina, Polygala arvensis, Polygonum barbatum, Polygonum donii, Polygonum glabrum, Polygonum lapathifolium, Polygonum plebejjum, Polygonum pulchrum, Potamogeton crispus, Potamogeton pectinatus, Ricinus communis, Rumex hastatus, Rungia pectinata, Saccharum spontanium, Saussurea heteromala , Setaria tomentosa, Silene conoidea, Solanum nigrum, Sonchus arvensis, Sorghum halepense, Spilanthes paniculata, Stellaria media, Strobilanthes atropurpurens, Strobilanthes dalhousianus, Thalictrum reniforme, Trichodesma indicum, Tridax procumbens, Trifolium resupinatum, Trigonella pubescence, Urena lobata, Urtica dioica, Vernonia cinerea, Veronica anagalis-aquatica, Vicia sativa*

Young Branches — *Hedera helix, Pteris cretica*

REFERENCES

1. Abraham, Z. 1997. Further ethnobotanical study of the Todas and Kotas of the Nilgiris: 249-254. *In* Jain, S. K. (*eds.*): *Contribution to Indian Ethnobotany*, Vol. 3. Sci. Publ., Jodhpur (India).

2. Agami, M.S., Beer & Waisel, Y. 1986. The morphology and physiology of turions in *Najas marina* L. in Israel. *Aquatic. Bot.* **26**: 371-376.

3. Agarwal, V.S. 2003. *Directory of Indian Economic Plants.* Bishen Singh Mahendra Pal Singh, Dehradun (India).

4. Aggarwal, V.S. 1986. *Economic Plants of India.* Kailash Prakashan, Calcutta.

5. Ahluwalia, K.S. 1952. Medicinal plants of Kangra valley. *Indian For.* **78**(4): 181-194.

6. Ahmad, N. & Younus, M. 1979. Aquatic plants of Lahore. *Pakistan Assoc. for Advancement Sci.* **12**: 1-24.

7. Ahmed, A. A. & Borthakur, S. K. 2005. *Ethnobotanical Wisdom of Khasis of Meghalaya.* Bishen Singh Mahendra Pal Singh, Dehradun (India).

8. Ahuja, P.S. 2001. Current status of propagation of medicinal plants in Indian Himalaya: 207-230. *In* Samant, S.S., Dhar, U. & Palni, L.M.S. (*eds.*): *Himalayan Medicinal Plants-Potential and Prospects.* Gyanodaya Prakashan, Nainital.

9. Alagesaboopathi, C. & Balu, S. 2000. Ethnobotany of Indian *Andrographis* Wallich *ex* Nees: 29-32. *In* Maheshwari, J.K. (*eds.*): *Ethnobotany and Medicinal Plants of Indian Subcontinent.* Sci. Publ., Jodhpur (India).

10. Alagesaboopathi, C., Dwarkan, P. & Balu, S. 2000. Plants used as medicine by tribals of Shevaroy Hill, Tamilnadu: 391-393. *In* Maheshwari, J.K. (*eds.*): *Ethnobotany and Medicinal Plants of Indian Subcontinent.* Sci. Publ., Jodhpur (India).

11. Ambasta, S.P. 1986. *The Useful Plants of India.* Publication and Information Directorate, C.S.I.R., New Delhi.

12. Aminuddin & Girach, R.D. 1991. Ethnobotanical studies on Bondo Tribe of district Karaput (Orissa), India. *Ethnobotany* **3**: 15-19.

13. Aminuddin & Girach, R.D. 1993. Observations on ethnobotany of the Bhuja-a-tribe of Sonabera plateau. *Ethnobotany* **5**: 83-86.

14. Ancibor, E. 1979. Systematic anatomy of vegetative organs of the Hydrocharitaceae. *J. Linn. Soc. Bot.* **78**: 237-286.

15. Anderson, T. 1919. The vegetative morphology of *Pistia* and Lemnaceae. *Proc. Roy. Soc. London* **19b**: 96-103

16. Anonymous, 1948-76a. *The Wealth of India: Raw Materials*. Vols. I-IX. C.S.I.R., New Delhi.

17. Anonymous, 1976b. *Medicinal Plants of India*, Vol. I. Cambridge Printing Press, New Delhi (India).

18. Anonymous, 1989. *A Conservation of Wetlands in India*. Govt. of India, Min. Envr. & For., New Delhi.

19. Anonymous, 1991. *Census of India, Series - 19*, (Directorate of Census Operations, Bhubneswar).

20. Anonymous, 1994. *Ethnobiology in India- A Status Report Ministry of Environment and Forest*. Govt. of India.

21. Anonymous, 2007. *Gebiedsrappostage KRW: Waterlichaam Buureserbeck 2007. Waterschap Rijn and IJS-SEL*, Doetinchem.

22. Ansarali, K.C. & Sivdasan, M. 2009. Ethnobotanical investigations in Lakshadweep Islands, India. *Ethnobotany* **21**: 18-24.

23. Ansari, A.A. 1991. Ethnobotanical notes on some plants of Khirsy, Pauri Garhwal, U.P. *Ethnobotany* **3**: 105-106.

24. Ansari, A. A. 1997. Less known wild edible plants of Shevory and Kolly hills of South India: 243-245. *In* Jain, S. K. (*eds.*): *Contribution to Indian Ethnobotany*, Vol. 3. Sci. Publ., Jodhpur (India).

25. Arber, A. 2003. *Water Plants: A Study of Aquatic Angiosperms*. Bishen Singh Mahendra Pal Singh, Dehradun.

26. Arora, R.K. 1995. Ethnobotany and its role in domestication and conservation of native plant genetic resources: 79-86. *In* Jain, S.K. (*eds.*): *A Manual of Ethnobotany*. Sci. Publ., Jodhpur (India).

27. Arora, R.K. 1997. Native food plants of the tribals in North-Eastern India: 137-152. *In* Jain, S. K. (*eds.*): *Contribution to Indian Ethnobotany*, Vol. 3. Sci. Publ., Jodhpur (India).

28. Arora, R.K., Maheshwari, M.L., Chandel, K.P.S. & Gupta, R. 1980. Mano (*Inula racemosa*): little known aromatic plant of Lahaul valley, India. *Econ. Bot.* **34**(2): 174-180.

29. Arora, R.K. & Pandey, A. 1996. *Wild Edible Plants of India-Diversity, Conservation and Use*. NBPGR. New Delhi.

30. Arya, K. R. & Goel, A. K. 2000. Ethnobotanical study of a remote tribal area of Almora District - A survey report- Part I: 247-252. *In* Maheshwari, J.K. (*eds.*): *Ethnobotany and Medicinal Plants of Indian Subcontinent*. Sci. Publ., Jodhpur (India).

31. Arya, K.R & Parkash, V. 2000. Ethnobotanical study of a remote tribal area of Almora District. *J. Econ. Taxon. Bot.* **23**(2): 247-252.

32. Asolkar, L.V., Kakkar & Chakae, P. 1992. *Second Supplement to Glossary of Indian Medicinal Plants with Active Principles, Part-I (A-K)*. C.S.I.R., New Delhi.

33. Aston, H. I. 1973. *Aquatic Plants of Australia Melbourne*. Univ. Press, Carlton

34. Aswal, B.S. 1996. Conservation of ethno-medicinal plants diversity of Garhwal Himalaya in India: 133-135. *In* Jain, S. K. (*eds.*): *Ethnobiology in Human Welfare*. Deep Publ., New Delhi.

35. Aswal, B.S. & Mehrotra, B.N. 1994. *Flora of Lahaul Spiti (A Cold Desert in North West Himalaya)*. Bishen Singh Mahendra Pal Singh, Dehradun (India).

36. Atkinson, E.T. 1882. *Economic Botany of the Himalayan Region*. Cosmo Publ., New Delhi.

37. Augustine, J. & Sivadasan, M. 2005. Ethnomedicinal plants of Periyar Tiger reserve, Kerela, India. *Ethnobotany* **16**: 44-49.

38. Awasthi, A.K. & Goel, A.K. 2000. Ethnobotanical studies on the tribe Onge from Little Andaman Island: 569-575. *In* Maheshwari, J.K. (*eds.*): *Ethnobotany and Medicinal Plants of Indian Subcontinent*. Sci. Publ., Jodhpur (India).

39. Baburaj, D., Britto, S., John, S., Mathew, G.K. & Rajan, S. 2000. Cultivated medicinal plants useful in homoeopathy found in Nilgiri District, Tamil Nadu: 234-245. *In* Maheshwari, J.K. (*eds.*): *Ethnobotany and Medicinal Plants of Indian Subcontinent*. Sci. Publ., Jodhpur (India).

40. Badola, H.K. 2001. Endangered medicinal plant species in Himachal Pradesh. *Curr. Sci.* **83**: 797-798.

41. Bajpayee, K.K. & Dixit, G. 1996. Ethnobotanical studies on food-stuffs of tribals of Terai region, Uttar Pradesh: 128-133. *In* Maheshwari, J.K. (*eds.*): *Ethnobotany in South Asia*. Sci. Publ., Jodhpur (India).

42. Balasubramanian, P. & Prasad, S. Nagendra. 1996. Medicinal plants among the Irulars of Attappady and Boluvampatti forests in the Nilgiri Biosphere reserve 253-259. *In* Maheshwari, J.K. (*eds.*) *Ethnobotany in South Asia*. Sci. Publ., Jodhpur (India).

43. Balick, M.K. & Cox, P.A. 1996. *Plants People and Culture: The Science of Ethnobotany*. Sci. American Libr., New York.

44. Ballabh, B. & Chaurasia, O.P. 2006. Ethnobotanical studies on Boto tribe in Ladakh. *Ethnobotany* **18**: 87-95.

45. Balu, S., Alagesaboopathi, C., & Madhavan, S. 2000. Botanical remedies for diabetes from the Cauvery Delta of Tamil Nadu: 359-362. *In* Maheshwari, J.K. (*eds.*): *Ethnobotany and Medicinal Plants of Indian Subcontinent*. Sci. Publ., Jodhpur (India).

46. Bandya, S.D., Agtey, V. V., & Kulkarni, D. K. 1990. Nutritional composition and the fruits and Doum palms (Hyphane) from the west coast of India. *Principes* **34**(1): 21-23

47. Banerjee, A. 2000. Ethnobotany of few plant species in the eroded soil of Birbhum, West Bengal: 537-530. *In* Maheshwari, J.K. (*eds.*): *Ethnobotany and Medicinal Plants of Indian Subcontinent*. Sci. Publ., Jodhpur (India).

48. Banerjee, L. K. & Bhattacharya, K. 1996. New distributional records from Orissa Coast. *Bull. Bot. Surv. India.* **14**(1-4): 1984-1986.

49. Banerjee, R.K. & Verma, U. 1994. Wetland Plant Resources for Conservation. *Curr. Sci.* **82**(11): 1336 – 1345.

50. Banerjee, R.N. & Ghora, C. 1996. On the domestic use of some unreported plants of west Dinapur district (West Bengal): 325-329. *In* Maheshwari, J.K. (*eds.*): *Ethnobotany in South Asia*. Sci. Publ., Jodhpur (India).

51. Barford, A.S. & Kvist, L.P. 1996. *Comparative Ethnobotanical Studies of the American Groups in Coastal Equador*. Royal Danish Acad. Sci. & Lrs., Copenhegen.

52. Barooah, C.J. & Mahanta, P. K. 2006. *Aquatic Angiosperm of Biswanath Chariali, Assam.*

53. Barua, K. N., Iswar, C. & Das, M. 2000. Ethnobotany of Rajbanshis of Assam: 609-614. *In* Maheshwari, J.K. (*eds.*): *Ethnobotany and Medicinal Plants of Indian Subcontinent*. Sci. Publ., Jodhpur (India).

54. Barua, P. & Sarma, G.C. 1984. Studies on the medicinal uses of plants by the Northeast tribes-III. *J. Econ. Taxon. Bot.* **11**: 71-76.

55. Beal, E.O. 1977. A manual of marsh and aquatic vascular plants of North Carolina with habitat data. *Agric. Exp. Stat. Tech. Bull.* **247**: 1-298.

56. Beal, E. O. & J. W. Thieret. 1986. *Aquatic and Wetland Plants of Kentucky.* Aquat. Kentucky.

57. Beg, M., Beg, M.Z. & Ali, S.J. 2006. Ethnomedicinal studies on sub-Himalayan forests of North Eastern U.P.: 259-298. *In* Trivedi, P.C. (*eds.*): *Medicinal Plants: Ethnobotanical Approach.* Agrobios, Jodhpur (India.).

58. Belal, Ahmed Esmat & Springuel, I. 1996. Economic value of plant diversity in arid environments. *Nature & Res.* **32**(1): 33-39.

59. Bellany, R. 1995. *Ethnobiology- Expedition Field Techniques.* Expedition Advisory Center, Royle Geogr. Soc., London.

60. Bennet, S.S.R., 1986. *Name Changes in Flowering plants of India and Adjacent Regions.* Triseas Publishers, Dehradun.

61. Bennet, S.S.R., Biswas, S., Chandra, V., Singh, P.J., Chandra, S. & Doayal, R. 1991. *Food From Forests.* I.C.F.R.E., Dehradun.

62. Bews, J.W. 1979. *The Worlds Grasses. Their Economics and Ecology.* Intl. Book Distr., Dehradun.

63. Bhalla, S., Patel, J.R. & Bhalla, N.P. 1996. Ethnomedicinal observations on some Asteraceae of Bundelkhand region, Madhya Pradesh: 175-178. *In* Maheshwari, J.K. (*eds.*): *Ethnobotany in South Asia.* Sci. Publ., Jodhpur (India).

64. Bhandary, M.J., Chandrashekar, K.R. & Kaverippa, K.M. 1996. Ethnobotany of Gowalis of Uttara Kannada district, Karnataka: 244-249. *In* Maheshwari, J.K. (*eds.*): *Ethnobotany in South Asia.* Sci. Publ., Jodhpur (India).

65. Bhargava, K.S. 1959. Usual and supplementary food plants of Kumaon. *J. Bomb. Nat. History Soc.* **56**: 26-31.

66. Bhat Gopalakrishna, K. & Nagendran, C. R. 2001. *Sedges and Grasses* (Dakshina Kannada and Udupi districts). Bishen Singh Mahendra Pal Singh, Dehradun, (India).

67. Bhat, S., Maheshwari, P., Kumar, S. & Kumar, A. 2002. *Mentha* species: *In vitro* regeneration and genetic transformation. *Mol. Biol Today* **3** (1): 11-23.

68. Bhatt, D.C., Mehta, S.K. & Mitaliya, K. D. 1999. Ethnomedicinal plants of Shetrunjaya Hills of Palitana, Gujrat. *Ethnobotany* **11**: 22-25.

69. Bhattacharjee, S.K, 2001. *Handbook of Medicinal Plants.* Pointer Publ., Jaipur (India).

70. Bhattacharyya, G. 1996. Medico-ethno-botanical value of Saurashtra weeds: 154-166. *In* Maheshwari, J.K. (*eds.*): *Ethnobotany in South Asia.* Sci. Publ., Jodhpur (India).

71. Bhattacharyya, G. 2000. Ethnobotanical wealth of the "Druk-Yul" (Bhutan): 94-98. *In* Maheshwari, J.K. (*eds.*): *Ethnobotany and Medicinal Plants of Indian Subcontinent.* Sci. Publ., Jodhpur (India).

72. Bhattarai, S., Chaudhary, R.P. & Taylor, R.S.L. 2008. Screening of selected ethnomedicinal plants of Manang District, Central Nepal for antibacterial activity. *Ethnobotany* **20**: 9-15.

73. Bhogaonkar, P.Y. & Ahmad, S.A. 2007. Folk herbal medicine in Unani system at Amravati District (Maharashtra): 222-235. *In* Sahu T. R. (*eds.*): *Indigenous Knowledge: An Application*. Sci. Publ., Jodhpur (India).

74. Bhogaonkar, P.Y. & Kanerkar, U. R. 2007. Indian ethnomedicine in human genetics: *In* Sahu T. R. (*eds*): 236-243. *Indigenous Knowledge: An Application*. Sci. Publ., Jodhpur (India).

75. Billore, K.V., Joseph, T.G. & Dave, S. K. 1998. Interesting folk remedies by the 'Lok Vaidyas' of Rajasthan for 'Swas Roga'. *Ethnobotany 10*: 42-45.

76. Bisset, N. G. & Vichtl, M. 2001. *Herbal Drugs and Pharmaceuticals*. Medpharm Sci. Publ., Stuttgart.

77. Biswas, A., Bari, M.A., Roy, M. & Bhadra, S.K. 2010. Inherited folk pharmaceutical knowledge of tribal people in the Chittagong Hill tracts. Bangladesh. *Indian J. Trad. Know. 9*: 77-89.

78. Biswas, A. K. & Calder, C. 1937. *Hand-Book of Common Water and Marshy Plants of India and Burma*. Govt. Press, Delhi.

79. Biswas, K. & Calder, C. 1984. *Hand-Book of Common Water and Marsh Plants of India and Burma*. Bishen Singh Mahendra Pal Singh, Dehradun (India).

80. Bondya, S.L., Khanna, K.K. & Singh, K.P. 2006. Ethnomedicinal uses of leafy vegetable from the tribal folk-lore of Achanakmar - Amarkantak Biosphere reserve (Madhya Pradesh and Chhattisgarh). *Ethnobotany 18*: 145-148.

81. Bor, N.L. 1973. *The Grasses of Burma, Ceylon, India and Pakistan* (excluding Bambuseae). Otto Koeltz antiquariat Koenigstein.

82. Bora, P. J. 2000. A study on ethnomedicinal uses of plants among the Bodo tribe of north east India: 583-589. *In* Maheshwari, J.K. (*eds.*): *Ethnobotany and Medicinal Plants of Indian Subcontinent*. Sci. Publ., Jodhpur (India).

83. Borman, S., Korth, R. & Temte, J. 1997. *Through the Looking Glass: A Field Guide to Aquatic Plants*. Wisconsin Lakes Partnership. Stevens Point, WI.

84. Borthakur, S.K. 1996. Wild edible plants in the market of Assam, India. An ethnobotanical investigation: 31-34. *In* Jain, S. K. (*eds.*): *Ethnobiology in Human Welfare*. Deep Publ., New Delhi.

85. Borthakur, S.K., Choudhury, B.T. & Gogoi, R. 2004. Folklore hepato – protective herbal recipes from Assam in Northeast India. *Ethnobotany 16*: 76-82.

86. Bose, K.C. 1984. *Pharmacopoea Indica*. Bishen Singh Mahendra Pal Singh, Dehradun (India).

87. Brahmam, M. 2000. Indigenous medicinal plants for modern drug development programme: revitalization of native health traditions. *Adv. Plant Sci. 13*(1): 1-10.

88. Brahmam, M., Dhal, N.K. & Saxena, H.O. 1996. Ethnobotanical studies among the Tanla of Malayagiri Hills in Dhenkanal district, Orissa, India: 393-396. *In* Jain, S. K. (*eds.*): *Ethnobiology in Human Welfare*. Deep Publ., New Delhi.

89. Brahmam, M. & Saxena, H.O. 1996. Ethnobotany of Gardamardan Hills-some noteworthy folk-medicinal uses. *Ethnobotany 2*: 71-79.

90. Brezy, O., Mehta, I. & Sharma, 1973. Studies on evapotranspiration of some aquatic weeds. *Weed Sci. 21*: 197-204.

91. Brickell, C. 1996. *Royal Horticultural Society: A-Z of Garden Plants*. Dorling Kindersley Ltd, London.

92. Bruggen, H.W.E. Van. 1968-1973. Revision of the genus *Aponogeton* (Apogetonaceae) I. The species of Madgasca. *Blumea* **16**: 243-263.

93. Brussel, D.E. 1998. *Potions, Poisons and Panaceas: An Ethnobotanical Study of Montserrat*. Southern Illinois Univ. Press, Carbondale.

94. Burkill I.H. 1935. *A Dictionnary of the Economic Products of the Malay Peninsula*. (*eds.*): Ministry of Agriculture (Malaysia). Crown Agents for the Colonies. London.

95. Bursch, E.M. 1971. *A Handbook of Water Plants*. Frederick. Warne & Co., Ltd., London.

96. Caius, J.F. 2003. *The Medicinal and Poisonous Plants of India*. Sci. Publ., Jodhpur (India).

97. Chadde, S.W. 1998. *A Great Lakes Wetland Flora*. Pocket Flora Press, Columet.

98. Chak, I.M & Sharma, J.N. 1965. Effect of asarone on experimentally induced conflict neurosin in rats. *Indian J. Exp. Biol.* **3**: 252.

99. Chakraborty, M.K., Bhattacharjee, A. & Pal, D.C. 2003. Plants used for thatching purpose by the tribals of Purulia district, West Bengal, India: 571-572. *In* Singh, V. & Jain, A.P. (*eds.*): *Ethnobotany and Medicinal Plants of India and Nepal*, Vol. 2. Sci. Publ., Jodhpur (India).

100. Chamanlal, 1997. The aquatic vegetation of Uttarkashi – Garhwal Himalaya region, U.P. *J. Econ. Taxon. Bot.* **5**(3): 565-583.

101. Chanderasekar, K. & Srivastava, S.K. 2003. Ethnomedicinal studies in Pin valley National Park, Lahaul-spiti, Himachal Pradesh. *Ethnobotany* **15**: 44-47.

102. Chandersekar, K. & Srivastva, S.K. 2005. New reports on aphrodisiac plants from Pin valley National Park, Himachal Pradesh. *Ethnobotany* **17**: 189-190.

103. Chandra, S. 1990. *Foundations of Ethnobotany (Pre-1900 Ethnobotany). A Review and Bibliography*. Deep Publ., New Delhi.

104. Chandra, V. 1997. *Edible Plants of Forestry Origin*. Indian Council of Forestry Research and Education, Dehradun.

105. Chatterjee, A. & Pakrashi, S.C. 1997. *The Treatise on Indian Medicinal Plants*. Vol. V. Publications and information directorate, C.S.I.R., New Delhi (India).

106. Chatterjee, D. 1947. Influence of east Mediterranean region flora on that of India. *Sci. Cult.* **13**: 9-11.

107. Chaudhury, D. & Neogi, B. 2000. Ethnobotany of Khasi and Chakma tribes of North East India: 583-589. *In* Maheshwari, J.K. (*eds.*): *Ethnobotany and Medicinal Plants of Indian Subcontinent*. Sci. Publ., Jodhpur (India).

108. Chauhan, N.S. 1974. Medicinal plants of Una forest division, Una district, H.P. lying in Shiwalik range. *Nagarjun.* **27**: 31-39.

109. Chauhan, N.S. 1983. *Some rare and interesting medicinal plants of H.P.* All India Summer Inst. on Agroforestry in the Western Himalaya, July 22-August 7, 1983.

110. Chauhan, N.S. 1988. Ethnobotanical study of medicinal plants of Himachal Pradesh: 187-198. *In* Kaushik, P. (*eds.*): *Indigenous Medicinal Plants Including Microbes and Fungi*. Today's and Tommorow's Printers & Publi., New Delhi.

111. Chauhan, N.S. 1989a. Herbal wealth of cold desert area of Himachal Pradesh: 65-68. *Proc. Nat. Seminar-Cum-Workshop on Cold Desert Area in India.* Govt. Press, Shimla, H.P.

112. Chauhan, N.S. 1989b. Potential of aromatic plants flora in Himachal Pradesh. *Indian Perf.* **33**(2): 118-122.

113. Chauhan, N.S. 1989c. Endangered ayurvedic pharmacopoeial plant resources of H.P.: 199-205. *In* Kaushik, P. (*eds.*): *Indigenous Medicinal Plants Including Microbes and Fungi.* Today's and Tommorow's Printers & Publi., New Delhi.

114. Chauhan, N.S. 1990. Medicinal orchids of Himachal Pradesh. *J. Orchid Soc. India* **4**(1,2): 99-105.

115. Chauhan, N.S. 1996. *Ethno-medicobotany of Pabbar valley in district Shimla, Himachal Pradesh (India)*: 30. *In*: UHF-IUFRO International Workshop in Prospects of Medicinal Plants. Solan (H.P.), India.

116. Chauhan, N.S. 1999. *Medicinal and Aromatic Plants of Himachal Pradesh.* Indus Publi. Co., New Delhi.

117. Chauhan, N.S. 2003. Important medicinal and aromatic plants of Himachal Pradesh. *Indian For.* **129**(8): 979-998.

118. Chaurasia, O.P., Ahmed, Z. & Ballabh, B. 2001. Potential aromatic floras of Himalayan cold district: Ladakh & Lahul-Spiti. *J. Econ. Taxon. Bot.* **25**(1): 91-97.

119. Chaurasia, O.P., Singh, B., & Sareen, S.K., 2000. Ethnomedicinal plants of Arctic Desert, Ladakh used in veterinary practices: 163-166. *In* Maheshwari, J.K. (*eds.*): *Ethnobotany and Medicinal Plants of Indian Subcontinent.* Sci. Publ., Jodhpur (India).

120. Chevallier, A. 1996. *The Encyclopedia of Medicinal Plants.* Dorling Kinderslay (London, New York, Stutgart, Moscow).

121. Chhetri, D.R. 2005. Ethnomedicinal plants of the Khangchendzonga. National Park, Sikkim. India. *Ethnobotany* **17**: 96-103.

122. Chopra, H.N., Khan, M.L. & Tripathi, R.S. 2005. Ethnomedicinal plants in the sacred groves of Manipur. *Indian J. Trad. Know.* **4**(1): 21-32.

123. Chopra, R.N. 1933. *Indigenous Drugs of India.* Calcutta, Art press, Calcutta.

124. Chopra, R.N., Chopra, I.C. & Verma, B.C. 1969. *Supplement to Glossary of Indian Medicinal Plants.* C.S.I.R., New Delhi.

125. Chopra, R.N., Nayar, S.L. & Chopra, I.C. 1956. *Glossary of Indian Medicinal Plants.* C.S.I.R., New Delhi.

126. Choudhury, B.C. 2000. Conserving wetlands: emerging scenaria: 131-138. *In* A. Bhardwaj, R. Badola & B.M. Rathore (*eds.*): Proceedings of the workshop on the '*Conserving Biodiversity in the 21*[st] *Century, Through Integrated Conservation and Development Planning on a Regional Scale.*' LBSNA Missourie and WII Dehradun.

127. Chowdhery, H.J. 1996. Arunachal Pradesh, its people, ethnobotanical diversity and conservation: 323-328. *In* Jain, S. K. (*eds.*): *Ethnobiology in Human Welfare.* Deep Publ., New Delhi.

128. Chowdhery, H.J. & Wadhwa, B.M. 1984. *Flora of Himachal Pradesh*, Vol. 1-3. Bot. Surv. India, Calcutta.

129. Choudhury, M. D., Shil, S. & Chakraborty, G. 2008. Ethno-medico botanical studies on Dimara Kachari of Cachar district, Assam. *Ethnobotany* **20**:128-132.

130. Chowdhury, H.J., Alam, M.K., & Hassan, M.A. 1996. Some traditional folk formularies against dysentery and diarrhoea in Bangladesh. *J. Econ. Taxon. Bot. Addl. Ser.* **12**: 20-23.

131. Chun-Lin, L. & Jieru, W. 1995. Ethnobotany of Jinuo nationalality in Xishuangbanna, Southwest China: edible plants. *Ethnobotany* **7**: 39-49.

132. Clarke, C.B. 1973. *Illustrations of Cyperaceae and Authentic Cyperaceae of Linnaeus.* Pama Prmlane. The Chronica Botanica New Delhi-1.

133. Collett, H. 1902. *Flora Simlensis.* Thacker Spink and Co. Calcutta and Shimla, Reprinted 1971. Bishen Singh Mahendra Pal Singh, Dehradun (India).

134. Cook, C.D.K. 1996. *Aquatic and Wetland Plants of India.* Oxford Univ. Press, London.

135. Cowrdin L.M., Carter V., Golet F.C., & LaRaoe E.T., 1979. *Classification of Wetlands and Deepwater Habitats of United States. US Fish and Wildlife Service,* Washington D.C.

136. Craford, G.W. 1983. *Palaeoethnobotany of the Kameda Peninsula.* Jamen Univ., Michigan, Ann - Abor.

137. Dagar, H.S. & Dagar, J.C. 1999. *Ethnobotany of Aborigines of Andaman-Nicobar Islands.* Surya Intl. Publ., Dehradun.

138. Dager, J.C. & Dager, H.C. 1996. Ethnobotanical studies of the Nicobarese of Chowra Island of Nicobar Group of islands. Surv. in India, Port Blair. Andaman & Nicobar Islands, India. *J. Econ. Taxon. Bot. Addl. Ser.* **12**: 381-388.

139. Dam, D.P. & Hajra, P. K. 1997. Observations on ethnobotany of the Monpas of Kamang district, Arunachal Pradesh: 153-159. *In* Jain, S.K. *(eds.)*: Contribution to Indian Ethnobotany, Vol. 3. Sci. Publ., Jodhpur (India).

140. Dandiya, P.C. & Sharma, J.D. 1961. Pharmacological effects of α-asarone and β- asarone. *Nature* **192**: 1299.

141. Dandiya, P.C. & Sharma, J.D. 1962. Studies on *Acorus calamus* V. Pharmacological action of α-asarone and β- asarone. *Indian Med. Res.* **50**: 46.

142. Das, A.K. & Sharma, G.D. 2003. Ethno-medicinal uses of plants by Manipuri and Barmab communities of Cachar District, Assam: 421-429. *In* Singh, V.P. & Jain, A.P. *(eds.)*: *Ethnobotany and Medicinal Plants of India and Nepal,* Vol. 1. Sci. Publ., Jodhpur (India).

143. Das, D. 1999. Wild food plants of Midnapore, West Bengal, during drought and flood. *J. Econ. Taxon. Bot. Addle. Ser.* **12**: 306-313.

144. Das, D. 2000. Wild food plants of Midnapore, West Bengal, during drought and food: 539-543. *In* Maheshwari, J.K. *(eds.)*: *Ethnobotany and Medicinal Plants of Indian Subcontinent.* Sci. Publ., Jodhpur (India).

145. Das, D. & Agarwal, V.S. 1991. *Fruit Drug Plants of India.* Kalyani Publ., New Delhi.

146. Das, H.S., Panda, P.C. & Patnaik, S.N. 1996. Traditional uses of wetland plants of eastern Orissa: 306-312. *In* Maheshwari, J.K. (eds.): *Ethnobotany in South Asia.* Sci. Publ. Jodhpur, India.

147. Dasdas, S.K. & Singh, N.P. 2001. Taxonomic studies on Eriocaulaceae in Karnataka. *J. Econ. Taxon. Bot.* **25**: 449-484.

148. Dasdas, S.K. & Singh, N.P. 2006. *Alternanthera philoxeroides* (Mart.) Griseb. (Amaranthaceae) - further extended distribution in Maharashtra. *J. Econ. Taxon. Bot.* **30**(1): 198-200.

149. Dash, P.K., Dhal, N.K. & Rout, N.C. 2007. Phyto-therapeutic uses of mangroves for primary health-care among the local inhabitants of Bhitarkanika wildlife Sanctuary, Orissa, India. *Ethnobotany* **19**: 49 - 54.

150. Dash, S.S. & Misra, M.K. 1999a. Taxonomic survey and systematic census of economic plants of Narayanapatna Hills of Koraput District, Orissa. *Ethnobotany* **11**: 21-24.

151. Dash, S.S. & Misra, M.K. 1999b. Tribal uses of plants from Narayanapatna region of Koraput district, Orissa. *Anc. Sci. Life* **15**: 230-237.

152. Dash, S.S. & Misra, M.K. 2000. Taxonomic survey and systematic census of economic plants of Narayanapatna Hills of Koraput District, Orissa: 473-498. *In* Maheshwari, J.K. (*eds.*): *Ethnobotany and Medicinal Plants of Indian Subcontinent.* Sci. Publ., Jodhpur (India).

153. Dastur, J.F. 1970. *Medicinal Plants of India and Pakistan.* B. D. Taraporevala Sons & Co. Pvt. Ltd., Bombay.

154. Deokota, R. & Chhetri, R. B. 2009. Traditional medication in Bhedetar of district Sunsari (Nepal). *Ethnobotany* **21**: 112-115.

155. Deshmukh, S.V. & Balaji, V. 1994. *Conservation of Mangrove Forest Genetic Resources- A Training Manual* (JTTO-CRSARD Project), M.S. Swaminathan Res. Foundation, Chennai.

156. Deshmukh, V.R., Muratkar, G.D. & Rothe, S.P. 2000. Preliminary observation on the medicinal and economically important leguminous plant species from Amravati Tehsil: 283-289. *In* Maheshwari, J.K. (*eds.*): *Ethnobotany and Medicinal Plants of Indian Subcontinent.* Sci. Publ., Jodhpur (India).

157. Devi, M. 2003. Wild edible plants of Sonipal district, Assam: 396-409. *In* Singh,V.P. & Jain, A.P. (*eds.*): *Ethnobotany and Medicinal Plants of India and Nepal,* Vol. 2. Sci. Publ., Jodhpur (India).

158. Dey, K. L. 1973. *The Indigenous Drugs of India.* The Chronica Bot., New Delhi.

159. Dey, K. L. & Bahadur, R. 1973. *The Indigenous Drugs of India.* Pama Primlane, New Delhi.

160. Dhaliwal, D.S. & Sharma, M. 1999. *Flora of Kullu District* (*Himachal Pradesh*). Bishen Singh Mahendra Pal Singh, Dehradun (India).

161. Dhiman, D.R. 1976. *Himachal Pradesh Ki Vanoshdhiya Sampada.* Imperial Printing Press, Dharamsala, H.P.

162. Dhyani, S.K. & Sharma, R.K. 1987. Exploration of socio-economic plant resources in Vyasi Valley in Tehri Garhwal. *J. Econ. Taxon. Bot.* **9**: 299-310.

163. Dixit, R.D. & Singh, S. 2004. Medicinal pteridophytes- An overview: 269-297. *In* Trivedi, P.C. (*eds.*): *Medicinal Plants, Conservation and Utilization.* Aviskar Publishers, Distributers, Jaipur, India.

164. Dixit, R.D. & Vohra, J.N. 1984. *A Dictionary of the Pteridophytes of India.* BSI, Department of environment, Howrah.

165. Dobriyal, R.M., Singh, G.S., Rao, K.S. & Saxena, K.G. 1997. Medicinal plant resources in Chhakinal watershed in north-western Himalaya. *J. Herbs Spices & Med. Plants* **5**:15-27.

166. Drake, J.A., Mooney, H.A., Castri, F., Groves, R.H., Kruger, F.J., Rejmanek, M. & Williamson, M. (*eds.*) 1989. *Biological Invasions: A Global Perspective.* John Wiley & Sons, New York.

167. Duhoon, S.S., Kopper, M.N. & Chandra, U. 1996. Seabuckthorn (*Hippophae* spp) - A less known wonder plant of ethno-medico-botanical importance in cold desert of India. *J. Econ. Taxon. Bot. Addl. Ser.* **12**: 43-45.

168. Duke, J.A. 1968. *Darine Ethnobotanical Dictionary*. (3rd edition). Sci. Publ., Jodhpur (India).

169. Duke, J.A. 1986. *An Isthmian Ethnobotanical Dictionary*. Sci. Publ., Jodhpur (India).

170. Duke, J.A., Godwin, M., Ducellier, J., Ann, P. & Duke K. 2002. *Handbook of Medicinal Herbs*. CRC press London.

171. Duthie J.F. 1973. *Flora of the Upper Gangatic Plain and of the Adjacent Shivalik and Sub-himalayan Tracts*. Vol-1. Bishan Singh Mahendra Pal Singh, Dehradun (India).

172. Duthie, J.F. 1888. *The Fodder Grasses of Northern India*. Sci. Publ., Jodhpur (India).

173. Dwarakan, P. & Ansari, A.A. 1996. Less known uses of plants of Kollimalai (Salem DT., Tamilnadu) in South India: 284-287. *In* Maheshwari, J.K. (*eds.*): *Ethnobotany in South Asia*. Sci. Publ., Jodhpur (India).

174. Dwarkan, P. & Alagesaboopathi, C. 2000. Traditional crude drug resources used for human and livestock diseases in Salem District, Tamilnadu: 421-424. *In* Maheshwari, J.K. (*eds.*): *Ethnobotany and Medicinal Plants of Indian Subcontinent*. Sci. Publ., Jodhpur (India).

175. Dwivedi, A.P.1993. *Forests-The Non-Wood Resources*. Intl. Book Distributors, Dehradun.

176. Dwivedi, S.N. 2003. Ethnobotanical studies and conservational strategies of wild and natural resources of Rewa district of Madhya Pradesh: 233-244. *In* Singh,V.P. & Jain, A.P. (*eds.*): *Ethnobotany and Medicinal Plants of India and Nepal*, Vol. 1. Sci. Publ., Jodhpur (India).

177. Farooq, S. 2005. *555 Medicinal Plants: Field and Laboratory Manual*. Intl. Book Distr., Dehradun (India).

178. Fassett, N. C. 1940. *A Manual of Aquatic Plants*. New York and London.

179. Faulks, P.J. 1958. *An Introduction to Ethnobotany*. Moredale, London.

180. Felger, R.S. & Moser, M. B. 1985. *People of the Desert and Sea: Ethnobotany of the Serj Indians*. Univ. Arizon Press, Tucson.

181. Ganai, K.A. & Nawachoo, I.A. 2003. Traditional treatment of toothache by the the Gujjar and Bakerwal tribes of Kashmir in India: 105-107. *In* Singh, V. & Jain, A.P. (*eds.*): *Ethnobotany and Medicinal Plants of India and Nepal*. Vol.-1. Sci. Publ., Jodhpur (India).

182. Ganglee, H.C. 1985. *Handbook of Indian Mosses*. Amerind Publ. New Delhi.

183. Gangwar, A.K. & Ramkrishnan, P.S. 1990. Ethnobiological notes on some tribes of Arunachal Pradesh, Northeastern India. *Econ. Bot.* **44**: 94-105.

184. Gaur, R.D., Rawat, D.S. & Dangwal, L.R. 1993. Some little known aquatic plants from Garhwal Himalaya (Dalsera Lake, Garhwal, Uttar Pradesh, Himalayas) *J. Indian Bot. Soc.* **90**: 135-145

185. Gaur, R.D., Semwal, J.K & Tiwari, J.K. 1983. A survey of high altitude medicinal plants of Garhwal Himalaya. *Bull. Medico-Ethnobot. Res.* **4** (3-4): 102-116.

186. Geetha, S. Lakshmi, G. & Ranjithakani, P. 1996. Ethnobotanical review: wild fibre yielding plants of Kolli Hills, Tamil Nadu. *J. Econ. Taxon. Bot. Addl. Ser.* **12**: 250-252.

187. Gefu, Jerome O., Abdu, Paul A. & Alawa, Clement B. (*eds.*) 2000. *Ethnoveterinary Practices, Research and Development*. Publ. by National Animal Production Research Institute, Shika Ahmadu Bello University, PMB 109, Zaria, Nigeria.

188. Ghora, C. 1996. On the domestic use of some unreported plants of West Dinapur district (West Bengal). *J. Econ. Taxon. Bot.* **12**: 325-328.

189. Ghosh, A. 2004. *Traditional Commercial Practices in Sustainable Development and Conservation of Man and Wetland:* 429-435. The 3rd IUCN World Cons. Cong. Bankok, Thailand.

190. Gill, L.S., Idu, M., & Ogbor, D.N. 1997. Folk medicinal plants; practices and beliefs of the Bini People in Nigeria. *Ethnobotany 4:* 53-66.

191. Gill, L.S., Nyawuame, H.G.K., Esezobor, E.I., & Osagie, I.S. 1993. Nigerian folk medicine: practices and beliefs of Esan people. *Ethnobotany 5:* 129- 142.

192. Girach, R.D. 1992. Medicinal plants used by Kondh tribe of district Phulbani (Orissa) in Eastern India. *Ethnobotany 4:* 53-66.

193. Girach, R.D. & Aminuddin 1995. Ethnomedicinal uses of plants among the tribals of Singhbum district, Bihar, India. *Ethnobotany 7:* 103-107.

194. Girach, R.D., Aminuddin, Ahmad, M. Brahmam, M. & Mishra, M.K. 1996. Native phytotherapy among rural population of district Bhadrak, Orissa: 162-164. *In* Jain, S. K. (*eds.*): *Ethnobiology in Human Welfare.* Deep Publ., New Delhi.

195. Girach, R.D., Aminuddin, Brahmam, M. & Mishra, M.K. 1997. Observations on ethnomedicinal plants of Bhadrak district, Orissa, India. *Ethnobotany 9:* 44-46.

196. Girach, R.D., Singh, S., Brahmam, M., Misra, M.K., 1999. Traditional treatment of skin diseases in Bhadrak District, Orissa. *J. Econ. Taxon. Bot. 23:* 499-504.

197. Girach, R.D., Singh, S., Brahmam, M. & Mishra, M.K. 2000. Traditional treatment of skin diseases in Bhadrak District, West Bengal, India: 535-538. *In* Maheshwari, J.K. (*eds.*): *Ethnobotany and Medicinal Plants of Indian Subcontinent.* Sci. Publ., Jodhpur (India).

198. Goel, A.K. & Tripathi, S. 2009. Ethnobotanical spectrum of Indian flora: an overview during the past 20 years. *Ethnobotany 21:*131-134.

199. Goel, G.V. & Rajendran, A. 2000. Cross-cultural ethnobotanical studies of Santhal Pargana (Eastern India) and Western Ghats (Southern Ghats): 147-150. *In* Maheshwari, J.K. (*eds.*): *Ethnobotany and Medicinal Plants of Indian Subcontinent.* Sci. Publ., Jodhpur (India).

200. Gogoi, R. & Borthakur, S.K. 1991. Plants in religio-cultural beliefs of Tai Khamtis of Assam (India). *Ethnobotany 3:* 89-95.

201. Gogoi, R. & Borthakur, S.K. 2001. Notes on herbal recipes of Bodo tribe in Kamrup district, Assam. *Ethnobotany 13:* 15-23.

202. Gogoi, R., Bokolial, D. & Hazarika, D. 2003. Preliminary observation of medicinal plants of Chandrapur area of Kamrup district, Assam: 410-415. *In* Singh, V. & Jain, A.P. (*eds.*): *Ethnobotany and Medicinal Plants of India and Nepal,* Vol. 1. Sci. Publ., Jodhpur (India).

203. Gogoi, R. & Das, M.K. 2003. Observations on some weeds of medicinal importance in Brahmputra valley of Assam: 434-441. *In* Singh, V. & Jain, A.P. (*eds.*): *Ethnobotany and Medicinal Plants of India and Nepal,* Vol. 1. Sci. Publ., Jodhpur (India).

204. Good, R. 1964. *The Geography of the Flowering Plants.* Longman Group Ltd. London.

205. Gopal, 1990. *Ecology and Management of Aquatic Vegetation in the Indian Subcontinent.* Kluwer Acad. Pub., Dordrecht.

206. Goud, S.P., Murthy, K.S., Rani, S.S. & Pullaiah, T. 2000. Non-timber forest resources in the economy of tribals of Nallamalais, Andhra Pradesh. *J. Non-Timber For. Prod.* **4**: 99-102.

207. Goud, S.P. & Pullaiah, T. 1996. Folk veterinary of Kurnool district, Andhra Pradesh. *Ethnobotany* **8**: 71-74.

208. Grotenhermen, F. & Russo, E. (*eds.*) 2002. *Cannabis and Cannabinoids Pharmcology, Toxicology and Therapeutic Potential.* The Haworth Press, Inc., New York.

209. Guha Bakshi, D.N., SenSarma, P. & Pal, D.C. 1999. *A Lexican of Medicinal Plants in India.* Vol. I. Naya Prakash, Calcutta, India.

210. Guha Bakshi, D.N., Sensarma, P. & Pal, D.C. 2001. *A Lexicon of Medicinal Plants in India.* Vol. II. Naya Prakash, Calcutta (India).

211. Guha, R. & Mondal, M.S. 2003. A new species of *Caldesia Pasl* (Alismataceae) from lower Bengal. *J. Econ. Taxon. Bot.* **27**: 1102-1106.

212. Gupta, R. 1961. Flora of Lal Dal. *Indian For.* **87**(5): 316-324.

213. Gupta, R. 1964. Survey record of medicinal and aromatic plants of Chamba forest division, Himachal Pradesh. *Indian For.* **90**:454-468.

214. Gupta, R. 1971. Medicinal and aromatic plants of Bhandal ranges, Churah forest division, Chamba district, Himachal Pradesh. *J. Bomb. Nat. History Soc.* **68**: 791-803.

215. Gupta, R. 1981. Plants in folk medicine of the Himalaya: 83 - 90. *In* Jain, S.K. (*eds.*): *Glimpses of Indian Ethnobtany.* Oxford and IBH Publ. Co., New Delhi.

216. Gupta, S.P. 1997. Native medicinal uses of plants by the asurs of Netarhat plateau (Bihar): 103-116. *In* Jain, S. K. (*eds.*): *Contribution to Indian Ethnobotany*, Vol. 3. Sci. Publ., Jodhpur (India).

217. Hajra, P.K. & Baishya, A.K. 1997. Ethnobotanical notes on the Miris (Mishings) of Assam plains: 161-168. *In* Jain, S. K. (*eds.*): *Contribution to Indian Ethnobotany*, Vol. 3. Sci. Publ., Jodhpur (India).

218. Hajra, P.K. & Das, B.K. 1982. Vegetation of Gangtok with special reference to exotic plants. *Indian For.* **107**: 554-566.

219. Harborne, J.B. & Baxter, H. 2001. *Chemical Dictionary of Economic Plants.* John Wiley and Sons Ltd, Chichester, New York.

220. Harshberger, 1896. Ethnobotany – the use of plants by aboriginal peoples.

221. Hartwell, J.L. 1967-71. Plants used against cancer: survey. *Lloydia* **30**: 379-436; **31**:71-170; **32**:79-107; 153-205; 247-296; **33**:97-104; 280-392; **34**:103-106.

222. Henery, A.N., Hosagoudar, V.B. & Ravikumar, K. 1996. Ethno-medico-botany of the Southern Western Ghats of India: 173-180. *In* Jain, S.K. (*eds.*): *Glimpses of Indian Ethnobotany.* Oxford & IBH Publ. Co. New Delhi, Bombay, Calcutta (India).

223. Henry, G.G. 1999. *Materia Medica.* Sci. Publ., Jodhpur (India).

224. Hooker, J.D. (1872-1897). *The Flora of British India*, Vols. 1-7. Reeve & Co. Ltd., Kent.

225. Hosagoudar, V.B. & Henry, A.N. 1996a. Ethnobotany of Kadars, Malasars and Muthuvans of the Anamalais in Coimbatore District, Tamil Nadu. India: 260-268. *In* Maheshwari, J.K. (*eds.*): *Ethnobotany in South Asia.* Sci. Publ., Jodhpur (India).

226. Hosagoudar, V.B. and Henry, A.N. 1996b. Ethnobotany of Soligas in Biligiri Rangana Betta. Karnataka, Southern India: 272-284. *In* Maheshwari, J.K. (*eds*.): *Ethnobotany in South Asia*. Sci. Publ., Jodhpur (India).

227. Hudidrom, B.K.S. 1996. Plants used in medico-sexual purposes by Meitei community in Manipur state, India: 364-367. *In* Maheshwari, J.K. (*eds*.): *Ethnobotany in South Asia*. Sci. Publ., Jodhpur (India).

228. Idu, M. 2009. Ethnobotany in Nigeria: Retrospects and Prospects. Ethnobotany *21*: 25-31.

229. Idu, M. Timothy, O. & Oghinan, O.M. 2008. Plants used for ethnomedicine in esan north east local government area of edo states in Nigeria. *Ethnobotany 20:* 85-90.

230. Imperato, F. 1994. Chemical constituents of *Pteris cretica* L. *Phytochemistry* **37**: 589-590.

231. Indra, K. & Venkataraju, R. R. 1998. Note on taxonomy and distribution of two rare and little known aquatic angiosperm from Andhra Pradesh. *J. Indian Bot. Soc.* **77**: 135 – 137.

232. Islam, M. 1989. *Aquatic Weeds of North East India*. Intl. Book Distri., Dehradun.

233. Islam, M. 1996. Ethnobotany of certain underground parts of plants of North-Eastern region, India: 338-344. *In* Maheshwari, J.K. (*eds*.): *Ethnobotany in South Asia*. Sci. Publ., Jodhpur (India).

234. Jadhav, D. 2009. Ethnomedicinal plants used in leaf therapy in Ratlam district (Madhya Pradesh). *Ethnobotany* **21**: 84-90.

235. Jain, A., Roshnibala, S., Kanilal, P.B., Singh, R.S. & Singh, B.K. 2007. Aquatic/ Semi aquatic plants used in herbal remedies in the wetlands of Manipur, Northeastern India. *Indian J. Trad. Know.* **6**: 346-351.

236. Jain, D.L., Baheti, A.M., Jain, S.R. & Khandelwal, K.K. 2010. Use of medicinal plants among tribes in Satpuda region of Dhule and Jalgaon district of Maharashtra-an ethnobotanical survey. *Indian J. Trad. Know.* **9**: 152-157.

237. Jain, S.K. 1964. Wild food plants of the tribal of Bastar, Madhya Pradesh. *Quart. J. Mythic Soc.* **54**: 73-87.

238. Jain, S.K. 1968. *Medicinal Plants*. Nat. Book Trust, New Delhi.

239. Jain, S.K. 1981. Observations on ethnobotany of the tribals of Central India: 193-198. *In* Jain, S.K. (*eds*.): *Glimpses of Indian Ethnobotany*. Oxford & IBH Publ., New Delhi, Bombay, Calcutta (India).

240. Jain, S.K. 1987. *A Manual of Ethnobotany*. Sci. Publ., Jodhpur.

241. Jain, S.K. 1991. *Dictionary of Indian Folk Medicine and Ethnobotany*. Deep Publ., New Delhi.

242. Jain, S.K. 2002. *Bibliography of Indian Ethnnobotany*. Sci. Publ., Jodhpur (India).

243. Jain, S.K. 2003. Notable foreign medicinal uses for some plants of Indian tradition. *Indian J. Trad. Know.* **2**: 321.

244. Jain, S.K. 2006. Ethnobotany in the New Millennium - some thoughts on future direction in Indian Ethnobotany. *Ethnobotany* **18**: 1-3.

245. Jain, S.K. 2008. *Number Therapy*. IVV Publishing House.

246. Jain, S.K. 2010. Ethnobotany in India: some thoughts on future work. *Ethnobotany* **22**: 1-4.

247. Jain, S.K. 2011. *A peep into folk knowledge in Moghul period – deep understanding of plant life. Ethnobotany* **22**: 158.

248. Jain, S.K. & Goel, A.K. 2005. Some Indian plants in Tibetan traditional medicine -1. *Ethnobotany* **17**: 127-136.

249. Jain S.K., Hajra, P.K. & Shanpru, R. 1977. Survey of edible plants in bazaar of Meghalaya. *Bull. Meghalaya Sci. Soc.* **2**: 29-34.

250. Jain S.K., Mudgal V., Banerjee D.K., Guha A., Pal D.C. & Das D. 1984. *Bibliography of Ethnobotany*. Bot. Surv. India., Calcutta.

251. Jain, S.K., Ranjan, V., Sikarwar, R.K.S., & Saklani, A. 1994. Botanical distribution of psychoactive plants of India. *Ethnobotany* **6**: 65-75.

252. Jain, S.K. & Rao, R.R. (*eds.*) 1977. *A Handbook of Field and Herbarium Methods*. Today's and Tomorrow's Printers and Publ., New Delhi.

253. Jain, S.K. & Saklani, A. 1992. Cross-cultural ethnobotanical studies in North-East India. *Ethnobotany* **4**: 25-28.

254. Jain, S.K. & Sharma, P.P. 2000. A review of parallel or unique indigenous uses of twenty five plants in India and some African countries. *Ethnobotany* **12**: 51-55.

255. Jain, S.K., Sinha, B.K. & Gupta, R.C. 1991. *Notable Plants in Ethnomedicine of India*. Deep Publ., New Delhi.

256. Jain, S.K. & Srivastava, S. 2001. Indian ethnobotanical literature in last two decades – a graphic review and future directions. *Ethnobotany* **13**: 1-8.

257. Jain, S.K. & Tarafdar, C.R. 1970. Medicinal plantlore of the Santals. A revival of P.O. Boddling's work. *Econ. Bot.* **24**: 241-278.

258. Jain, S.P. 1996. Ethno-medico-botanical survey of Chaibara, Singhbhum district, Bihar: 403-408. *In* Maheshwari, J.K. (*eds.*): *Ethnobotany in South Asia*. Sci. Publ., Jodhpur (India).

259. Jain, S.P., 1997. Observations on ethnobotany of the tribals of central India: 61-65. *In* Jain, S. K. (*eds.*): *Contribution to Indian Ethnobotany*, Vol. 3. Sci. Publ., Jodhpur (India).

260. Jain, S.P., Abraham, Z. & Shah, N.C. 1997. Herbal remedies among 'Ho' tribe in Bihar: 117-121. *In* Jain, S. K. (*eds.*): *Contribution to Indian Ethnobotany*. Vol. 3. Sci. Publ., Jodhpur (India).

261. Jain, S.P. & Puri, H.S. 1994. An ethno-medico-botanical survey of Parvati valley in Himachal Pradesh (India). *Ethnobotany* **2**: 11-18.

262. Jain, S.P. & Singh, S.C. 1997. An ethno-medico-botanical survey of Ambikapur district, M.P.: 83-91. *In* Jain, S. K. (*eds.*): *Contribution to Indian Ethnobotany*, Vol. 3. Sci. Publ., Jodhpur (India).

263. Jain, S.P., Singh, S.C., Srivastava S., Singh, J., Mishra N.P. & Prakash, A. 2010. Hitherto unreported ethnomedicinal uses of plants of Betul district of Madhya Pradesh. *Indian J. Trad. Know.* **9**: 522-525.

264. Jamir, N.S. 1995. Studies on wild edible fruits in Nagaland State, India. *J. Non-wood Forest Prod.* **2** (192): 79-82.

265. Jamir, N.S., Limasemba & Jamir, N. 2008. Ethnomedicinal plants used by Konyak Naga tribes of Mon district in Nagaland. *Ethnobotany* **20**: 48-53.

266. Janaki Ammal, E.K. 1955. *An Introduction to the Subsistence Economy of India: 16-22. Changing the face of the Earth*. Princeton Inn. Preinceton, New Jeresy.

267. Jha, R.R. & Varma, S.K. 1996. Ethnobotany of Sauria Paharias of Santhal Pargana, Bihar: 1. *Ethnobotany* **8**: 31-35.

268. Jha, U. N. 1965. Hydrophytes of Ranchi. *Trop. Ecol.* **6**: 98 – 105.

269. Jha, V. Mishra, S. Kargupta, A.N. & Jha, A. 1996. Leaves and flowers utilized as supplementary vegetables in Darbhanga (North Bihar) and their ethnobotanical significance: 305-402. *In* Maheshwari, J.K. (*eds.*): *Ethnobotany in South Asia*. Sci. Publ., Jodhpur (India).

270. Joshi, A.C. 1952. Aquatic vegetation of Lahoul. *The Palaeobotanist* **1**: 277-280.

280. Joshi, K. 2008. *Swertia* L. (Gentianaceae) in Nepal: Ethnobotany and agenda for sustainable management. *Environ. Leaflets* **12**: 1-6.

281. Joshi, K. 2009. Indigenous uses of wetland plant diversity of two valleys (Kathmandu and Pokhara) in Nepal. *Ethnobotany* **21**: 11-17.

282. Joshi, P. 1995. *Ethnobotany of the Primitive Tribes in Rajasthan*. Rupa Books Pvt. Ltd., Jaipur.

283. Kak, A. M. 1984a. Aquatic known weeds of Kashmir Himalayas. *J. Bombay Nat. History Soc.* **5**: 514 - 518.

284. Kak, A. M. 1984b. Aquatic Cyperaceae in the N.W. Himalaya. *J. Bombay Nat. History Soc.* **5**: 687 - 703.

285. Kak, M. M. 1990. Aquatic and wetland vegetation of the Kashmir Himalaya. *J. Econ. Taxon. Bot.* **14**: 1 - 14.

286. Kala, C.P. 2003. *Medicinal Plants of Indian Trans-Himalaya*. Bishen Singh Mahendra Pal Singh, Dehradun (India).

287. Kala, C.P. 2004. *The Valley of Flowers Myth and Reality*. Intl. Book Distr., Dehradun.

288. Kala, C.P. & Rawat, G.S. 2001. Human use and conservation status of wild medicinal herbs in the Bhyundar valley, Western Himalaya: 547-559. *In* Pande, P.C. & Samant, S.S. (*eds.*): *Plant Diversity of the Himalaya*. Gyanodaya Publ., Nainital (India).

289. Kamal, R. & Mathur, N. 1991. Histamine a biogenic amine from *Parthenium hysterophorus* Linn. *J. Pharm. Res.* **4**: 213-214.

290. Kamble, S.Y., Patil, S.R., Sawant, P.S., Sawant, S. Pawar, S.G. & Singh, E.A. 2010. Studies on plants used in traditional medicine by Bhilla tribe of Maharashtra. *Indian J. Trad. Know.* **9**: 591-598.

291. Kapoor, A., Kanwar, P. & Gupta, R. 2010. Traditional recipes of district Kangra of Himachal Pradesh. *Indian J. Trad. Know.* **9**: 282-288.

292. Kapur, S.K. 1985. Observations on the floristic composition of Kangra valley (Himachal Pradesh). *New Bot.* **12**:151-165.

293. Kapur, S.K. 1993. Ethnomedico plants of Kangra valley (Himachal Pradesh). *J. Econ. Taxon. Bot.* **17**(2): 395 - 408.

294. Kapur, S.K. & Singh, P. 1996. Traditionally important medicinal plants of Udhampur district (Jammu Province)- Part I: 75-82. *In* Maheshwari, J.K. (*eds.*): *Ethnobotany in South Asia*. Sci. Publ., Jodhpur (India).

295. Kapur, S.K. & Nanda, S. 1996a. Traditionally important medicinal plants of Bhaderwah hills (Jammu Province)-Part II: 56-61. *In* Maheshwari, J.K. (*eds.*): *Ethnobotany in South Asia*. Sci. Publ., Jodhpur (India).

296. Kapur, S.K. & Nanda, S. 1996b. Traditionally important medicinal plants of Bhaderwah hills (Jammu Province)- Part III: 70-75. *In* Maheshwari, J.K. (*eds.*): *Ethnobotany in South Asia.* Sci. Publ., Jodhpur (India).

297. Kapur, S.K. & Srivastava, T.N. 1996. Traditionally important medicinal plants of Udhampur district (Jammu Province)- Part II: 82-89. *In* Maheshwari, J.K. (*eds.*): *Ethnobotany in South Asia.* Sci. Publ., Jodhpur (India).

298. Kapur, S.K., Nanda, S. & Srivastava, T.N. 1996. Ethnobotanical uses of RRL-Herbarium-III: 50-56. *In* Maheshwari, J.K. (*eds.*): *Ethnobotany in South Asia.* Sci. Publ., Jodhpur (India).

299. Karnick, C.R. 1994. *Pharmacopocial Standards of Herbal Plants.* Vol. I. Sri Satguru Publ., Delhi (India).

300. Kashyap, S. R. 1993. *Liverworts of the Western Himalayas and the Punjab Plain.* Chron. Bot., New Delhi.

301. Katewa, S.S., & Arora, A., 1997. Some plants in folk medicines of Udaipur district (Rajasthan). *Ethnobotany* **9**: 48-51.

302. Katewa, S.S., Nag, Ambika & Guria, B. 2000. Ethnobotanical studies on wild plants for food from the Aravali Hills of South-East Rajasthan: 259-264. *In* Maheshwari, J.K. (*eds.*): *Ethnobotany and Medicinal plants of Indian Subcontinent.* Sci. Publ., Jodhpur (India).

303. Kaul, M.K. 1996. Strategies for sustainable use of threatened medicinal plant resources in western himalaya: 284-286. *In* Jain, S. K. (*eds.*): *Ethnobiology in Human Welfare.* Deep Publ., New Delhi.

304. Kaul, M.K. 1997. *Medicinal Plants of Kashmir and Ladakh Temperate and Cold Arid Himalaya.* Indus Publ. Co., New Delhi.

305. Kaul, M.K. 2010. High altitude botanicals in integrative medicine-case studies from Northwest Himalaya. *Indian J. Trad. Know.* **9**:18-25.

306. Kaur, S. 1989. Pteridophytes in science and technology: 61-68. *In* Bir S.S. & Saggo, M.L.S. (*eds.*): *Perspectives in Plant Sciences in India.* Today and Tomorrow's Printer Publ., New Delhi.

307. Kaushik , P. & Dhiman, A.K. 1984. Common Medicinal Pteridophytes. *Indian Fern Jl.* **12**: 139-45.

308. Kaushik, P. 1998. Ethnobotanical *Importance of Ferns of Rajasthan: Indigenous Medicinal Plants.* Today and Tommorrow's Printers and Publ., New Delhi.

309. Kaushik, P. & Dhiman, A.K. 2000. *Medicinal Plants and Raw Drugs of India.* Bishen Singh Mahendra Pal Singh, Dehradun.

310. Kayal, R.N., Saha, S. & Ghoshal, P.P. 2003. A floristic account of Indian medicinal plants (originated, cultivated, naturalized or introduced) used in homoepathic drug prepration: 60-67. *In* Singh, V. & Jain, A.P. (*eds.*): *Ethnobotany and Medicinal Plants of India and Nepal,* Vol. 1. Sci. Publ., Jodhpur (India).

311. Keller, H.A. & Ghillean, P.T. 2008. Plants associated with fish by the Guaranies of misiones, Argentina. *Ethnobotany* **20**: 1-8.

312. Khan, A.A. 2001. Need for an index to compute the relative reliability of ethnobotanical claims. *Ethnobotany* **13**: 84-86.

313. Khan, A.A. 2005. Ethnobotany in twenty-first century in India. *Ethnobotany* **17**: 71-78.

314. Khan, I. A. & Khanum, A. 2005. *Role of Biotechnology in Medicinal and Aromatic Plants.* Ukaaz Publ., Hyderabad, Andhra Pradesh.

315. Khan, S.S. 2003. *Vistas in Ethnobotany*. Vol. I. Indian Journal of Applied and Pure Biology.

316. Khanna, K.K., Mudgal, V., Shukla, G. & Srivastava, P.K. 1996. Unreported ethnomedicinal uses of plants from Mirzapur district. Uttar Pradesh. *J. Econ. Taxon. Bot. Addl. Ser.* *12*: 112-117.

317. Khare, C.P. 2004. *Indian Herbal Remedies: Rational Western Therapy, Ayurvedic and Other Traditional Usage, Botany*. Springer, New York.

318. Khare, P.B. 1996. Ferns and fern allies- their significance and fantacies. *App. Bot. Abstr.* NBRI, Lucknow *16*(1): 50-63.

319. Khare, P.K. & Khare, L.J. 2000. Plants used in rheumatism by rural people of Chhatarpur District, Madhya Pradesh, India: 301-304. *In* Maheshwari, J.K. (*eds.*): *Ethnobotany and Medicinal Plants of Indian Subcontinent*. Sci. Publ., Jodhpur (India).

320. Kharkongor, P. & Joseph, J. 1997. Folklore medico-botany of rural Khasi and Jaintia tribes in Meghalaya: 195-207. *In* Jain, S. K. (*eds.*): *Contribution to Indian Ethnobotany*, Vol. 3. Sci. Publ., Jodhpur (India).

321. Khullar, S.P. 1984. The ferns of Western Himalayas. A few additions, corrections and annotations. *Indian Fern. J.* **1**: 89-95.

322. Khullar, S.P. 1994. *An Illustrated Fern Flora of West Himalaya*, Vol I. Intl. Book Distr., Dehradun.

323. Khullar, S.P. 2000. *An Illustrated Fern Flora of West Himalaya*, Vol II. Intl. Book Distr., Dehradun.

324. Khuroo, A. A., Rashid, L., Reshi, Z., Dar, G.H. & Wafai, B.A. 2006. The alien flora of Kashmir Himalaya. *Biol. Invasions* **9**: 269-292.

325. Kindscher, K. 1987. *Edible Plants of the Prame: An Ethnobotanical Guide*. Univ. Press Kanas, Lawrence.

326. Kirtikar, K.R. & Basu, B.D. 1984. *Indian Medicinal Plants*. Vols. I-IV. Bishen Singh Mahendra Pal Singh, Dehradun (India).

327. Koelz, W.N. 1979. Notes on the ethnobotany of Lahaul, a province of the Punjab. *Quar. J. Crude Drug. Res.* **17**: 1-56.

328. Kothari, M.J. 2001. A revision of family Potamogetonaceae of India. *Bull. Bot. Surv. India* **43**(1-4): 151-194.

329. Kowarik, I. 2003. Human agency in biological invasions: secondary releases foster naturalisation and populatin expansion of alian plant species. *Biol. Invasions* **5**: 293-312.

330. Krishna, B. & Singh, S. 1987. Ethnobotanical observations in Sikkim. *J. Econ. Bot.* **9**: 1-7.

331. Kshirsagar, R.D. & Singh, N.P. 2000. Less – known ethnomedicinal uses of plants in Coorg district of Karnataka state, Southern India. *Ethnobotany* **12**: 12-16.

332. Kshirsagar, R.D. & Singh, N.P. 2007. *Ethnobotany of Mysore and Coorg, Karnataka State*. Bishen Singh Mahendra Pal Singh, Dehradun. (India).

333. Kulkarni, D.K. & Kumbhojkar, M.S. 1996. Plant conservation in Maharashtra during 17th century. *Ethnobotany* **12**: 39-38.

334. Kumar, A. 1996. Some ethnomedicinal plants of the Murias of the Indravati tiger reserve. Bastar (Madhya Pradesh): 201-205. *In* Maheshwari, J.K. (*eds.*): *Ethnobotany in South Asia*. Sci. Publ., Jodhpur (India).

335. Kumar, G.M. & Naqshi, A.R. 1990. Ethnobotany of Jammu – II Banihal. *J. Econ. Taxon. Bot.* **14**(1): 67 - 74.

336. Kumar, J. 2006. Cybernetic management of wetlands (A model for North Bihar). Flood plain wetlands of Bihar: 50-55. Proceedings of the National Seminar on *"The Environmental Status and Appropriate Use of Flood Plain Wetlands of Bihar"* held at B.N. College, Patna on 14 & 15 May, 2005. Naward, Patna (India).

337. Kumar, P.S. & Banerjee, L.K. 1999. *Potamogeton* L. the common pond weed in eastern India. *Envis No.* **6**: 12-14.

338. Kumar, R. Vijaya & Pullaiah, T. 2003. A survey of wild edible plants of Chenchu tribes of Pakistan district (Andhra Pradesh): 146-199. *In* Khan, S.S. (*eds.*): *Vistas in Ethnobotany*, Vol. I. Indian Journal of Applied and Pure Biology.

339. Kumar, S. 2002. *The Medicinal Plants of North-East India*. Sci. Publ. Jodhpur (India).

340. Kumar, S. & Nagiyan, P. 2006. Assessment and conservation and medicinal plant wealth of Haryana: 146-199. *In* Trivedi, P.C. (*eds.*): *Medicinal Plants: Ethnobotanical Approach*. Agrobios, Jodhpur (India).

341. Kumar, S. & Narain, S. 2010. Herbal remedies of wetlands and macrophytes in India. *Indian J. Pharma. & Bio Sciences* **1**(2): 1-12

342. Kumar, S. S. 1995. *Recent Studies on Indian Bryophytes*. Bishen Singh Mahendra Pal Singh, Dehradun (India).

343. Kumar, V. & Jain, S. K. 1998. A contribution of ethnobotany of Surguja district in Madhya Pradesh, India. *Ethnobotany* **10**: 89-96.

344. Kurian, J.C. 1999. *Plants That Heal.* Oriental Watchman Publ. House, Pune.

345. Lairel, S.A. 2002. *Biodiversity and Traditional Knowledge – Equitable Partenership in Practice*. Earthscan Publ. Ltd., London.

346. Lal, B., Vats, S.K., Singh, R.D. & Gupta, A.K. 1996. Plants used as ethnomedicine and supplementary food by Gaddis of Himachal Pradesh, India: 384-387. *In* Jain, S. K. (*eds.*): *Ethnobiology in Human Welfare*. Deep Publ., New Delhi.

347. Lal, C., Negi, B.S. Kukreti, M., Singh, M. E. & Ghidiyal, J.C. 1997. The aquatic vegetation of Uttarkashi-Garhwal Himalayan region, U.P. *J. Econ. Taxon. Bot.* **21**: 79-82.

348. Lalramnghinglova, H. 1999. Ethnobotanical and agroecological studies on genetic resources of food plants in Mizoram State. *J. Econ. Taxon. Bot.* **23**(2): 637-644.

349. Lalramnghinglova, H. 2002. Ethnobotanical study on the edible plants of Mizoram. *Ethnobotany* **14**: 23-33.

350. Lalramnghinglova, H. 2003. *Ethnomedicinal Plants of Mizoram*. Bishen Singh Mahendra Pal Singh, Dehradun. (India).

351. Lalramnghinglova, J. H. 1996. Ethnobotany of Mizoram –A preliminary survey: 439-459. *In* Maheshwari, J.K. (*eds.*): *Ethnobotany in South Asia*. Sci. Publ., Jodhpur (India).

352. Lindley, J. 1981. *Flora Medica; A Botanical Account of All the More Important Plants Used in Medicine*. Ajay Book Service. New Delhi., India.

353. Mahajan, K.K. 1989. *Indian Wetlands: An Overview and Their Management. Wetland Conservation.* Environ. Comm. Centre, Udaipur, India.

354. Mahato, A.K. & Mahato, P. 1996. Ethnobotanical wealth of Chhotanagpur plateau-IV. Some medicinal plants used against intestinal worms: 389-391. *In* Maheshwari, J.K. (*eds.*): *Ethnobotany in South Asia.* Sci. Publ., Jodhpur (India).

355. Mahato, R.B. & Chaudhary, R.P. 2005. Ethnomedicinal plants of Palpa district, Nepal. Ethnobotany **17**: 152-163.

356. Maheshwari, J.K. 1992. *Ethnobotany in India.* Sci. Publ. Jodhpur.

357. Maheshwari, J.K. 1996. *Ethnobotany in South Asia.* Sci. Publ. Jodhpur (India).

358. Maheshwari, J.K. 2003. *Ethnobotany and Medicinal Plants of Indian Subcontinent.* Sci. Publ., Jodhpur (India).

359. Maheshwari, J.K. & Paul, S.R. 1975. The exotic flora of Ranchi. *J. Bombay Nat. History Soc.* **72**(1): 158-188.

360. Maheshwari, J.K. & Singh, J.P. 1984. Contribution to the ethnobotany of Bhoxa tribe of Hijnar and Pauri Garhwal districts, U.P. *J. Econ. Taxon. Bot.* **5**: 251-259.

361. Maheshwari, J.K. & Singh, H. 1992. Plants used by Bhora P.T.G. of Dehradun (U.P.). *Vanyajati* **40**: 1-8.

362. Maheshwari, J.K., Singh, K.K. & Saha, S. 1996. *Ethnobotany of tribals of Mirzapur District, Uttarpradesh, Economic Botany Information Service.* NBRI, Lucknow.

363. Maliya, S. D. 2006. The Aquatic and Wetland Flora of Mainpuri district, U.P., India. *J. Econ. Taxon. Bot.* **30**(3): 533 - 546.

364. Maliya, S.D. 2009. Ethnomedicinal practices among tribals and indigenous people of Shravarti and Bahraich districts, (UP). *Ethnobotany* **21**: 121-123.

365. Maliya, S. D. & Singh, S. M. 2004. Diversity of aquatic and wetland macrophytes vegetation of Uttar Pradesh (India). *J. Econ. Taxon. Bot.* **28**(4): 935-975

366. Malla, B. & Chhetri, R.B. 2009. Observations on some ethnomedicinal plants in Kavrepalanchowk district, Nepal. *Ethnobotany* **21**: 41-45.

367. Mamgain, M.D. 1975. Himachal Pradesh District Gazetteers: Lahaul and Spiti. 135.

368. Manandhar, N. P. 1987. An ethnobotanical profile of Manang Valley, Nepal. *J. Econ. Taxon. Bot.* **10**: 207- 213.

369. Manandhar, N.P. 1996a. Traditional practice for oral health care in Nepal: 408-411. *In* Maheshwari, J.K. (*eds.*): *Ethnobotany in South Asia.* Sci. Publ., Jodhpur, India.

370. Manandhar, N.P. 1996b. Ethnobotanical observations on ferns and fern allies of Nepal: 414-422. *In* Maheshwari, J.K. (*eds.*): *Ethnobotany in South Asia.* Sci. Publ., Jodhpur (India).

371. Mandal, R. N., Saha, G. S., Das, K. M. & Choudhury, B. P. 2006. Eco- floristic survey of aquatic macrophytes in Cita Form - a case study. *J. Econ. Taxon. Bot.* **30**(4): 776 - 782.

372. Mandal, S.K. & Basu, S.K. 1996. Ethnobotanical studies among some tribals of Nilgiri district, Tamil Nadu: 268-271. *In* Maheshwari, J.K. (*eds.*): *Ethnobotany in South Asia.* Sci. Publ., Jodhpur (India).

373. Manilal, K.S. 1989. Linkages of ethnobotany with other sciences and discipline. *Ethnobotany* **1**: 15-24.

374. Mao, A.A. 1993. A preliminary report note on the folklore botany of Mao Naga of Manipur (India). *Ethnobotany* **5**: 143-147.

375. Martin, M. 1972. *Introduction a 1' Ethnobotanique du Cembodge*. Paris.

376. Martin, M., Mathias, Evelyn & Censtance (*eds.*) 2001. *Ethnoveterinary Medicine – An Annonated Bibliography of Community Animal Healthcare*. ITDG Publishing, London.

377. Martin, P. & Retnam, R.K. 2005. *Ethnomedicinal Plants*. Agrobios. India.

378. Masih, S.K. 2003. Diversity of ethnomedicinal wealth in Amar Kantak plateau region (India): 204-214. *In* Khan, S. S. *Vistas in Ethnobotany* Vol. I. Indian Journal of Applied and Pure Biology.

379. Matthew, K.M. 1969. Exotic flora of Kodaikanal and Palni Hills. *Rec. Bot. Surv. India.* **20**(1): 1-241.

380. May, L.W. 1978. The economic uses and associated folklore of ferns and fern allies. *Bot. Rev.* **44**: 491-528.

381. Maya, S., Sarojini M.V. & Sreekandan, N.G. 2003. Ethnobotanical notes on the flora of sacred tanks of Kerala. *Ethnobotany* **15**: 55-59

382. Meena, K.L. & Yadav, B.L. 2010. Some traditional ethnomedicinal plants of Southern Rajasthan. *Indian J. Trad. Know.* **9**: 471-474.

383. Megoneitso, K. & Rao, R. 1983. Ethnobotanical studies in Nagaland-4. Sixty two medicinal plants used by Angami-Nagas. *J. Econ. Taxon. Bot.* **4**: 167-172.

384. Mehra, P. N. 1982. *Cytology of East Indian Grasses*. P. Kapur Rajbandhu Industrial Co., New Delhi.

385. Mehrotra, B.N. 1990. Quality control requirement of medicinal plants used in traditional medicines. *Ethnobotany* **2**: 19-24.

386. Mehrotra, S. & Mehrotra, B.N. 2005. Role of traditional and folklore herbals in the development of new drugs. *Ethnobotany* **17**: 104-111.

387. Minnis, P. (*ed.*) 2000. *Ethnobotany: A Reader*. Univ. Oklahoma Press. Norman.

388. Mishra, D.K., Samanta, G. & Mishra, T.K. 1996. Ethnomedicine of tribe Kharia of Midnapore district, West Bengal: 329-331. *In* Maheshwari, J.K. (*eds.*): *Ethnobotany in South Asia*. Sci. Publ., Jodhpur (India).

389. Mishra, D. P., & Sahu, T. R., 1984. Euphorbiaceous plants used in medicine by the tribes of Madhya Pradesh, India. *J. Econ. Taxon. Bot.* **5**: 791-793.

390. Mishra, A. 2008. *Studies on Aquatic and Marshy Angiospermic Plants of Eastern Uttar Pradesh*. Univ. of Allahabad, Allahabad.

391. Mitsch, W.I. & Gosselink, I. G. 1986. *Wetlands*. Van Nostrand-Reinhold, New York.

392. Mohanty, R.B. & Padhy, S.N. 1996. Traditional phytotherapy for diarrhoeal diseases in Ganjam and Phulbami districts of south orissa, India. *Ethnobotany* **8**: 60-66.

393. Molla, H.A. & Roy, B. 1996. Some ethnomedicinal claims from Jalpaiguri district of West Bengal: 322-324. *In* Maheshwari, J.K. (*eds.*): *Ethnobotany in South Asia*. Sci. Publ., Jodhpur (India).

394. Mudaliyar, C.T. 1921. *A Handbook of Some South Indian Grasses*. Printed by the Suptd. Govt. Press, Madras.

395. Mudgal, V. 1987. Literature on ethnobotany – Indian and foreign: 45-57. *In* Jain, S.K. (*eds.*): *A Manual of Ethnobotany*. Sci. Publ., Jodhpur (India).

396. Mudgal, V., Pal, D.C., Kayal, R.N. & Saha, S. 1999. *Ethnobotany of Totopara*. Bishen Singh Mahendra Pal Singh, Dehradun (India).

397. Muenscher, W.C. 1944. *Aquatic Plants of the United States*. Cornell Univ. Press.

398. Mukerjee, A.K. 1984. *Flora of Pachmarhi and Bori Reserves*. Howrah.

399. Murakami, T., Maehashi, H., Tanaka, N., Stake, T., Kurashi, H. Komazawa, Y., Saiki, Y. & Chem, C.M. 1986. Chemical constituents of *Pteris cretica* L. *Carcinogenesis* **105**: 640-648.

400. Muratkar, G.D. & Rothe, S.P. 2000. Preliminary observation on the medicinal and economically important leguminous plant species from Amravati Tehsil: 283-289. *In* Maheshwari, J.K. (*eds.*): *Ethnobotany and Medicinal Plants of Indian Subcontinent*. Sci. Publ., Jodhpur (India).

401. Murthy, S.R., Rani, S.S. & Pullaiah, T. 2003. Wild edible plants of Andhra Pradesh, India: 613-630. *In* Singh, V & Jain, A.P. (*eds.*): *Ethnobotany and Medicinal Plants of India and Nepal,* Vol. 2. Sci. Publ., Jodhpur (India).

402. Nadkarni, K.M., 1976. *Indian Materia Medica,* Vols. I–II. Popular Prakashan Private Ltd. (Popular Press), Bombay.

403. Nag, A., Galav, P., Katewa, S.S., & Swarnkar, S. 2009. Traditional herbal veterinary medicines from Kotda tehsil of Udaipur district (Rjasthan). *Ethnobotany* **21**: 103-106.

404. Nair, N.B. 1989. A spatial study of the Neyyer River in the light of the river-continuum-consept. *Trop. Ecol.* **30**(1): 101-110.

405. Nair, N.C. 1977. *Flora of Bashahr Himalaya*. Int. Biosciences Publ., Hissar.

406. Nandan, K.B. & Singh, C.B. 2004. Useful macrophytes in Kawar lake, North Bihar, India. *Indian J. For.* **27**(3): 241-244.

407. Narain, S. 2006. Additions to the aquatic and marshy plants of Hamirpur and Mahoba district (U.P.) India. *J. Phytol. Res.* **191**(1): 135-137.

408. Narain, S. & Singh, M. 2008. Aquatic and marshy angiosperms of Sarsainawar wetland of Etawah district, Uttar Pradesh, India. *J. Indian Bot. Soc.* **87**:157-161.

409. Nath, S.C. & Bordoloi, D.N. 1989. Ethnobotanical observation on some medicinal folklore of Tirap district, Arunachal Pradesh. *J. Econ. Taxon. Bot.* **13**: 321-325.

410. Nautiyal, A.R., Nautiyal, M.C. & Purohit, A. N. 1997. *Harvesting Herbs-2000. Medicinal and Aromatic Plants-An Action Plan for Uttarkhad*. Bishen Singh Mahendra Pal Singh, Dehradun. (India).

411. Nautiyal, S. 1981. Some medicinal plants of Garhwal Hills- A traditional use. *J. Sci. Res. Plate No. Med.* **2**: 12-17.

412. Nautiyal, S., Maikhuri, P.K., Rao, K.S. & Saxena, K.G. 2003. Ethnobotany of the Thlchha Bhotiya tribe of the buffer zone villages in Nanda Devi Biosphere Reseve, India: 119-142. *In* Singh, V. & Jain, A.P. (*eds.*) *Ethnobotany and Medicinal Plants of India and Nepal,* Vol. 1. Sci. Publ., Jodhpur (India).

413. Nayar, E.R. 2010. Indian plant genetic resources: role of ethnobotany in determining priorities and issues. *Ethnobotany* **22**: 86-96.

414. Nayar, M.P. 1977. Changing patterns of the Indian flora. *Bull. Bot. Surv. India* **19**: 145-154.

415. Negi, K.S. & Gaur, R.D. 1991. A contribution to the edible wild fruits of Uttar Pradesh hills. *Bull. Bot. Surv. India* **43**: 233-266.

416. Negi, K. S., Gaur, R.D. & Tiwari, J.K. 1999. Ethnobotanical notes on the flora of Har-ki-Doon (District Uttarkashi), Garhwal Himalaya, Uttar Pradesh. India. *Ethnobotany* **11**: 9-17.

417. Negi, K.S., Tiwari, J.K. & Gaur, R.D. 1985. A contribution to the flora of Dodotal- a high altitude lake in the Garhwal Himalaya (Utterkashi), U.P. *J. Bombay Nat. History Soc.* **82**: 258-272.

418. Negi, P.S. & Hajra, P.K. 2007. Alien flora of Doon valley, North West Himalaya. *Curr. Sci.* **92** (7): 968-978.

419. Negi, P.S. & Subramani, S.P. 2002. Ethnobotanical study in the village Chhitkul of Sangla valley, district Kinnaur, Himachal Pradesh. *J. Non-Timber Forest Prod.* **9**(3-4): 113-120.

420. Negi, Y.S. & Bhalla, P. 2002. Collection and marketing of important medicinal and aromatic plants in tribal areas of Himachal Pradesh. *Indian For.* **128**(6): 641-649.

421. Neuwinger, H.D. 1996. *African Ethnobotany- Poisons and Drugs*. Chapman & Hall.

422. Neuwinger, H.D. 2000. *African Traditional Medicine: A Dictionary of Plant Uses and Applications*. Medphrm. Sci. Pub. Sturtgart, Germany.

423. Newall, C.A., Anderson, L. A. & Phillipson, J.D. 1996. *Herbal Medicine: A Guide for Health Care Professionals*. The Pharmaceutical Press, London.

424. Noumi, E. 2010. Ethno-medico-botanical survey of medicinal plants used in the treatment of asthma in the Nkongsamba region, Camroon. *Indian J. Trad. Know.* **9**: 491-495.

425. Nunez, D.R. & Castro, C.O. 1995. Medicinal plants and a multipurpose complex sold in the market of Funchal (Island of Madeira, Portugal). *Ethnobotany* **7**: 75-82.

426. Painuli, R.M. & Maheshwari, J.K. 1996. Some interesting ethnomedicinal plants used by Sahariya tribe of Madhya Pradesh. *J. Econ. Taxon. Bot. Addl. Ser.* **12**: 179-185.

427. Pal, D.C. 2000. Some plants in traditional and homeopathic systems of medicine in India: 79-83. *In* Maheshwari, J.K. (*eds.*): *Ethnobotany and Medicinal Plants of Indian Subcontinent*. Sci. Publ., Jodhpur (India).

428. Pal, D.C. & Srivastava, J.N. 1976. Preliminary notes on ethnobotany of Singhbhum district Bihar. *Bull. Bot. Surv. India* **18**: 247- 250.

429. Panda, A. & Misra, K.M. 2011. Ethnomedicinal survey of some wetland plants of south Orissa and their conservation. *Indian J. Trad. Know.* **10**(2): 296-303.

430. Panda, P.C. & Das, P. 2000. Medicinal plant lore of the tribals of Baliguda sub-Division, Phulbani district, Orissa: 515-521. *In* Maheshwari, J.K. (*eds.*): *Ethnobotany and Medicinal Plants of Indian Subcontinent*. Sci. Publ., Jodhpur (India).

431. Panda, S. 1996. Plant use - a recent prespective from Eastern Himalayas: 332-337. *In* Maheshwari, S.K. (*eds.*): *Ethnobotany in South Asia*. Sci. Publ., Jodhpur (India).

432. Pande, P.C., Tiwari, L. & Pande, H.C. 2006. *Folk-Medicine and Aromatic Plants of Uttaranchal*. Bishen Singh Mahendra Pal Singh, Dehradun. (India).

433. Pandey, A.K., Bora, H.R. & Deka, S.C.1996. An ethno-medico botanical study of Golghat district, Assam: native plant remedies for jaundice. *J. Econ. Taxon. Bot.* **12**: 344-349.

434. Pandey, D.N. 1998. *Ethnoforestry (Local Knowledge for Sustainable Forestry and Livelihood Security)*. Himanshu Publ., Udaipur, New Delhi.

435. Pandey, H., Prakash & Chauhan, S.K. 2000. Antiseptic property of *Solanum surattense* Burm f.: 41-42. *In* Maheshwari, J.K. (*eds.*): *Ethnobotany and Medicinal Plants of Indian Subcontinent*. Sci. Publ., Jodhpur (India).

436. Pandey, R.K. & Pandey, C. 2009. Medicinal value of aquatic and wetland plants of Varanasi district. *Indian J. Trop. Biodiversity* **17**(2): 141-150.

437. Pandey, R.P. 2004. *Enhydra fluctuans* Lour. (Asteraceae). A new record from Rajasthan India. *J. Econ. Taxon. Bot.* **28**(3): 527-528.

438. Pandey, R.P. & Parmar, P.J. 1994. The exotic flora of Rajasthan. *J. Econ. Taxon. Bot.* **18**(1): 105-121.

439. Panthi, M. P. & Chaudhary, R. P. 2003. Ethnomedicinal plant resources of Arghakhanchi district, West Nepal. *Ethnobotany* **15**: 71-86.

440. Parabia, M. & Pathak, S. 2007. Use of herbal resources in primary healthcare and the knowledge feedback mechanism as an interactive system for tribal empowerment. *Ethnobotany* **20**: 106-110.

441. Pareek, A. 1994. Preliminary ethnobotanical notes on the plants of aquatic habitats of Rajasthan. *J. Phytol. Res.* **7**: 73-76.

442. Parkash, V. & Aggarwal, A. 2010. Traditional uses of ethnomedicinal plants of lower foot-hills of Himachal Pradesh-I. *Indian J. Trad. Know.* **9:** 519-521.

443. Parmar, C. & Kaushal, M.K. 1982. *Wild Fruits of the Sub-Himalayan Region*. Kalyani Publ., New Delhi.

444. Parrotta, J.A. 2001. *Healing Plants of Peninsular India*. CABI Publishing House.

445. Patil, D.A. 2000. Sanskrit plant names in an ethnobotanical perspectives. *Ethnobotany* **12**: 60-64.

446. Patil, D.A. 2009. Significance of some vernacular plants names in Buldhana district (Maharashtra) vis-à-vis their origin. *Ethnobotany* **21**: 91-94.

447. Patil, P.S., Dushing, Y.A. & Patil, D.A., 2007. Observations on plantlore in Buldhana district of Maharashtra. *Ancient Sci. Life* **17**(1): 43-49.

448. Payee, Gabriell DeBear (*ed.*) 2000. *Cultural Uses of Plants: A Guide to Learning About Ethnobotany*. New York Bot. Gdn. Press, New York.

449. Perrings, C. Dehnen-Schmutz, K. Touza, J. & Williamson, M. 2005. How to manage biological invasions under globalization. *Trends Ecol. Evol.* **20**: 212- 215.

450. Perry, F. 1938. *Water Garden*. London.

451. Pieroni, A. 1999. Gathered wild food plants in the upper valley of the Serchio River (Garfagnana), Central Italy. *Econ. Bot.* **53**(3): 327-341.

452. Polunin, O. & Stainton, A. 1984. *Flowers of the Himalaya*. Oxford Univ. Press, Delhi.

453. Polunin, O. & Stainton, A. 1987. *Concise Flowers of the Himalaya*. Oxford Univ. Press, Delhi.

454. Powers, 1874. Aboriginal Botany – the study of all forms of vegetation that aboriginal people used for commodities such as medicine, food, textiles and ornaments.

455. Prajapati, N.D., Purohit, S.S., Sharma, A.K. & Kumar, T. 2006. *A Hand book-Medicinal Plants - A Complete Source Book*. Agrobios, Jodhpur (India).

456. Prakasha, H.M., Krishnappa, M., Krishnamurthy, Y.L. & Poornima, S.V. 2010. Folk medicine of NR Puna taluk in Chikmagalur district of Karnataka. *Indian J. Trad. Know.* **9**(1): 55-60.

457. Prasad, B.N. 1996. Leucoderma in ancient Indian medicine. *British J. Dermat.* **78**: 355.

458. Prasad, V. K., Rajagopal, T., Kanit, Y. & Badrinath, K.V.S. 2002. Food plants of Konda Reddis of Rampa Agency. East Godavari district, Andhra Pradesh- a case study. *Ethnobotany* **11**: 92-96.

459. Punekar, S.A. Malpure, N.V. & Lakshinaraayan 2003. Five new species of *Eriocaulon* L. (Eriocaulaceae) from the western Ghats. *India Rheedea* **13**: 19-27.

460. Punjani, B.L. 2002. Plants used in contact therapy by tribals of Sabarkantha (Gujrat). *Ethnobotany* **14**: 57-59.

461. Pushpangadan, P. & Kumar, B. 2005. Ethnobotany, CBD, WTD and the biodiversity act of India. *Ethnobotany* **17**: 2-12.

462. Rahman, M. A. 2000. Ethno-medico-botanical knowledge among tribals of Bangladesh: 89-93. *In* Maheshwari, J.K. (*eds.*): *Ethnobotany and Medicinal Plants of Indian Subcontinent*. Sci. Publ., Jodhpur (India).

463. Rai, M.K. & Upadhyay, S.K. 1996. Medicinal plants of Chhindwara district: retrospect and prospect: 174-188. *In* Rajak, R.C. & Rai, M.K. (*eds.*) *Herbal Medicines Biodiversity and Conservation Strategies*. Intl. Book Distributors, Dehradun, India.

464. Rajendran, A. & Sikarwar, R.L.S. 2003. Intra-cultural ethnobotanical studies on the tribe Kadars (Tamilnadu) and Sahariyas (Madhya Pradesh), India: 662-664. *In* Singh, V. & Jain, A.P. (*eds.*): *Ethnobotany and Medicinal Plants of India and Nepal*, Vol. 2. Sci. Publ., Jodhpur (India).

465. Rajendran, S.M. & Aswal, B.S. 2000. Some flowering plants used as cosmetics among tribals of Nilgiris, Tamil Nadu, India: 425-430. *In* Maheshwari, J.K. (*eds.*): *Ethnobotany and Medicinal Plants of Indian Subcontinent*. Sci. Publ., Jodhpur (India).

466. Rajendran, S.M., Sekar, K.C. & Sudaresan, V. 2002. Ethnomedicinal lore of Valaya tribals in seithur hills of Virudunagar district, Tamil Nadu, India. *Indian J. Trad. Know.* **1**: 59

467. Raju, R.A. 2000. *Wild Plants of Indian Sub-Continent and Their Economic Use*. CBS Publ. & Distr., New Delhi.

468. Rama Rao, N. & Henery, A. N. (*eds.*) 1996. *Ethnobotany of Eastern Ghats in Andhra Pradesh, India*. B.S.I., Calcutta.

469. Rama Rao N., Seetharami Reddy, B.V.A., & Prasanthis, T.V.V. 2008. Folk herbal remedies for rheumatoid arthritis in Srikakulam district of Andhra Pradesh. *J. Econ. Taxon. Bot.* **1**: 157-162.

470. Ramdas, S.R., Ghotge, N.S., Ashalata, S., Mathur, N.P., Broome, V.G. & Rao, S. 2000. Ethnoveterinary remedies used in common surgical conditions in some districts of Andhra Pradesh and Maharashtra, India. *Ethnobotany* **12**: 100-112.

471. Rana, J.C. 2004. *Lesser Known Food Plants of Himachal Pradesh.* N.B.P.G.R., Regional Station, Phagli, Shimla (H.P.).

472. Rana, S., Datt, B. & Rao, R.R. 2000. Strategies for sustainable utilization of plant resources by the Tribals of Tons valley, Western Himalaya: 90-104. *In* Maheshwari, J.K. (*eds.*): *Ethnobotany and Medicinal Plants of Indian Subcontinent.* Sci. Publ., Jodhpur (India).

473. Rana, T .S., Datt, B. & Rao, R.R. 2003. *Flora of Tons Valley.* Bishen Singh Mahendra Pal Singh, Dehradun. (India).

474. Ranjan, P. 1999. A contribution to some of the medicinal plants of Indo-Nepal border area adjoining the districts of Madhubani and Sitamarhi. *Ethnobotany* **11**: 135-137.

475. Ranjan, P. 2000. A contribution to some of the medicinal plants of Indo-Nepal border area adjoining the districts of Madhubani and Sitamarhi: 651-659. *In* Maheshwari, J.K. (*eds.*): *Ethnobotany and Medicinal Plants of Indian Subcontinent.* Sci. Publ., Jodhpur (India).

476. Ranjan, P. 2003. A contribution to some ethnobotanically important plants of Nepal: 345-353. In Singh, V. & Jain, A.P. (*eds.*): *Ethnobotany and Medicinal Plants of India and Nepal,* Vol. 1. Sci. Publ., Jodhpur (India).

477. Ranjan, S., Gupta, H.C. & Kumar, S. 2000. Exotic medicinal plants of Lucknow: 205-222. *In* Maheshwari, J.K. (*eds.*): *Ethnobotany and Medicinal Plants of Indian Subcontinent.* Sci. Publ., Jodhpur (India).

478. Rao, B. U.V.U. 2006. Collection and documentation of ethno-medicinal and non-timber forest produts (N.T.F.P.). Information from Royalseema region of Andhra Pradesh , India. *Ethnobot. Leaflets* **10**: 294-304.

479. Rao, N.R. & Henry, A.N. 1995. *The Ethnobotany of Eastern Ghats in Andhra Pradesh, India.* B.S.I., Calcutta.

480. Rao, N.R., Rajendran, A. & Henery, A.N. 2000. Phyto-zootherapy of the tribes of Andhra Pradesh: 331-335. *In* Maheshwari, J.K. (*eds.*): *Ethnobotany and Medicinal Plants of Indian Subcontinent.* Sci. Publ., Jodhpur (India).

481. Rao, R.R. 1996. Traditional knowledge and sustainable development: key role of ethnobiologists. *Ethnobotany* **8**: 14-27.

482. Rao, R.R & Jamir, N.S. 1982. Ethnobotanical studies in Nagaland II. 54 medicinal plants used by Nagas. *J. Econ. Taxon. Bot.* **36**: 176-181.

483. Rao, R.R. & Negi, B. 1980. Observations on the ethnobotany of Khasi and Garo tribes in Meghalaya (India). *J. Econ. Taxon. Bot.* **1**: 157-162.

484. Rao, V.V. & Rajasekharan, P. E. 2002. Threatened medicinal plant resources and conservation needs. *The Botanica* **52**: 53-63.

485. Rao, G. R., Subhashchandran, M. D. & Ramachandra, T. V. 2011. Wetland flora of Uttara Kannada, In Environment Education for Ecosystem. *Curr. Sci.* **95**(5): 678-679.

486. Rastogi, M.A. 1960. Medicines from the wild. A case study of the Great Himalayan Park. *People & Cult.*: 74-75.

487. Rastogi, R.P. & Mehrotra, B.N. 1990. *Compendium of Medicinal Plants.* Vol. I. (1960-69) Central Drugs Research Institute, Lucknow and National Institute of Science Communication, New Delhi (In-

dia).488. Rastogi, R.P. & Mehrotra, B.N. 1991. *Compendium of Medicinal Plants*. Vol. II. (1970-79) Central Drugs Research Institute, Lucknow and National Institute of Science Communication, New Delhi (India).

489. Rastogi, R.P. & Mehrotra, B.N. 1993. *Compendium of Medicinal Plants*. Vol. III. (1980-84) Central Drugs Research Institute, Lucknow and National Institute of Science Communication, New Delhi (India).

490. Rastogi, R.P. & Mehrotra, B.N. 1995a. *Compendium of Medicinal Plants*. Vol. IV. (1985-89) Central Drugs Research Institute, Lucknow and National Institute of Science Communication, New Delhi (India).

491. Rastogi, R.P. & Mehrotra, B.N. 1995b. *Compendium of Medicinal Plants*. Vol. V. (1990-94) Central Drugs Research Institute, Lucknow and National Institute of Science Communication, New Delhi (India).

492. Rau, M.A. 1961. A recent survey of medicinal plant resources of Lahaul valley (Panjab India). *Ind. Res. SPlate No. Nr. Intl. Symp.* **4**: 201-205.

493. Ravindran, K.C., Venkatesan, K., Balakrishnan, V., Chellappan, K.P. & Balasubramanian, T. 2005. Ethnomedicinal studies of Pichavaram mangroves of East Coast, Tamilnadu. *Indian J. Trad. Know.* 5: 201-205.

494. Rawat, M.S. & Chaudhury, S. 1998. Ethno-medico-botany of Arunachal Pradesh (Nyshi and Apatani tribes). Bishen Singh Mahendra Pal Singh, Dehradun. (India).

495. Reddy, C.S. 2008. Catalogue of invasive alien flora of India. *Life Sci. Jour.* **5**(2): 84-89.

496. Reddy, C.S & Raju, V.S. 2002. Additions to the weed flora of Andhra Pradesh. *J. Econ. Taxon. Bot.* **26**: 195-198.

497. Reddy, C.S. & Reddy, K.N. 2004. *Cassia rotundifolia* Pers. (Caesalpiniaceae): a new record for India. *J. Econ. Taxon. Bot.* **28**: 73-74.

498. Reddy, C.S., Bhanja, M.R. & Raju, V.S. 2000. *Cassia uniflora* Miller: a new record for Andhra Pradesh, India. *Indian. J. For.* **23**(3): 324-325.

499. Reddy, K.N., Madhuri, V., Subbaraju, G.V. & Hemadri, K. 2005. Ethnotherapeutics of certain Ayurvedic medicinal Plants of Kodapalli fort, Andhra Pradesh. *Indian For.* **131**: 442-448.

500. Reddy, S. & Raju, V.S. 2000. Folklore biomedicine for common veterinary diseases in Nalgonda district, Andhra Pradesh, India. *Ethnobotany* **12**: 113-117.

501. Redford, K.H. & Mansur, J.A. 1997. *Traditional People and Biodiversity Conservation in Large Tropical Landscapes*. America Verde Publ., Arlington, Virginia, USA.

502. Reid, G.K. 1961. *Ecology of Inland Waters and Estuaries*. Reinhold Publ. Co., New York.

503. Renfrew, J.M. 1973. *Paleoethnobotany- The Prehistoric Food Plants of the Near East and Europe*. Columbia Univ., New York.

504. Retnam, R.K. & Martin, P. 2006. *Ethnomedicinal Plants*. Agrobios, India.

505. Rosakutty P. S, Roslin A. S., & Ignacimuthu S, 1999. Some traditional folklore medicinal plants of Kanyakumari district. *J. Econ. Taxon. Bot.* **23**: 369–375.

506. Rosakutty, P.S., Roslin, A. Stella & Ignarimuthu, S. 2000. Some traditional folklore medicinal plants of Kanyakumari District (Tamil Nadu): 369-375. *In* Maheshwari, J.K. (*eds.*): *Ethnobotany and Medicinal Plants of Indian Subcontinent*. Sci. Publ., Jodhpur (India).

507. Rout, S.D. & Panda, S.K. 2010. Ethnomedicinal plant resources of Mayurbhanj district, Orissa. *Indian J. Trad. Know.* **9**: 68-72.

508. Roy, A. Banerjee, L.K. & Mukherjee, K. 1999. *Sagittaria montevidensis* Cham. and Schlecht. (Alismtaceae)- A new record for India. *Rheedea* **9**(1): 85-88.

509. Roy, B., Halder, A.C. & Pal, D. C. 1998. *Plants for Human Consumption in India*. B. S.I., Calcutta.

510. Roy, G.P. 1984. *Flora of India. Grasses of Madhya Pradesh*. B.S.I., Calcutta.

511. Sahu, S.C., Dhal, N.K. & Mohanty, R.C. 2009. Ethnobotanical study of Deogarh district (Orissa) with respect to plants used for treating skin diseases. *Ethnobotany* **21**: 46-50.

512. Said, D.C. 1997. *Hamdard Pharmacopoeia of Eastern Medicine*. Sri Satguru Publ., Delhi.

513. Saini, D.C.1996a. Ethnobotany of Basti district, Uttar Pradesh. *J. Econ. Taxon. Bot. Addl.Ser.* **12**: 138-153.

514. Saini, D.C. 2008. Traditional uses of Pteridophytes among Baiga tribes of Amarkantak, Anuppur district, M.P. *Ethnobotany* **20**: 65-69.

515. Saini, D.C., Singh, S.K. & Rai, K. 2010. *Biodiversity of Aquatic and Semi-aquatic Plants of Uttar Pradesh (with special reference to Eastern Uttar Pradesh)*. Gomti Nagar, Lucknow.

516. Saini, V. K.1996b. Plants in the welfare of tribal woman and children in certain areas of Central India: 140-144. *In* Jain, S. K. (*eds.*): *Ethnobiology in Human Welfare*. Deep Publ., New Delhi.

517. Saklani, A. & Jain, S.K. 1994. *Cross-Cultural Ethnobotany of North-East India*. Deep Publ., New Delhi (India).

518. Samant, S. 1999. Prioritization of biological conservation sites in India wetland: 155-167. *In* Singh, S., Sastry, A.R.K., Mehta, R. & Uppal, V. (*eds.*): *Setting Biodiversity Conservation Priorities For India*. World Wide Fund for Nature, India.

519. Samant, S.S., Dhar, U. & Rawal, R.S. 1996. Natural resource use by some natives within Nanda Devi Biosphere reserve in West Himalaya. *Ethnobotany* **8**: 40-50.

520. Samant, S.S., Dhar, U. & Rawal, R.S. 2001a. Diversity and distribution of wild edible plants Indian Himalaya: 421-482. *In* P.C. Pandey & Samant, S.S. (*eds.*): *Plant Diversity of the Himalaya*. Gyanodaya Prakashan, Nainital. India.

521. Samant, S.S., Dhar, U. & Rawal, R.S. 2001b. Diversity, disrtibution, and indigenous uses of threatented medicinal plants of Askot wildlife sanctuary in West Himachal: Causes and management - perspectives: 167-184. *In* Samant, S.S., Dhar, U. & Palni, M.S. (*eds.*): *Himalayan Medicinal Plants, Potential and Perspects*. Gyanodaya Prakasan, Nainital, India.

522. Samant, S.S., Dhar, U. & Palni, L.M.S. 1998. *Medicinal Plants of Indian Himalaya: Diversity, Distribution Potential Values*. Himavikas Publ. No. 13, Gyanodaya Prakasan., Nainital, India.

523. Samant, S.S. & Pangtey, P.S. 1994. Aquatic, marshy and wetland flora of district Pithoragarh (Kumaon Himalaya). *J. Econ. Taxon. Bot.* **18**(1): 47-54.

524. Samanta, A. K. & Das, D.C. 2003. Ethnobotanical studies on *Typha elephantina* Roxb. (Typhaceae) in the southern parts of West Bengal, India. *J. Econ. Taxon. Bot.* **27**: 576-9.

525. Samwatsar, S. & Diwanji, V.B. 1996. Plants used in snake, scorpion and incect bites/stings by Adibasis of Jhabua (M.P.) India. *J. Econ. Taxon. Bot. Addl. Ser.* **12**: 199-200.

526. Samwatsar, S. & Diwanji, V.B. 1999. Plants used for rheumatism by the tribals of Western M.P. *J. Econ. Taxon. Bot.* **23**: 369–375.

527. Samwatsar, S. & Diwanji, V.B. 2000. Plants used for rheumatism by the tribals of Western M.P.: 305-314. *In* Maheshwari, J.K. (*eds.*): *Ethnobotany and Medicinal Plants of Indian Subcontinent.* Sci. Publ., Jodhpur (India).

528. Sanilkumar, M.G., & Thomas, K. J. 2007. Indigenous medicinal usages of some macrophytes of the Muriyad wetland in Vembanad – Kol, Ramser site, Kerela. *Indian J. Trad. Know.* **6**(2): 365-367.

529. Sanyal, D. 1994. *Vegetable Drugs of India.* Bishen Singh Mahendra Pal Singh, Dehradun (India).

530. Saren, A.M., Sen, R. & Pal, D.C. 2000. A contribution to the ethnobotany of Bankura District, West Bengal: 545-555. *In* Maheshwari, J.K. (*eds.*): *Ethnobotany and Medicinal Plants of Indian Subcontinent.* Sci. Publ., Jodhpur (India).

531. Sarin, Y.K. 1967. A survey of vegetable raw resources of Lahaul. *Indian For.* **93**(7): 489-499.

532. Sarin, Y.K. 1990. Some less known but effective folk remedies in North-West Himalayan region. *Ethnobotany* **2**: 39-43.

533. Sarin, Y.K. & Chopra, C.L. 1985. Medicinal and aromatic plant resources of Himachal Pradesh, present status and future prospects. *Workshop Sci. & Tech., Shimla,* Oct. 29-30

534. Sarkar, N., Rudra, S. & Basu, S.K. 2000. *Ethnobotany of Bangriposi, Mayurbhanj.* Orissa: 509-514. *In* Maheshwari, J.K. (*eds.*): *Ethnobotany and Medicinal Plants of Indian Subcontinent.* Sci. Publ., Jodhpur (India).

535. Sarma, S.K., Bhattacharya & Devi, B. 2002. Traditional use of herbal medicines by Madahi tribe of Nalbari district of Assam. *Ethnobotany* **14**: 103-111.

536. Sarma, S.K. & Saikia, M. 2010. Utilization of wetland resources by the rural people of Nagaon district, Assam. *Indian J. Trad. Know.* **5**: 145-151.

537. Sasidharan, N. & Swarupnandan, K. 2000. Ethnobotanical studies on the hill tribes in the Shola forests of high ranges Kerala, South India: 451-466. *In* Maheshwari, J.K. (*eds.*): *Ethnobotany and Medicinal Plants of Indian Subcontinent.* Sci. Publ., Jodhpur (India).

538. Satapathy, K. B. 2008. Interesting ethnobotanical uses from Juang, Kolha and Munda tribes of Keonjhar district, Orissa. *Ethnobotany* **20**: 99-105.

539. Satapathy, K.B. & Brahmam, M. 1996. Some medicinal plants used by tribals of Sundergarh district, Orissa, India: 153-158. *In* Jain, S. K. (*eds.*): *Ethnobiology in Human Welfare.* Deep Publ., New Delhi.

540. Satapathy, K.B. & Brahmam, M. 2000. Some interesting phytotherapeutic claims of tribals of Jaipur district (Orissa), India: 467-472. *In* Maheshwari, J.K. (*eds.*): *Ethnobotany and Medicinal Plants of Indian Subcontinent.* Sci. Publ., Jodhpur (India).

541. Satya, V. & Solanki, C.M. 2008. Clans of Bhils and their role in conservation. *Ethnobotany* **20**: 80-84.

542. Saxena, H.O. 1986. Observations on the Ethnobotany of Madhya Pradesh. *Bull. Bot. Surv. India* **28**: 149 - 156.

543. Saxena, H.O., Brahmam, M. & Dutta, P.K. 1997. Ethnobotanical studies in Orissa: 125-135. *In* Jain, S.K. (*eds.*): *Contribution to Indian Ethnobotany*. Sci. Publ., Jodhpur (India).

544. Saxena, K.G. 1991. Biological invasion in the Indian sub-continent: review of invasion by plants: 53-73. *In* Ramakrishnan, P.S. (*eds.*): *Ecology of Biological Invasion in the Tropics*. Intl. Sci. Publ., New Delhi.

545. Schoen, Allen M. & Wynu, Susan, G. (*eds.*) 1998. *Complementary and Alternative Veterinary Medicine: Principles and Practice*. Mosby. Inc. St. Louis. Miss. USA.

546. Schultes, R.E. 1962. Tapping our heritage of ethnobotanical lore. *Econ. Bot.* **8**: 224-227.

547. Schultes, R.E. 1967. The place of ethnobotany in the ethnopharmacologic search for psychotomimetic drugs: 35-57. *In* Effron, D. (*eds.*): *Ethnopharmacologic Search for Psychoactive Drugs*. Publ. No. **1645**, Washington.

548. Schultes, R.E. 1979. Amazonia as a source of new economic plants. *Econ. Bot.* **32**: 259-266.

549. Schultes, R.E. 1983. Richard Spruce: An early ethnobotanist and explorer of the northwest Amazon and northern Andes. *J. Ethnobiol.* **3**: 139-147.

550. Schultes, R.E. 1986. The reason for ethnobotanical conservation. *Bull. Bot. Surv. India* **28**: 203-324.

551. Schultes, R.E. & Raffauf, R.F. 1990. *The Healing Forest*. Dioscorides Press, Portland, Oregon.

552. Schultes, R.E. & Reis, S.V. (*ed.*) 1997. *Ethnobotany-Evolution of a Discipline*. Dioscorides Press. Portland, Oregon.

553. Sculthorpe, C.D. 1967. *The Biology of Aquatic Vascular Plants*. London.

554. Sebastian, M.K. 1984. Plants used as veterinary medicines, galactagogues and fodder in the forest areas of Rajasthan. *J. Econ. Taxon. Bot.* **5**: 785.

555. Seetharam, Y.N., Chalageri, G., Haleshi, C. & Vijay. 1999. Folk Medicine and ethnomedicine of North – Eastern Karnataka. *Ethnobotany* **11**: 32-37.

556. Seitel, Peter (*ed.*) 2001. *Safeguarding Traditional Cultures: A Global Assessment. Centre for Folk Life and Cultural Heritage*, Smithsonian Institution, Washington, DC.

557. Semwal, D.P., Pardha Saradhi, P., Kala, C.P. & Sajwan B. S. 2010. Medicinal plants used by local Vaidyas in Ukhimath block, Uttarakhand. *Indian J. Trad. Know.* **9**: 480-485.

558. Sen, J.K. & Behera, L.M. 2009. Traditional use of herbal medicines against headache and migrane by the tribals of Bangladesh district of Orissa. *Ethnobotany* **21**: 127-130.

559. Sen, S. & Batra, A. 1997. Ethno-medico-botany of household remedies of Phagi tehsil of Jaipur district (Rajasthan). *Ethnobotany* **9**: 122-128.

560. Seshavatharam, V. 1990. Traditional uses and the problem of noxious growth: 201-218. *In* Gopal, B. (*eds.*): *Ecology and Management of Aquatic Vegetation in Indian Subcontinent*. Kluwer Acad. Publi. Dordrecht, Netherlands.

561. Seth, M. K. & Kumar, S. 2007. Pteridophytic flora of Bilaspur, Himachal Pradesh. Part I. *In* Prashar, I.B. & Ahluwalia, A.S.: *A General Account of Bilaspur District and 13 Most Common Pteridophytes*. Bishen Singh Mahendra Pal Singh, Dehradun. (India).

562. Seth, M.K., Kumari, A., Bhandari, A. & Khullar, S.P. 2002. Common pteridophytes of Shimla (Himachal Pradesh): Families Adiantaceae, Hemionitidaceae, Davalliaceae and Aspleniaceae. *J. Econ. Taxon. Bot.* **1***:* 57-64.

563. Shah, N.C. 1997. Ethnobotany of *Cannabis sativa* in Kumaon Region, India. *Ethnobotany* **9**: 117-121.

564. Shah, N.C. 2008. Conservation of wild medicinal plants: need for a comprehensive strategy endangered species. *Kurukshetra* **46** (3)*:* 15-17.

565. Sharma, B. & Maheshwari, S. 2005. Traditional medicinal practices of Gaddi tribes in Kangra district, Himachal Pradesh. *Indian J. Trad. Know.* **4**(2): 173-188.

566. Sharma, B.D. 1984. *Exotic Flora of of Allahabad*. B.S.I., Dehradun.

567. Sharma, B.D. & Malhotra, S.K. 1984. A contribution to the ethnobotany of tribals areas in Maharashtra. *J. Econ. Taxon. Bot.* **5**: 533-537.

568. Sharma, B.D. & Rana, J.C. 1999. Traditional medicinal uses of plants of Himachal Hills. *J. Econ. Taxon. Bot.* **23**(3): 237-240.

569. Sharma, B.D. & Rana, J.C. 2000. Traditional medicinal uses of plants of Himachal Hills: 173-176. *In* Maheshwari, J.K. (*eds.*): *Ethnobotany and Medicinal Plants of Indian Subcontinent*. Sci. Publ., Jodhpur (India).

570. Sharma, B.D. & Rana, J.C. 2005. *Plant Genetic Resources of Western Himalaya: Status and Prospectus*. Bishen Singh Mahendra Pal Singh, Dehradun. (India).

571. Sharma, K.R. & Sood, M. 1997. *Important Medicinal Plants of Himachal Pradesh*. Vol. I. Dr. Y.S. Parmar Univ. Hort. & Forestry, Nauni, H.P. (India).

572. Sharma, P.K., Chauhan, N.S. & Lal, B. 2004. Observation on the traditional phytotherapy among the inhabitants of Parvati valley in Western Himalayas, India. *J. Ethnopharmacol.* **92**: 167-176.

573. Sharma, P.K., Dhyani, S.K. & Shanker, V. 1979. Some useful and medicinal plants of the district Dehradun and Shiwalik. *J. Sci. Res. Plate No. Med.* **1**: 17-43.

574. Sharma, S. C. 1996. A medicobotanical study in relation to veterinary medicines of Shahjahanpur district (Uttar Pradesh). *J. Econ. Taxon. Bot. Addl. Ser.* **12:** 123-127.

575. Sharma, S.K. 2000. Plants used as Henna dye by Bhills of Southern Rajasthan: 257. *In* Maheshwari, J.K. (*eds.*): *Ethnobotany and Medicinal Plants of Indian Subcontinent*. Sci. Publ., Jodhpur (India).

576. Sharma, U.K. 2003. Folk and herbal medicine among Nepalese of Assam: 599-603. *In* Maheshwari, J.K. (*eds.*): *Ethnobotany and Medicinal Plants of Indian Subcontinent*. Sci. Publ., Jodhpur (India).

577. Shiddamallayya, N., Yasmeen, Arza & Kumar, G. 2010. Medico-botanical survey of Kumar Parvattha Kukke Subramanya, Manglore. Karnataka. *Indian J. Trad. Know.* **9**: 96-99.

578. Shukla, G. & Verma, B.K. 1996. Roots – a vital plant part to cure body ailments among tribals/rural folklore of western Bihar: 392-395. *In* Maheshwari, J.K. (*eds.*): *Ethnobotany and Medicinal Plants of Indian Subcontinent*. Sci. Publ., Jodhpur (India).

579. Sikarwar, R.L.S. & Kaushik, J.P. 1997. Aquatic and semi-aquatic plants of Morana district (M.P.). *J. Econ. Taxon. Bot.* **21**(3): 639-647.

580. Silori, O.S., Dixit, A.M., Gupta, L. & Mistry, N. 2009. Observations on medicinal plants richness and associated conservation issues in district Kachchh, Gujarat: 137-180. *In* Trivedi, P.C. (*eds.*): *Medicinal Plants Utilization and Conservation.* Aavishkar Publ., Jaipur (India).

581. Singh, A.K. 2000. A contribution to the ethnobotany of sub- Himalayan region of eastern Uttar Pradesh: 237-246. *In* Maheshwari, J.K. (*eds.*): *Ethnobotany and Medicinal Plants of Indian Subcontinent.* Sci. Publ., Jodhpur (India).

582. Singh, A.K., Singh, R. N. & Singh, S.K. 1987. Some ethnobotanical plants of Terai region of Gorakhpur district, India. *J. Econ. Taxon. Bot.* **23**: 185-198.

583. Singh, B., Chaurasia, O.P., & Ballabh, B. 2001. Edible wild plants of Trans-Himalayan cold desert: 483-512. *In* Pandey, P.C. & Samant, S.S. (*eds.*): *Plant Diversity of the Himalaya.* Gyanodaya Prakashan, Nainital, India.

584. Singh, B., Chaurasia, O.P. & Jadhav, K.L. 1996a. An ethnobotanical study of Indus valley (Ladakh): 92-101. *In* Maheshwari, J.K. (*eds.*): *Ethnobotany in South Asia.* Sci. Publ., Jodhpur (India).

585. Singh, G., Singh, H.B. & Mukerjee, T.K. (*eds.*) 2003a. Ethnomedicine of North-East India. Proc. Nat. Sem. Trad. Knowledge on Herbal Med. Plant Res. north-east India: Protection, Utiliztion and Conservation. N.I.S.C.I.R., New Delhi.

586. Singh, G.S. 1991. A contribution to ethnomedicine of Alwar district of Rajasthan. *Ethnobotany* **11**: 97-99.

587. Singh, G.S. 2003a. Ethnobotanical inventory of Sariska National Park in Aravalli hills of eastern Rajasthan, India: 181-195. *In* Singh, V. & Jain, A.P. (*eds.*): *Ethnobotany and Medicinal Plants of India and Nepal,* Vol. 1. Sci. Publ., Jodhpur (India).

588. Singh, G.S. 2004a. Ethno-biological and bio-medicinal wealth of Western Himalaya, India. *J. Med. Arom. Plate No. Sci.* **26**(3):517-526.

589. Singh, G.S., 2004b. Indigenous knowledge and conservation practices. Tribal society of Western Himalaya: a case study of Sangla valley. *Stud. Tires. Tribals* **2**(1): 29-35.

590. Singh, G.S., Rao, K.S. & Saxena, K.G. 1998. Environment, ecological and socio-economic impact of introduced crops in western Himalaya: a case study of Kullu Valley. *J. Human Ecol.* **9**: 63-72.

591. Singh, H. 1988. Ethnobiological treatment of piles by Bhoxas of Uttar Pradesh. *Anc. Sci. Life.* **8**: 167-170.

592. Singh, H. & Sharma, M. 2006. *Flora of Chamba district (Himachal Pradesh).* Bishen Singh Mahendra Pal Singh. Dehradun.

593. Singh, H.B. 2003b. Medicinal viable pteridophytes of India: 418-420. *In* Chandra, S. & Srivastava, M. (*eds.*): *Pteridology in the New Millenium.* Golden Jublee Vol. Kluwar Acad. Publi.

594. Singh, H.B. & Arora, R.K. 1978. *Wild Edible Plants of India.* New Delhi, ICAR.

595. Singh, H.B. & Viswanathan, M.V. 1996. Useful pteridophytes of India- a gift of nature to human beings: 24-36. *In* Maheshwari, J.K. (*eds.*): *Ethnobotany in South Asia,* Sci. Publ., Jodhpur (India).

596. Singh, H.B. 2003c. Economically viable pteridophytes of India: 421-426. *In* Chandra, S. & Srivastava, M. (*eds.*): *Pteridology in the New Millenium.* Golden Jublee Vol. Kluwar Acad. Publi.

597. Singh, H.B., Singh, R.S. & Sandhu, J.S. (*eds.*) 2003b. *Herbal Medicine of Manipur: A Colour Encyclopedia.* Daya Publ. House, New Delhi.

598. Singh, J., Bhuyan, T.C. & Ahmed, A. 1996b. Ethnobotanical studies on the Mishing tribes of Assam with special reference to food and medicinal plants - I. *J. Econ. Taxon. Bot. Addl. Ser.* **12**: 350-356.

599. Singh, K.K. & Kumar, K. 2000a. *Ethnobotanical Wisdom of Gaddi Tribe of Western Himalayas.* Bishen Singh Mahendra Pal Singh, Dehradun.

600. Singh, K.K. & Kumar, K. 2000b. Observations in ethnoveterinary medicine among the Gaddi Tribe of Kangra valley, Himachal Pradesh. *Ethnobotany* **12**: 42-44.

601. Singh, K.K. & Maheshwari, J.K. 1983. Traditional phytotherapy amongst the tribals of Varanasi district, U.P. *J. Econ. Taxon. Bot.* **4**: 829-838.

602. Singh, K.K. & Prakash, A. 1994. Indigenous phytotherapy among the Gond tribe of Uttar Pradesh. India. *Ethnobotany* **6**: 37-41.

603. Singh, K.K. & Prakash, A. 1996. Observations on ethnobotany of the Kol tribe of Varanasi district Uttar Pradesh: 133-137. *In* Maheshwari, J.K. (*eds.*): *Ethnobotany in South Asia.* Sci. Publ., Jodhpur (India).

604. Singh, K.K. & Rao, R.R. 2003. Studies on some potential ethnomedicine among the tribals of Uttar Pradesh, India: 87-93. *In* Khan, S. S. (*eds.*): *Vistas in Ethnobotany* Vol. I. Indian Journal of Applied and Pure Biology.

605. Singh, K.K., Palvi, S.K., & Singh, H.B. 1980. Survey of some medicinal plants of Dharchula block-I Pithorgarh district of U.P. *Bull.Medico-ethnobot.Res.* **1**: 7.

606. Singh, N.P. 2001. *Flora of Maharashtra State, Dicotyledons.* Vol. 2. B.S.I., Kolkata.

607. Singh, P., Kumar, E., Vinita, Devi & Huidrom, B.K. 2000. Ethnomedicinal studies of some plants used to enhance vocalism by the traditional Meitei singers of Manipur: 629-635. *In* Maheshwari, J.K. (*eds.*): *Ethnobotany and Medicinal Plants of Indian Subcontinent.* Sci. Publ., Jodhpur (India).

608. Singh, P., Sisodia, S. & Shah, J. 2007. Angiospermous weeds in medicinal importance in Gujarat state: 17-24. *In* Chopra, A.K., Khanna, D.R., Prasad, G., Malik, D.S. & Bhutiani, R. (*eds.*): *Medicinal Plants: Conservation, Cultivation and Utilization.* Daya Publ. House, Delhi (India).

609. Singh, P.B. 1993. Medicinal plants of ayurvedic importance from Mandi district of Himachal Pradesh. *Bull. Medico-ethnobot. Res.* **14**: 126- 136.

610. Singh, P.B. 1999. *Illustrated Field Guide to Commercially Important Medicinal and Aromatic Plants of Himachal Pradesh (with special reference to Mandi District).* Published by Society for Herbal Medicine and Himalaya Biodiversity, Mandi, H. P.

611. Singh, P.B. & Aswal, B.S. 1992. Medicinal plants of Himachal Pradesh used in Indian pharmaceutical industry. *Bull. Medico-ethnobot. Res.* **13**: 172-208.

612. Singh, S.C. & Srivastava, G.N. 2000. Exotic medicinal plants of Lucknow District, U.P., India: 223-235. *In* Maheshwari, J.K. (*eds.*): *Ethnobotany and Medicinal Plants of Indian Subcontinent.* Sci. Publ., Jodhpur (India).

613. Singh, S.K. & Rawat, G.S. 2000. *Flora of Great Himalayan National Park – Himachal Pradesh.* Bishen Singh Mahendra Pal Singh, Dehradun.

614. Singh, V. 1996. Ethnomedicobotany of Dards tribe of Gurez valley in Kashmir Himalaya: 129-132. *In* Jain, S.K. (*eds.*): *Ethnobiology in Human Welfare.* Deep Publ. New Delhi (India).

615. Singh, V. & Pandey, R.P. 1980. Medicinal plantlore of the tribals of eastern Rajasthan. *J. Econ. Taxon. Bot.* **1**: 137-147.

616. Singh, V. & Pandey, R.P. 1998. A study of some wetland ecosystems of Udaipur and environs, Rajasthan. *J. Econ. Taxon. Bot.* **22**: 217-224.

617. Singh, V. & Chauhan, N.S. 2005. Traditional practices of herbal medicines in the Lahaul valleys. Himachal Himalayas. *Indian J. Trad. Know.* **4**(2): 208-220.

618. Singh, V. & Pandey, R. P. 1996. Ethnomedicinal plants used for venereal and gynaecological diseases in Rajasthan (India): 154-166. *In* Maheshwari, J.K. (*eds.*): *Ethnobotany in South Asia.* Sci. Publ., Jodhpur (India).

619. Singh, V. & Singh, P. 1981. Edible wild plants of eastern Rajasthan. *J. Econ. Taxon. Bot.* **2**: 197-207.

620. Singh, V.K. & Zaheer, A.A. 1996. Folk medicinal plants used for family planning in India: 184-186. *In* Jain, S.K. (*eds.*): *Ethnobiology in Human Welfare.* Deep Publ. New Delhi (India).

621. Sinha, B.K., Maina, V. & Padhye, P.M. 1996. Ethno-medicinal plants of Bay Islands for skin care: 375-378. *In* Jain, S. K. (*eds.*): *Ethnobiology in Human Welfare.* Deep Publ. New Delhi.

622. Sinha, R.K. 1986. *Ethnobotanical Study of Manipur.* Manipur Univ., Imphal.

623. Sinha, R.K. 1987. Ethnobotany – Medicinal plants. *Frontier Bot.* **1**: 123-152.

624. Sinha, R.K. 1996. Conservation of cultural diversity of indigenous people essential for protection of biological diversity: 280-283. *In* Jain, S.K. (*eds.*): *Ethnobiology in Human Welfare.* Deep Publ., New Delhi.

625. Sinha, R.K. & Sinha, S. 2001. Ethnobiology: role of indigenous and ethnic societies in biodiversity conservation, human health protection and sustainable development. Surabhi Publ. Jaipur.

626. Siwakoti, M. & Siwakoti, S. 1999. Ethnomedicianl uses of plants among Limbu of Morang district, Nepal. *Ecoprint.* **5**: 79-84.

627. Siwakoti, M. & Siwakoti, S. 2000. Ethnomedicinal uses of plants among the Satar tribe of Nepal: 99-108. *In* Maheshwari, J.K. (*eds.*): *Ethnobotany and Medicinal Plants of Indian Subcontinent.* Sci. Publ., Jodhpur (India).

628. Siwakoti, M. & Verma, S. K. 1996. Medicinal plants of the Terai of eastern Nagpur: 414-423. *In* Maheshwari, J.K. (*eds.*): *Ethnobotany and Medicinal Plants of Indian Subcontinent.* Sci. Publ., Jodhpur (India).

629. Solanki, C.M., Satya, V. & Dubey, S. 2007. Indigenous knowledge of Veterinary medicines among tribals of West Nimar: 46-53. *In* Sahu T.R. *Indigenous Knowledge: An Application.* Sci. Publ., Jodhpur (India).

630. Sood, S.K. & Kaushal, B. 2008. *Fibre and Fodder Resources of India.* Satish Serial Publ. House, Azadpur, Delhi.

631. Sood, S.K., Nath, R. & Kalia, D.C. 2001. *Ethnobotany of Cold Desert Tribes of Lahaul-Spiti (N.W. Himalaya).* Deep Publ., New Delhi.

632. Sood, S.K. & Prakash, V. 2007. *Edible Roots and Uderground Stems of Ethnic India.* Satish Serial Publ. House, Delhi.

633. Sood, S.K., Rawat, S. & Rawat, D. 2009a. *Dye, Masticatory and Fumitory Resources of India.* Intl. Book Distr., Dehradun.

634. Sood, S.K., Sharma, P., Sharma, R. & Dogra, K.S. 2010. *Healing Herbs for Arthritis and Rheumatism. Folk Medicines for the Cure of Arthritis and Rheumatism*. Lambert Acad. Publ., Germany.

635. Sood, S.K., Sharma, S.K., Kumar N. & Kumar, H. 2009b. *Ethnobotanical Studies on Trees, Shrubs and Climbers of Himalaya*. Satish Serial Publ. House, Delhi, India.

636. Sood, S.K. & Sudershana. 2008. *Root Drugs Plants of Ethnic India*. Satish Serial Publ. House, Azadpur, Delhi.

637. Sood, S.K. & Thakur, S. 2004. *Ethnobotany of Rewalsar Himalaya (Dist. Mandi, Himachal Pradesh, India)*. Deep Publ., New Delhi.

638. Srivastava, S.K. 2003a. Floristic composition of Pong Dam Sanctuary, Himachal Pradesh. *Phytotaxonomy* **3**: 156-160.

639. Srivastava, S.K. 2003b. Floristic diversity of Renuka lake wetland and its environs: 93-109. *In* Janarthanam, M.K. & Narasimhan, D. (*eds.*): *Plant Diversity, Human Welfare and Conservation*. Goa Univ. Goa.

640. Srivastava, S.K., Chowdhery, H. J. & Rawat, G.S. 1987. Some economically important plants from Beas-Sutlej link project area in Himachal Pradesh: 131-146. *In* Sharma, M.R. & Gupta, B.K. (*eds.*): *Recent Advances in Plant Sciences*. Bishen Singh Mahendra Pal Singh, Dehradun.

641. Srivastava, R.C., & Jain, S.K. 2000. Cumulative index to Ethnobotany, Vol. 1-12, 1989-2000, with analysis of region and themes covered. *Ethnobotany* **12**: 139-161.

642. Srivastava, R.C., Singh, V.P. & Singh, M.K. 2003. Medicinal plants of Jaunpur district (U.P.): 148-159. *In* Singh, V. & Jain, A.P. (*eds.*): *Ethnobotany and Medicinal Plants of India and Nepal*, Vol. 1. Sci. Publ., Jodhpur (India).

643. Srivastava, R.C. & Srivastava, C. 2007. Diversity and economic importance of wetland flora of Gorakhpur district (U.P.). *J. Econ. Taxon. Bot.* **31**(1): 70-77.

644. Stainton, A. 1988. *Flowers of Himalaya, A Supplement*. Oxford Univ. Press, Delhi.

645. Stubbendiek, J., Hatch, S.L. & Hisch, K.J. 1994. North American Range Plants. Nebraska Press.

646. Subramani, S.P., Vaneet Jishtu & R.S. Rawat 2005. Plant Diversity Conservation Needs. *The Botanica* **55**: 45 -59.

647. Subramaniam, A. 2000. A survey of medicinal plants from Chitheri Hills in Dharampuri district, Tamilnadu: 395-416. *In* Maheshwari, J.K. (*eds.*): *Ethnobotany and Medicinal Plants of Indian Subcontinent*. Sci. Publ., Jodhpur (India).

648. Subramanyam, A. K. 1962. *Aquatic Angiosperms: Systematic Account of common Indian Aquatic Angiosperms. Bot. Monogr.* **3** (I-VI): 1-190, CSIR, New Delhi.

649. Subudhi, H.N., Choudhury, B.P., & Acharya, 1992. Some potential medicinal plants of Mahanadi Delta in the state of Orissa. *J. Econ. Taxon. Bot.* **16**(2): 479-487.

650. Sudhakar, A. & Vedhavathy, S. 2000. Wild edible plants used by the tribals of Chotoor district (Andhra Pradesh), India: 321-329. *In* Maheshwari, J.K. (*eds.*): *Ethnobotany and Medicinal Plants of Indian Subcontinent*. Sci. Publ., Jodhpur (India).

651. Sunil, C. N.& Sivadasan, M. 2000. *Kyllinga polyphylla* Willd. *ex* Kunth (Cyperaceae): a new record for India. *Rheedea* **10**(1): 81-83.

652. Swami, A. & Gupta, B.K. 1996. A note of some commonly occuring medicinal weeds of Udhampur district (J.&K.): 89-91. *In* Maheshwari, J.K. (*eds.*): *Ethnobotany in South Asia*. Sci. Publ., Jodhpur (India).

653. Tarafder, C.R. & Chaudhuri, H.N. 1997. Less known medicinal uses of plants among the tribals of Hazaribagh district of Bihar: 93-101. *In* Jain, S. K. (*eds.*): *Contribution to Indian Ethnobotany*, Vol. 3. Sci. Publ., Jodhpur (India).

654. Tarafder, M., Qiang T., B S. Nathan, R. & Ragu-Nathan, T. S. 1997. The Impact of Technostress on Role Stress and Productivity. *J. Man. Infor. Sys.* **24**: 301-328.

655. Taylor, P.M. 1990. *The Folk Biology of the Tobelo People – A Study in Folk Classification*. Smith Inst. Press, Washington.

656. Teg, H., Das, A.K. & Loyi, H. 2007. *Anti-inflammatory Plants used by the Khamti Tribe of Lohit District in Eastern Arunachal Pradesh*. India.

657. Thakor, A. B. 2009. Ethnomedicinal plants of Vaarli tribe in Valsad district (Gujarat). *Ethnobotany* **21**: 116-120.

658. Thakur, R.S., Puri, H.S., & Husani, A. 1989. *Major Medicinal Plants of India*. Central Institute of Medicinal & Aromatic Plants, Lucknow, India.

659. Thomas, J. & Britto, A. J. De. 1999. Weeds of medicinal importance in Tirunelveli District in Tamilnadu. *Ethnobotany* **11**: 22-25.

660. Thomas, J. & Britto, A.J. De. 2000. Weeds of medicinal importance in Tirunelveli District in Tamilnadu: 363-367. *In* Maheshwari, J.K. (*eds.*): *Ethnobotany and Medicinal Plants of Indian Subcontinent*. Sci. Publ., Jodhpur (India).

661. Thomas, J. & Britto, A. John De. 2003. Ethnobotanical study in Wynad district in Kerela: 815-824. *In* Singh, V. & Jain, A.P. (*eds.*): *Ethnobotany and Medicinal Plants of India and Nepal*, Vol. 2. Sci. Publ., Jodhpur, India.

662. Tiner, R.W. 1992. *A Field Guide to Coastal Wetland Plants of the Northeastern United States*. America.

663. Tiwari, K.C. & Tiwari, V.P. 1996. Some important medicinal plants of the tropical, subtropical and temperate region of Siang, Subansiri and Tirap districts of Arunachal Pradesh: 359-364. *In* Maheshwari, J.K. (*eds.*): *Ethnobotany and Medicinal Plants of Indian Subcontinent*. Sci. Publ., Jodhpur (India).

664. Tiwari, L. & Pandey, P.C. 2006. Ethnoveterinary plants of Uttarakashi district, Uttaranchal, India. *Ethnobotany* **18**: 139-144.

665. Towle, M.A. 1961. *The Ethnobotany of Pre-Columbian Peru*. Aldine Publ., Chikago.

666. Tripathi, A., Dwivedi, G.K. & Singh, O.P. 2010. A survey of indigenous semi-aquatic and angiosperm biodiversity of district, Jaunpur, Uttar Pradesh. *Intl. J. Plant Sci.* **1**: 353-356.

667. Trivedi, G. N., Molla, H.A. & Pal, D.C. 1985. Some uses of plants from the tribal areas of Chotanagpur, Bihar. *Nagarjun.* **29**: 15-18.

668. Trivedi, P.C. 2004. *Medicinal Plants Conservation and Utilization*. Avishkar Publ., Jaipur.

669. Trivedi, P.C. 2009. *Medicinal Plants Conservation and Utilization* (IInd edition). Avishkar Publ., Jaipur.

670. Trivedi, P.C. 2010. *Ethnic Tribes and Medicinal Plants*. Pointer Publ., Jaipur.

671. Turner, N.J., Thompson, L.C., Thompson, M.T. & York, A.Z. 1990. *Thompson Ethnobotany*. Royal British Columbia Museum, Canada.

672. Udayakumar, M., Narayanswamy, Dh., Kanakashanthi, Aj. & Thangavel, Se. 2010. A floristic study in a perennial lake of Thiruvallur District, South India. *Webmed Central Ecol.* **1**(10): 137.

673. Uniyal, B. 2003. Utilization of medicinal Plants by the rural woman of Kullu, Himachal Pradesh. *Indian J. Trad. Know.* **2**(4): 366-370.

674. Uniyal, M.R. 1989. *Medicinal Flora of Garhwal Himalayas*. Shree Baidyanath Ayurved Bhawan, Nagpur.

675. Uniyal, M.R., Bhat, A.V. & Chaturvedi, 1973. Preliminary observations on medicinal plants of Lahaul Spiti division in Himachal Pradesh. *Bull. Medico-ethnobot. Res.* **3**(1): 1-26.

676. Uniyal, M.R. & Chauhan, N.S. 1971. Medicinal plants of Uhal valley in Kangra Forest Division. H.P. *J. Res. Ind. Med.* **6**(3): 287-299.

677. Uniyal, M.R. & Chauhan, N.S. 1973a. Traditionally important medicinal plants of Kangra Valley in Dharamshala For. Circle, H.P. *J. Res. Ind. Med.* **8**(1): 76-85.

678. Uniyal, S., Awasthi, A. & Rawat, G.S. 2002. Traditional and ethnobotanical uses of plants in Bhagirathi Valley, Western Himalaya. *For.* **25** (3): 310-318.

679. Uniyal, S.K., Singh, K.N., Jawal, P. & Lal, B. 2006. Traditional use of medicinal plants among the tribal communities of Chota Bhangal, Western Himalaya. *J. Ethnobiol. & Ethnomedicine* **2**(14):1-8.

680. Upadhye, A.S., Vartak, V.D. & Kumbhojkar, M.S. 1994. Ethno-medicobotanical studies in Western Maharastra, India. *Ethnobotany* **6**: 25-31.

681. Upadhye, R. & Singh, J. 2005. Ethno-medical uses of Plants from Tikri forest of Gonda district (U.P.). *Ethnobotany* **17**: 167-170.

682. Uphof, J.C. Th. 1968. *Dictionary of Economic Plants*. H.R. Eugelmann (J. Cramer), Weinheim.

683. Uphof, J.C. Th. 2001. *Dictionary of Economic Plants*. Bishen Singh Mahendra Pal Singh, Dehradun. (India). & A.R.G. Gartner Verlag Koman, Ruggell.

684. Vartak, V.D. & Suryanarayan, M.C. 1995. Enumeration of wild edible plants from Sundra Island, Mulshi Reservoir, Pune district. *J. Econ. Taxon. Bot.* **19**(3): 555-569.

685. Vasudeva, S.M. 1999. Economic importance of pteridophytes. *Indian Fern J.* **16**: 130-152.

686. Venu, P., T., J.S. Rajakumar, & Danial, P. 2003. On the identity of and sterlity in *Nymphaea rubra* Roxb. *ex* Andrews (Nympheaceae). *Rheedea* **13**: 29-34.

687. Verma, P., Khan, A.A. & Singh, K.K. 1995. Traditional phytotherapy among the Baiga tribe of Shadol district of Madhya Pradesh, India. *Ethnobotany* **7**: 69-73.

688. Verma, S. & Chauhan, N.S. 2006. Studies on ethno-medico-botany of Kunihar forest division, district Solan (H.P). *Ethnobotany* **18**: 10-165.

689. Verma, S. & Chauhan, N.S. 2007. Indigenous medicinal plants knowledge of Kunihar forest division, district Solan. *Indian J. Trad. Know.* **6**(3): 494-497.

690. Vijayan, L., Vijayan, V.S., Prasad, S.N. & Muralidharan, S. 2007. *Conservation of Wetlands in Tamil Nadu*. Coimbatore.

691. Vishwanathan, M.V. 1997. Ethnobotany of the Malyalis in North Arcot District, Tamil Nadu, India. *Ethnobotany* **9**: 77-79.

692. Vishwanathan, M.V. 2000. Edible and medicinal plants of Ladakh (J&K): 151-154. *In* Maheshwari, J.K. (*eds.*): *Ethnobotany and Medicinal Plants of Indian Subcontinent*. Sci. Publ., Jodhpur, India.

693. Viswanathal, M.B., Kumar, E. Harrison Prem & Ramesh, N. 2006. *Ethnobotany of the Kanis (Kalakkad- Mundanthurai. Tiger Reserve in Tirunelveli, District Tamilnadu. India).* Bishen Singh Mahendra Pal Singh, Dehradun.

694. Wadoodkhan, M. A. 2000. The genus *Scripus* L. in Maharashtra. *Rheedea* **10(1)**: 19-32.

695. Wadoodkhan, M. A., Chavan, D.P. & Solanki, S., 2006. *Scirpus articulatus* (S.L.)- a complex in Cyperaceae. *J. Econ. Taxon. Bot.* **30**(2): 438-448.

696. Wangchuk, P., Samten, U., Thinley, J. & Afaq, S.H. 2008. High altitude plants used in Bhutanese traditional Medicine. *Ethnobotany* **20**: 54-64.

697. Ward, H.M. 1980. *Grasses. A Hand Book for Use in the Field and Laboratory.* Sci. Publ., Jodhpur (India).

698. Watt, G. 1972. *A Dictionary of Economic Products of India,* Vols. I-VI. Periodical Experts, Delhi, India.

699. Wielgorskaya, Tatiana 1995. *Dictionary of Generic Names of Seed Plants*. Bishen Singh Mahendra Pal Singh, Dehradun.

700. Yadav, S., Yadav, J.P., Arya, V. & Panghal M. 2010. Sacred groves in conservation of plant biodiversity in Mahendergarh district of Haryana. *Indian J. Trad. Know.* **9**: 693 -700.

701. Yadav, S.R., Sardesai, M.M. & Gaikwad, S.P. 2000. Two new species of *Utricularia* L. (Lenibuariaceae) from peninsular India. *Rheedea* **10**(2): 107-112.

702. Yasodamma, N., Mehar, S.K. & Paramageetham, C. 2009. Threat assessment (IUCN Categorization) for ethnomedicinal plants used by Chenchu tribe of Gundlabramheswaram in Nallamalai hills in Andhra Pradesh. *Ethnobotany* **21**: 52-60.

703. Yoganaraosimhan, S.N. 1996. *Medicinal Plants of India.* Vol. I- Karnataka. Interline Publ., Bangalore.

704. Yonzone, G.S., Yonzone, A. & Bhujel, R.B. 1996. Contribution to the ethnobotany of Darjeeling district, India: 388-389. In Jain, S. K. (eds.): *Ethnobiology in Human Welfare*. Deep Publ., New Delhi.

M

Magnesium   66, 227

Magnesium phosphate   23

Maheshwari   17, 60, 102, 130, 131

Maize   6

Malaria   58, 71, 85, 182

Malaya   134

Malic Acid   171

Malvaceae   19, 22, 206, 226

Mamgain   5

Manandhar   24, 29

Mandi   6

Mangroves   1

Maranthaceae   125

Marchantiaceae   150

Marsileaceae   150

Marsilea minuta L.   150

Martynia annua L   151

Martyniaceae   151

Matricaria Ester   78

Matthew   17

Matwana Khad   29

Measles   113, 150

Medicinal plant families   275

Medicinal wetland plants   324

Meena   24

Menismine   71

Menispermaceae   70

Menstrual disorders   73, 149

Menstruation   71

Menstural colic   156

Mental diseases   156

Mental disorders   63

Mentha longifolia L.   155

Mentha piperita L   156

Methionine   110, 222

Methylehavicol   168

Methylmorphaline   63

Milk flow   227

Mirabilis jalapa L   158

Miraxanthins   158

Monochlamdos   273

Monocot   273

Monocot families   268

Monocotyledonous   237

Moraceae   115, 116

Mouth freshners   282

Mouth sores   135

Mudaliyar   130

Muenscher   2

Muscular pains   81

Myanmar   129

Myo-Inositol   130

Myricetin   134

Myricyl alcohol   173

N

Naiadaceae   163, 188, 189

Nair   3

N-Alkanes   77

N-Alkanols   77

Nanda   97

Naphtholglycosidase   134

Narcotic   58

Naringenin   49

Nariodin   166

Neoxanthin B   139

Nepal   32, 135, 213

Neriodorin   166

Nerium indicum Mill   165

Nervous disorder   96

N-Hexacosanoic Acid   109

N-Hexacosanol   106

Nicotinic acid   223

Night blindness   21

Nitrates   53, 105

N-Triacontanyl   116

PLATE 1

Ali Khad (Deoli)

Ali Khad (Ghagas)

Ali Khad (Toba Village)

Gamrola Khad

Ghamber Khad
(A Panormic View)

Ghamber Khad (from Bridge)

Govind Sagar Lake
(Chadol Village)

Govind Sagar Lake
(Chandpur Village)

Govind Sagar Lake
(Kungerhati Village)

Govind Sagar Lake
(Village Kothi)

Lanjhta Khad

Lyond Khad

PLATE 2

Marol Khad (Water Fall)

Marsand Khad

Raul Khad

Saryali Khad
(Naghiyar Village)

Saryali Khad
(Tarlai Village)

Satluj River (Sungan Village)

Sauli Khad (Kyana Village)

Sir Khad (Bhatrog Village)

Sir Khad (Patta Village)

Suker Khad (Barthin)

Suker Khad (Barthin Bridge)

Totoh Khad
(Tributary of Ali Khad)

PLATE 3

Jagatkhana Water Tank

Kothi Babdi

Kuala Water Tank

Riwalser Water Tank

Deoli Fish Farm

Cattle Grazing
Near Water Source

Bùffaloes (Grazing)

Bundles of Fuel Wood

Fodder Cutter

Women Cutting Fodder Grass

Women Carrying Fodder Grass

Locals Washing Clothes

PLATE 4

Men Holding 'Jungda' (Yoke)

Local Woman Making Ropes

Local Woman Making Brooms

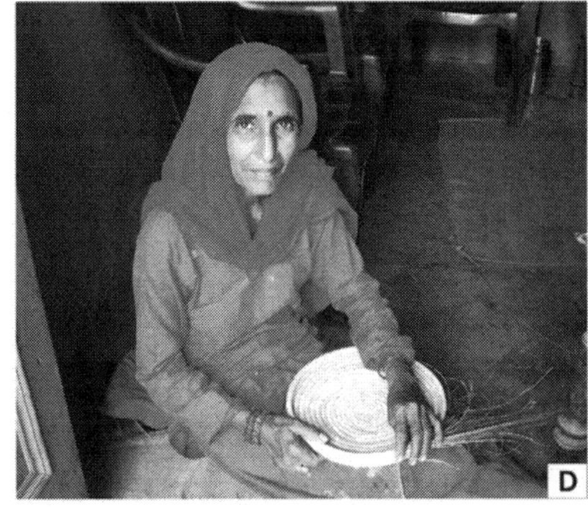

Local Woman Making Basket

View of Kitchen Showing 'Chulha'

Ropes for Tying Animals

PLATE 5

Local Dwellings

Pucca Dwellings Near Khad

Village School

'Kuchha' House

Temporary Dwellings Near Khad

Local Woman with her Children

PLATE 6

Offerings to Lord Shiva

Boat

Worship Place in Home

Young Lady Carrying Sand

Fisherman

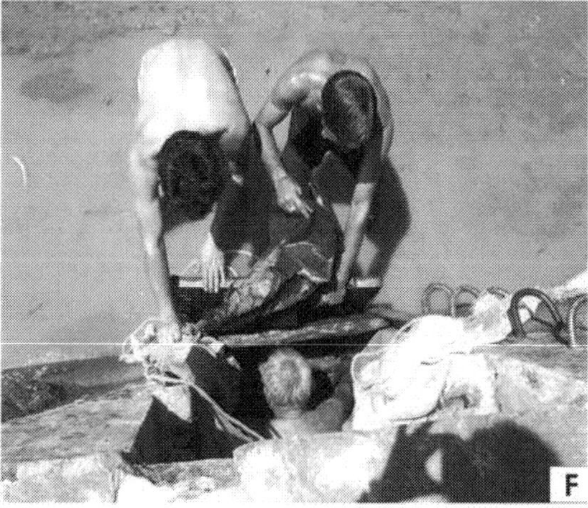

Fishermen with their Catch

PLATE 7

Abelmoschus crinitus

Abrus precatorius

Abutilon indicum

Achyranthes aspera

Acorus calamus

Adiantum capillus-veneris

PLATE 8

Adiantum incisum

Aerva sanguinolenta

Aeschynomene aspera

Ageratum conyzoides

Ajuga bracteosa

Amaranthus gangeticus

PLATE 9

Amaranthus paniculatus

Amaranthus tricolor var. *gangeticus*

Amaranthus viridis

Ampelopteris prolifera

Andrographis paniculata

Anisomeles indica

PLATE 10

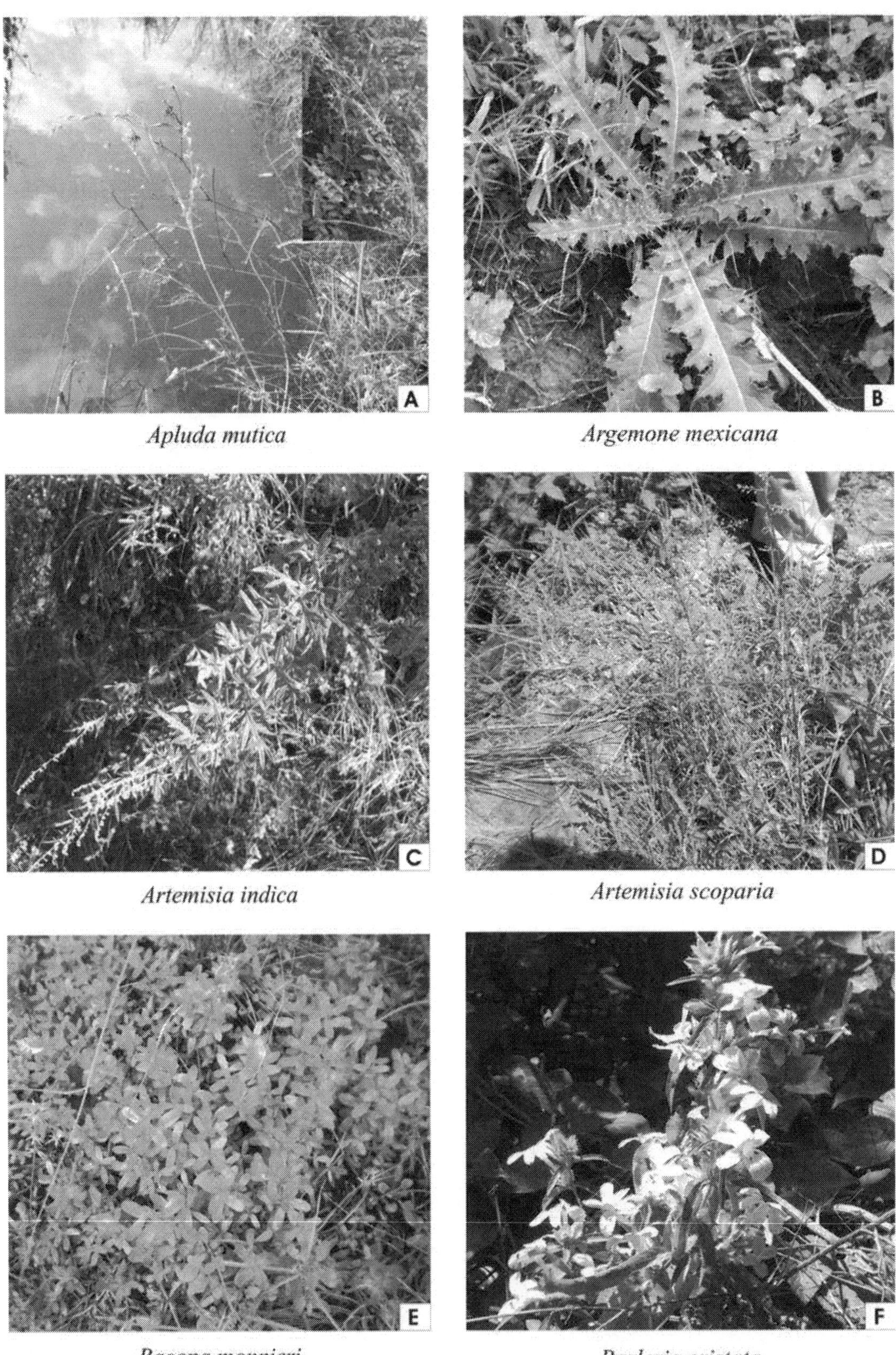

Apluda mutica

Argemone mexicana

Artemisia indica

Artemisia scoparia

Bacopa monnieri

Barleria cristata

PLATE 11

Begonia picta

Bidens pilosa

Blyxa auberti

Boehmeria platiphylla

Bromus catharticus

Bryophyllum calycinum

PLATE 12

Calamintha umbrosum

Canna indica

Cannabis sativa

Capsella bursa pastoris

Cardiospermum halicacobum

Cassia absus

PLATE 13

Cassia occidentalis

Centella asiatica

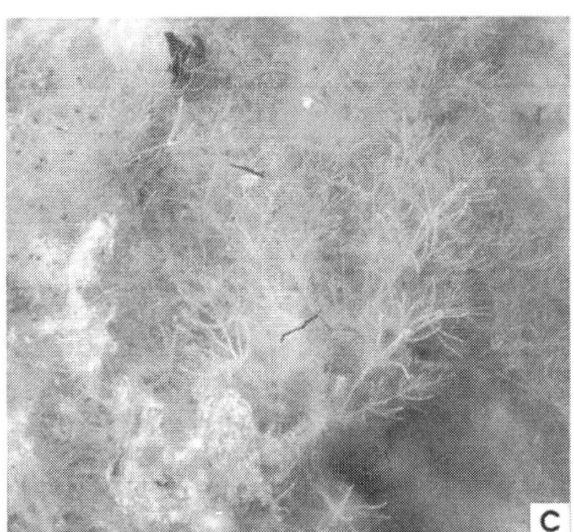

Ceratophyllum demersum

Cheilanthes bicolor

Chenopodium album

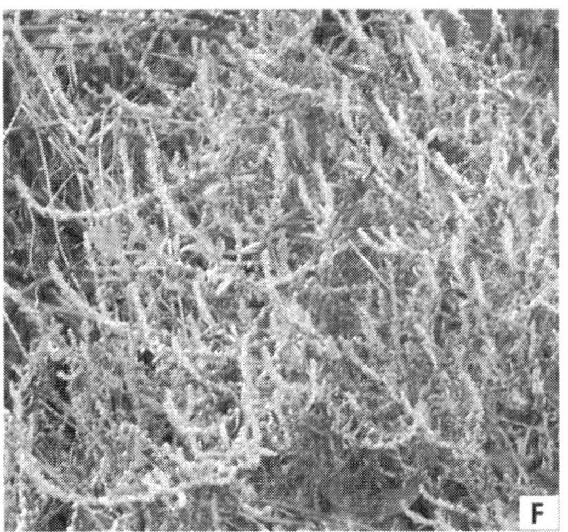

Chenopodium ambrosiodes

PLATE 14

Chenopodium murale

Christella dentata

Cissampelos pareira

Clematis gouriana

Coix lachryma-jobi

Colocasia antiquorum

PLATE 14

Chenopodium murale

Christella dentata

Cissampelos pareira

Clematis gouriana

Coix lachryma-jobi

Colocasia antiquorum

PLATE 15

Commelina diffusa

Commelina paludosa

Convolvulus arvensis

Conyza bonariensis

Conyza stricta

Costus speciosus

PLATE 16

Crotolaria alata

Crotolaria mysorensis

Cryptolepis buchanani

Cucumis pubescens

Curcuma longa

Cyathocline purpurea

PLATE 17

Cymbopogon citratus

Cymbopogon martinii

Cynodon dactylon

Cyperus compressus

Cyperus distans

Cyperus flabelliformis

PLATE 18

Cyperus iria

Cyperus rotundus

Datura stramonium

Debregeasia hypoleuca

Dichanthium annulatum

Dicliptera roxburghiana

PLATE 19

Digitaria griffithii

Dioscorea belophylla

Dioscorea bulbifera

Echinochloa frumentacea

Eleusine indica

Emilia sonchifolia

PLATE 20

Equisetum arvense

Equisetum debile

Eriophorum comosum

Eupatorium adenophorum

Euphorbia geniculata

Euphorbia helioscopia

PLATE 21

Euphorbia hirta

Euphorbia parviflora

Ficus hispida

Ficus roxburghii

Fragaria indica

Fragaria nubicola

PLATE 22

Fumaria indica

Galium aparine

Geranium nepalense

Girardinia heterophylla

Gloriosa superba

Gnaphalium pensylvanicum

PLATE 23

Gomphrena celosioides

Hedera helix

Hedychium spicatum

Heteropogon contortus

Hydrilla verticillata

Impatiens balsamina

PLATE 24

Ipomoea cairica

Ipomoea carnea

Ipomoea muricata

Ipomoea nil

Ipomoea pestigridis

Justicia simplex

PLATE 25

Lactuca dissecta

Lannea coromandelica

Lantana camara

Lathyrus aphaca

Leea crispa

Lepidagathis cuspidata

PLATE 26

Lespedeza sericea

Lindernia ciliata

Macrotyloma unifloruma

Marchantia palmata

Marsilea minuta

Martynia annua

PLATE 27

Medicago denticulata

Melothria heterophylla

Mentha longifolia

Mentha piperita

Micromeria biflora

Mirabilis jalapa

PLATE 28

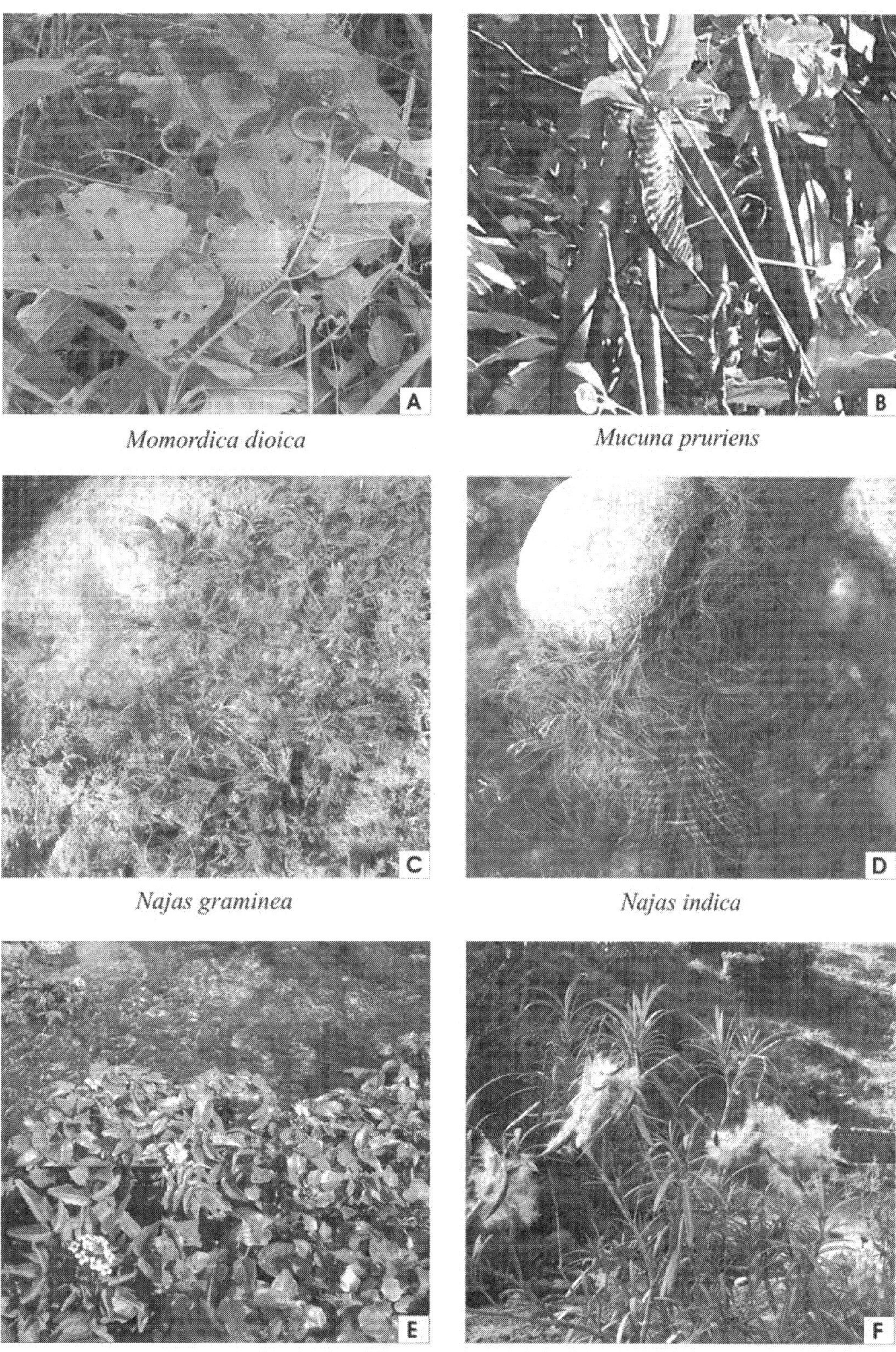

Momordica dioica

Mucuna pruriens

Najas graminea

Najas indica

Nasturtium officinale

Nerium indicum

PLATE 29

Ocimum basilicum

Oroxylum indicum

Oxalis corniculata

Parthenium hysterophorus

Paspalum distichum

Pennisetum lanatum

PLATE 30

Phragmites karka

Phyllanthus urinaria

Physalis longifolia

Physalis minima

Plectranthus coetsa

Plumbago zeylanica

PLATE 30

Phragmites karka

Phyllanthus urinaria

Physalis longifolia

Physalis minima

Plectranthus coetsa

Plumbago zeylanica

PLATE 31

Poa supina

Pogostemon plectranthoides

Polygala arvensis

Polygonum barbatum

Polygonum barbatum sub sp. *gracile*

Polygonum donii

PLATE 32

Polygonum glabrum (A)

Polygonum lapathifolium (B)

Polygonum minus (C)

Polygonum plebejjum (D)

Polygonum pulchrum (E)

Polygonum serrulatum (F)

PLATE 33

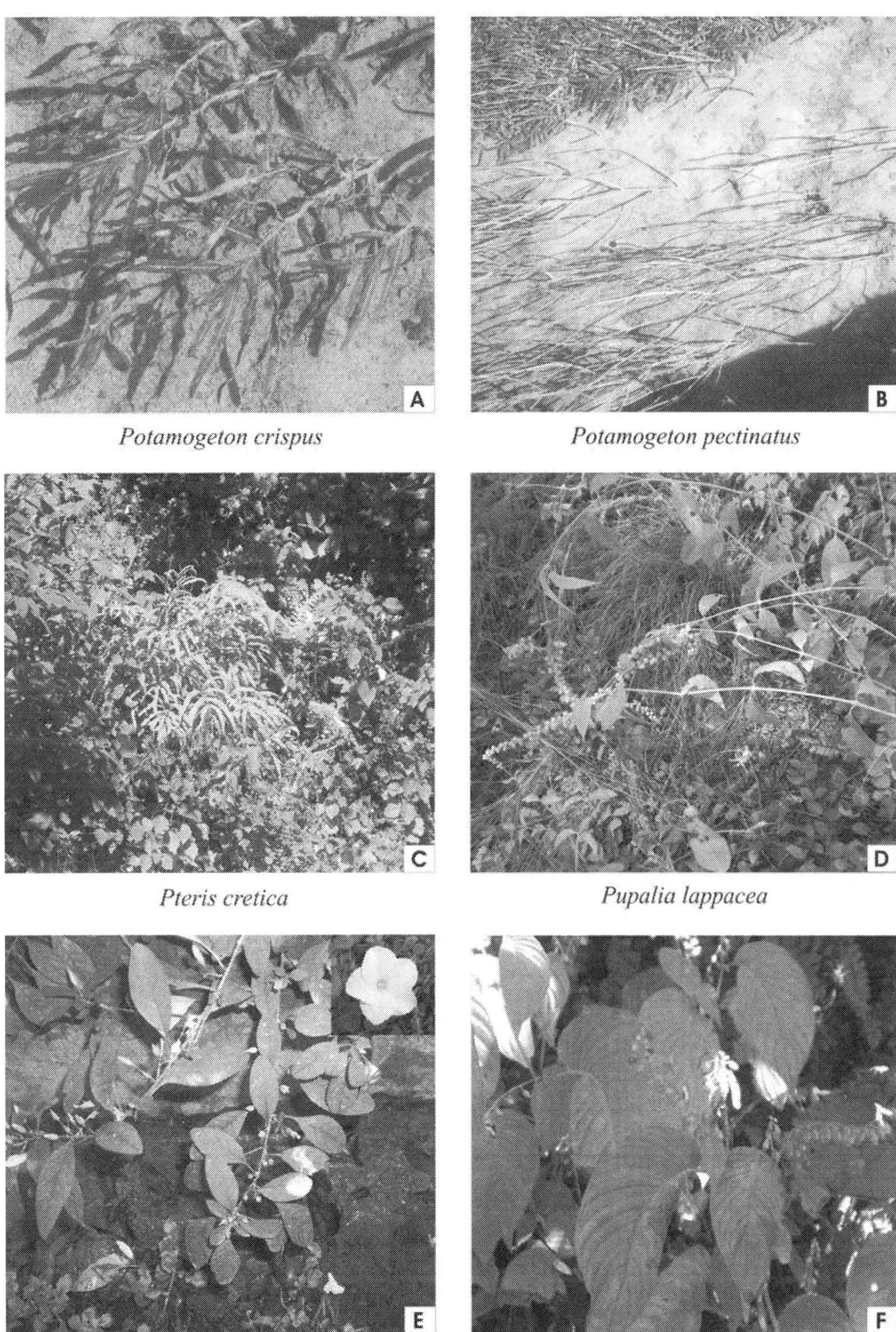

Potamogeton crispus

Potamogeton pectinatus

Pteris cretica

Pupalia lappacea

Reinwardtia indica

Rhynchoglossum obliquum

PLATE 34

Ricinus communis

Roylea cinerea

Rubus ellipticus

Ruellia patula

Rumex hastatus

Rumex nepalensis

PLATE 35

Rungia pectinata

Saccharum munja

Saccharum spontanium

Salix oxycarpa

Saussurea heteromala

Selaginella chrysocaulos

PLATE 36

Setaria tomentosa

Sida acuta

Silene conoidea

Solanum nigrum

Solanum xanthocarpum

Sonchus arvensis

PLATE 37

Sonchus asper

Sorghum halepense

Spilanthes acmella

Spilanthes paniculata

Stellaria media

Strobilanthes atropurpurens

PLATE 38

Strobilanthes dalhousianus

Taraxacum officinale

Thalictrum reniforme

Thevetia neriifolia

Trichodesma indicum

Tridax procumbens

PLATE 39

Trifolium resupinatum

Trigonella pubescense

Uraria picta

Urena lobata

Urtica dioica

Vernonia cinerea

PLATE 40

Veronica anagalis-aquatica

Veronica anagalis-aquatica; Flowering Shoot

Vicia sativa

Viola pilosa

Withania somnifera

Xanthium strumarium

Made in the USA
Monee, IL
07 July 2026

56549925R00266